Plumbing 401

PHCC EDUCATIONAL FOUNDATION PLUMBING APPRENTICE & JOURNEYMAN TRAINING COMMITTEE

Australia • Brazil • Japan • Korea • Mexico • Singapore • Spain • United Kingdom • United States

Plumbing 401
Plumbing-Heating-Cooling-Contractors–
National Association Educational Foundation

Vice President, Technology and Trades
Professional Business Unit: Gregory L. Clayton
Product Development Manager: Ed Francis
Product Manager: Vanessa L. Myers
Editorial Assistant: Nobina Chakraborti
Director of Marketing: Beth A. Lutz
Executive Marketing Manager: Taryn Zlatin
Marketing Manager: Marissa Maiella
Production Director: Carolyn Miller
Production Manager: Andrew Crouth
Content Project Manager: Kara A. DiCaterino
Art Director: Benjamin Gleeksman
Production Technology Analyst: Thomas Stover

Library of Congress Control Number: 2008929255
ISBN-13: 978-1-4180-6536-2
ISBN-10: 1-4180-6536-6

Delmar
5 Maxwell Drive
Clifton Park, NY 12065-2919
USA

Cengage Learning is a leading provider of customized learning solutions with office locations around the globe, including Singapore, the United Kingdom, Australia, Mexico, Brazil and Japan. Locate your local office at: **international.cengage.com/region**

Cengage Learning products are represented in Canada by Nelson Education, Ltd.

Visit us at **www.InformationDestination.com**

For more learning solutions, please visit our corporate website at **www.cengage.com.**

Printed in the United States
2 3 4 5 6 7 XX 16 15 14 13

Table of Contents

About PHCC and the PHCC Educational Foundation

THE PHCC EDUCATIONAL FOUNDATION

The Plumbing–Heating–Cooling–Contractors—National Association Educational Foundation was incorporated in 1986 as a 501(c)(3) tax-exempt, charitable organization. The purpose of the foundation is to help shape the industry's future by developing and delivering educational programs that positively impact every aspect of the plumbing–heating–cooling contractor's business. Major programs include apprentice, journeyman, and business management training.

The foundation relies on contributions from many sources. Gifts to the Foundation Endowment Fund are noninvadable, ensuring the foundation's financial integrity for generations to come. Donors include contractors, manufacturers, suppliers, and other industry leaders who are committed to preparing plumbing–heating–cooling contractors and their employees to meet the challenges of a constantly changing marketplace.

The PHCC Educational Foundation mission statement sets forth this vision:

> The Plumbing–Heating–Cooling–Contractors—National Association Educational Foundation develops apprentice, journeyman, and business management training programs for the success of plumbing and HVACR contractors, their employees, and the future of the industry.

ABOUT PHCC

The Plumbing–Heating–Cooling–Contractors—National Association (PHCC) has been an advocate for plumbing, heating, and cooling contractors since 1883. As the oldest trade organization in the construction industry, approximately 4,000 member companies nationwide put their faith in the association's efforts to lobby local, state, and federal government; provide forums for networking and educational programs; and deliver the highest quality of products and services.

PHCC's mission statement is the guiding principle of the association:

> The Plumbing–Heating–Cooling–Contractors—National Association is dedicated to the promotion, advancement, education, and training of the industry for the protection of the environment, and the health, safety, and comfort of society.

A complete account of PHCC's history is included in the PHCC Educational Foundation publication *A Heritage Unique*.

Acknowledgements

A special thank you is extended to Merry Beth Hall for her time and dedication as the *Plumbing 401* Project and Subject Matter Expert Committee Coordinator. Merry Beth is the Director of Apprentice & Journeyman Training with the PHCC Educational Foundation.

The following subject matter experts provided their time and expertise to the writing of this book:

Kirk Alter, Purdue University, West Lafayette, Ind.
Todd A. Aune, CTO, Inc., Harlingen, Tex.
Charles L. Chalk, Maryland PHCC, Ellicott City, Md.
Larry W. Howe, Howe Heating & Plumbing, Inc., Sioux Falls, S.Dak.
Eric L. Johnson, Quality Plumbing & Mechanical, Kodak, Tenn.
Michael J. Kastner, Jr., Kastner Plumbing & Heating, West Friendship, Md.
Richard R. Kerzetski, Universal Plumbing & Heating Co., Las Vegas, Nev.
Robert Kordulak, The Arkord Company, Belmar, N.J.
Frank R. Maddalon, F. R. Maddalon Plumbing & Heating, Hamilton, N.J.
Robert Muller, John J. Muller Plumbing & Heating, Matawan, N.J.
Larry Rothman, Roto Rooter Services Company, Cincinnati, Ohio
James S. Steinle, Atomic Plumbing, Virginia Beach, Va.
Orville Taecker, Andor, Inc., Watertown, S.Dak.

Credit for authorship of the original PHCC *Plumbing Apprentice Manuals* is extended to:

Ruth H. Boutelle
Patrick J. Higgins
Charles R. White
Richard E. White

About the Authors

Robert G. Konyndyk

Robert G. Konyndyk is Chief of the Plumbing Division within the Bureau of Construction Codes, Department of Labor and Economic Growth, State of Michigan. Mr. Konyndyk plans, organizes, directs, and controls a statewide license program that encompasses more than 15,000 professionals. In addition Mr. Konyndyk manages state field inspectors, oversees plumbing code development and administration, and supervises product acceptance. Prior to state employment he owned and operated a plumbing contracting firm for 10 years, serving as its licensed master plumber and licensed mechanical contractor. Mr. Konyndyk started from the ground up, so to speak, when he was "buried alive" during an airport excavation on his first day in a national union apprentice training program.

Mr. Konyndyk also served on active duty in the Air Force as a computer repairman after working in his father's contracting business. He has several licenses and certifications and focused his formal education on the plumbing profession with its technical merits. He has served on numerous plumbing code committees and in associations such as the American Society for Testing Materials International, International Code Council, and National Sanitation Foundation International, dealing with public health and plumbing issues since 1985. Mr. Konyndyk feels his greatest accomplishment was being one of three representatives chosen to initially develop the *International Plumbing Code* (first draft, *Joint Model Plumbing Code*) in 1994 as the Building Officials and Code Administrators representative. He enjoys teaching code as a state administrator and has traveled as an instructor to several states for different code groups. His goal has been to contribute to greater uniformity and stability in national code issues.

Edward T. Moore

Edward T. Moore is the author of the *Residential Construction Academy's Plumbing Video Set*. He holds a Bachelor of Science degree in Mechanical Engineering. Mr. Moore is currently the Department Manager and an Instructor for the Building Construction Trades and Air Conditioning Program at York Technical College in Rock Hill, South Carolina. He has also served as an instructor for the Industrial Maintenance and Welding programs. A licensed Master Plumber for South Carolina, Mr. Moore is also the owner of Moore Plumbing and Cabinetry. He has earned the NATE certification in Air Conditioning and Heat Pumps and is currently working on a Master of Science degree in Manufacturing Engineering. Mr. Moore recently acquired his HERS certification (Home Energy Rating Services), which allows him to give Energy Star ratings to homes.

Mr. Moore currently resides in Clover, South Carolina, with his wife and three children.

Preface

This is the fourth in a series of four plumbing textbooks suitable for training in an industry or academic setting. The content has been selected and organized to suit students and trainees who have had at least three years of formal or on-the-job training. The subject matter progresses on the assumption that students are applying what they learn in a school shop or field environment.

This edition has been written to reflect the latest methods, components, codes, and standards. Illustrations have been included to help readers grasp concepts more completely. Examples show a logical, step-by-step process. Finally, there are field applications and extensive safety information.

The curriculum that supports this book and the content itself have been developed and approved by the PHCC Educational Foundation Plumbing Apprentice & Journeyman Training Committee. The committee has endeavored to respond to industry demands for training material best suited to a field-oriented program. It is the combination of classroom and field experience that the committee wishes to promote throughout the industry.

CHAPTER 1

Service Professionalism, Leaks, and Drainage Problems

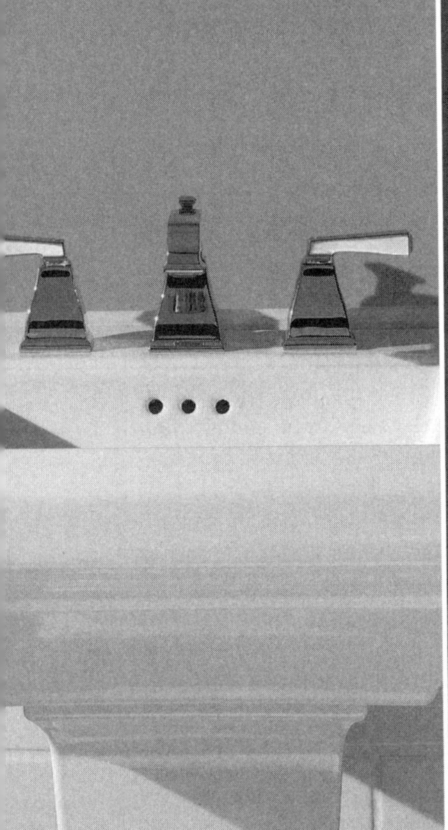

LEARNING OBJECTIVES

The student will:

- Discuss the professional attitude necessary for a successful service technician.
- Describe the importance of the customer–employee relationship.
- Explain when to make replacements rather than repairs.
- Describe the necessary steps to repair leaks and remove stoppages in drainage systems.

PROFESSIONAL UNDERSTANDING AND ATTITUDE

This chapter and the next ten chapters in *Plumbing 401* examine the repair and servicing of fixtures, fittings, appliances, appurtenances, piping, and other facets of the plumbing system that may require maintenance operations.

It is extremely important that each technician understand that he is a professional, trained specialist having not only mechanical skills with proper service techniques but also a strong emphasis on good public and customer relations. Mechanical skill is only a part of best practices during service and repair procedures. The service technician must be aware that he represents the employer in the eyes of the customer. Therefore, the technician must be courteous, use appropriate language, and be attentive to his personal appearance. Professionalism aids a company in its reputation and growth, which translates to continued employment, increased compensation, possible advancement, and most important, a sense of satisfaction.

The best service technicians recognize that service and repair work offers the greatest number of challenges in the plumbing profession. Maintaining an attitude of facing the challenge and final accomplishment requires a great deal of mental understanding with the best mechanical skills. Mental understanding recognizes that your success is dependent upon technical knowledge, mechanical skills, an ability to communicate with customers, and often sales ability. Your ability to communicate with customers is heavily based upon your attitude. Communication skills are centered on listening!

THE CUSTOMER–EMPLOYEE RELATIONSHIP

Service and repair work is unique in that, often, you as a single employee are interacting directly with the individual who is contracting with your employer for company services, unlike general new construction work. Service technicians are often aggressive, independent individuals with a great sense of pride and ability. Those qualities aid an employee in being self-motivated. But your technical skills may play a less important role than your skills as a listener, communicator, and analyst. If you are presently a service technician, there is an extremely good chance your employer has recognized that you have the skills previously mentioned. Customer relationship skills are similar to installation skills in that they can and must be improved through a concerted effort.

Many service companies use a policy manual to ensure proper employee performance and service procedures so that everyone knows what is expected of an employee. The manual describes the performance actions required for employees and addresses various situations that may be encountered on the job. If your company has such a manual, you should study it and conform to its requirements.

In the Field

1. Treat the customer and his property with respect.
2. Consider the customer's attitude, needs, and wants.
3. Focus on the service task. Customers should never overhear personal conversations as they may believe you are not working on the task at hand.
4. Do not criticize any product or existing installation.
5. If you have made a time commitment, be prompt. If you find that you will be delayed, call all affected customers.
6. All can be summed up as follows: Conduct yourself in a thoroughly professional manner at all times.

BASIC SERVICE AND REPAIR PRINCIPLES

Service and repair procedures with general laundry-type lists can be of great value to a technician in conducting service work. Following these lists increases the technician's ability to finish the service task by reducing time, adhering to safety principles, and allowing greater attention to detail. Service technicians understand that attention to detail is one of the most important factors in bringing a repaired system back on line without unnecessary system failures.

This list is provided as a very basic aid to enable you to develop an improved list or to challenge you to consider other operations to improve your performance:

Basic Guidelines for a Service and Repair Plumber

1. Be proud of your work and your employer. Your attitude provides confidence to the customer.
2. Conduct yourself in a professional manner at the site. Personal cell phone conversations and inappropriate language are a few of the many things that offend customers.
3. Organize your work schedule to minimize travel time and maximize the time on the job.
4. Ask for help from your supervisor if you need it. Moving heavy equipment by yourself may cause damage to the customer's surroundings or may injure you.
5. Keep the work area safe while working and secure it when you leave.
6. Use drop cloths if conditions require them and maintain a neat work area during the service call. A clean area reduces the time required to clean up after completion of the work.
7. Have a clean pair of shoes or shoe covers for particularly messy situations. It is very common for customers to monitor your trips to and back from the service truck to see how it affects their "castle."
8. Keep the trips to the service vehicle to a minimum. Customers often judge your abilities on your being organized.
9. Wear protective clothing, gloves, and eye and face protection, especially when working on any drainage system, device, or problem.
10. Keep your service vehicle clean and organized.
11. Protect and secure your truck inventory and tools from damage or theft.
12. Take time to maintain your tools so that they are clean and in proper working order. Replace worn tools that cannot be maintained in like-new condition.

While no suggestions list can be truly complete, the hints listed above can help you form attitudes and set your own personal practice policies in becoming a professional service and repair technician.

> **In the Field**
>
> Two important factors should be considered when working on drainage systems. Harmful chemicals may have been used by a customer prior to your service call and may be splashed or drained on you during the cleaning process. Second, in medical facilities including medical gas installations, bacterial or infectious diseases may be present in the systems during service operations.

> **In the Field**
>
> Use only the best techniques and proper tools, especially when working on finished surfaces.

BASIC PRINCIPLES AND REPAIR VERSUS REPLACEMENT DECISIONS

The growth and development of an apprentice or student plumber to the status of journeyman requires the student to understand and practice the general principles of drainage, venting, and water supply. Further, the service and repair technician must know how to apply those principles to products and systems encountered in the field. Your previous training and course materials have provided basic principles and applications to new installations. In this last year of training, students will define and refine their problem-solving skills.

It is extremely important to understand the internal workings of the products installed in order to service and repair them. A technician will develop awareness of the equipment to restore a plumbing fixture or appliance to normal service after a breakdown.

Many skills are required to analyze the problem and then to produce a satisfactory repair. Replacement may be the most economical repair in many cases, and good judgment in deciding whether to repair or replace defective products is of great importance. Skills in advising the customer of later costs are critical as an employee and will be appreciated by the customer at a later date. Several other considerations must also be discussed with the customer. For example, the customer

> **In the Field**
>
> Keep abreast of your industry. Read trade magazines and attend trade meetings, seminars, and shows.

In the Field

1. Maintain a notebook for difficult service calls. Use this information to organize your efforts in obtaining greater information in preparation for similar types of calls. This will also aid you in stocking your service vehicle in the future.
2. A good technician strives to be better. The most critical factor in making a repair-versus-replacement decision is for you to take time to determine what the problem is in order to solve what is really wrong.
3. Fix it right the first time.

In the Field

Your introduction to the service aspect of plumbing service and repair must consider that your company's service department is the backbone of the business, and when new construction is declining, your company's competition will increase because of their reduced charges. Quality work based upon honest business practices will benefit your career.

may require the system to be operable temporarily, even with diminished capabilities, to allow for basic living functions or the conduct of business. Often, however, repair parts are not immediately available, and the technician may have to improvise to make the system operate temporarily.

SERVICING FIXTURE AND DRAINAGE SYSTEM PROBLEMS AND REPAIRING LEAKS

The plumber who specializes in service and repair work must be able to identify a variety of fixtures, appliances, and appurtenances during the course of his career, as well as understand the operating principles of the equipment. In this chapter we examine the leakage or drainage problems for fixtures and describe service and repair work first for residential applications and then for commercial settings.

Plumbing service and repair technicians are often called upon to resolve leakage or drainage problems for fixtures. Table 1–1 through Table 1–4 will provide troubleshooting tips to aid in your diagnosis and resolutions for residential fixtures. The similarities in corrective actions for residential and commercial systems are a matter of common sense.

Troubleshooting Residential Fixtures

Table 1–1 Bathtub or Shower

Fixture Problem	Probable Cause	Test	Solution
Fixture does not drain	Stoppage in drain line Stoppage in trap or in waste line	Visual observation Visual observation	Clear the drain Snake or "plunge" the line. (When using a plunger on the line, cover overflow opening. Be alert for the presence of drain cleaning solutions when a plunger is used. Always wear eye protection.)

Table 1–2 Kitchen or Bar Sink

Fixture Problem	Probable Cause	Test	Solution
Fixture does not drain properly or fixture drains slowly	Stopped up or constricted drain (usually caused by grease build-up)	Visual observation Listen for gurgling noise or bubbling in drain	Clear drain mechanically, or, if absolutely necessary, use a chemical.

Table 1–3 Laundry Tray

Fixture Problem	Probable Cause	Test	Solution
Fixture does not drain or drains slowly	Stopped up or constricted line (usually lint, strings, etc.)	Visual observation	Clear the drain line.

Table 1–4 Water Closet

Fixture Problem	Probable Cause	Test	Solution
Leaks on floor at base	Bad seal from bowl to flange Casting hole in fixture Plugged sewer	Test to see if fixture "rocks" or leaks at the base. Visual inspection Flush water through system	Replace seal. Firm up uneven areas. Replace fixture. Clear stoppage.
Water running into bowl continuously	Leaking flush valve and/or tank ball Water from ballcock refill tube discharging into the overflow Water rising above overflow level Cracked overflow tube threads (may occur on brass tubes) Float hanging up against tank or other parts Float has insufficient buoyancy Siphon on refill tube	Place dye in tank and see if it runs into bowl. Visual inspection when the ballcock should be off Visual inspection Dye test, conducted after ballcock and tank bell are tested as sound Visual observation Remove float and shake, listen for water. Plastic float may be waterlogged. If refill tube is below the water line, it may act as a siphon leg	Replace tank ball or flush valve. Replace ballcock. Replace ballcock or lift guide assembly. Replace overflow tube. Align float rod. Replace float. Raise the level of the refill tube.

Troubleshooting Commercial Drainage Problems

Numerous system drainage and leakage problems are identified along with their remedies in the following text with insights that will apply to all residential and commercial structures.

Drainage Waste and Vent Lines

These systems include both branches and stacks in the drainage system and are referred to by plumbers collectively as the DWV system. While the problems may be simple, such as poor drainage caused by obstructions, the resulting consequences can be very serious health hazards. The indications of improper operation are slow or inoperable drains, visual signs of leakage, and odors. A first consideration is that the odors could be a sign of deterioration of trap seal conditions or could be an indication of leaks in the DWV venting system, involving small numbers of piping or fittings. The solution for obstructions is cleaning the lines with drain cleaning equipment. Replacing the lines may also be an option based on economics and a discussion with the customer or his representative. Prior to beginning the work, assess the area in which you will be working in order to minimize damage to the building environment as drain pipe removal will become a serious cleanup concern.

In the Field

Exercise common sense when selecting the pipe, fittings, and connection methods for your repair. Often, bands, clamps, and other repair connections that are easier to install will not meet code requirements. Refer to your code book tables for a list of the acceptable pipes and fittings. Most codes have sections that list the approved methods for making joints and connections categorized by the type of piping material.

Sanitary Building Drains and Sewers

The sanitary main systems include the building drain in the structure and the building sewer outside the structure that connects to the public sewer or individual sewage disposal system such as a septic tank and drain field. One indication of problems in these primary systems is slow drainage in the surrounding building drain stacks and branches. The obvious problem would be an inoperable system virtually shutting down the building and creating a health hazard. These primary line difficulties are generally the result of obstructions or could be caused by a faulty backwater valve. The solution for obstructions is cleaning the lines with drain cleaning equipment. Prior to using mechanical cleaning equipment, a service and repair technician has to analyze the system and identify any interceptors or backwater valves, which of course cannot be involved in cable cleaning. Replacing the building drain and sewer are generally the last choice in dealing with these problems due to costs.

New technology related to public sewer utilities now offers the reconstruction of lines with relining equipment procedures. The terminology referring to these new methods is "trenchless construction." Old systems are cleaned, analyzed with video camera equipment, and relined. The relining product can consist of several different materials, which are commonly dipped or impregnated with a sealer that is later hardened. This is a very specialized process that can take differences in piping size into consideration. The material is pulled from one opening to another and then cured in place. The curing may include several methods, such as steam or hot air. Part of the process is to send equipment through the system to open any leads or branches entering the main. Local codes or practices often require added completion requirements such as a water test and final video analysis. Making a final video is a very wise move because any indications of water left in the system after a water test could indicate a water pocket due to backfall or a line settling due to poor backfill conditions. These problems of course can be corrected only by excavation and replacement. Figure 1–1 shows the sewer relining process.

Building Storm Systems

Building storm systems start with the roof drains and move through the systems to the conductors, building storm drain, building storm sewer, and finally the place of disposal, which can be a municipal storm system or point of disposal on the property such as a holding pond or reservoir. Another storm system classified as building subsoil piping will be discussed in later chapters. The building storm system can only receive clear water waste and must not have waste connections from the building.

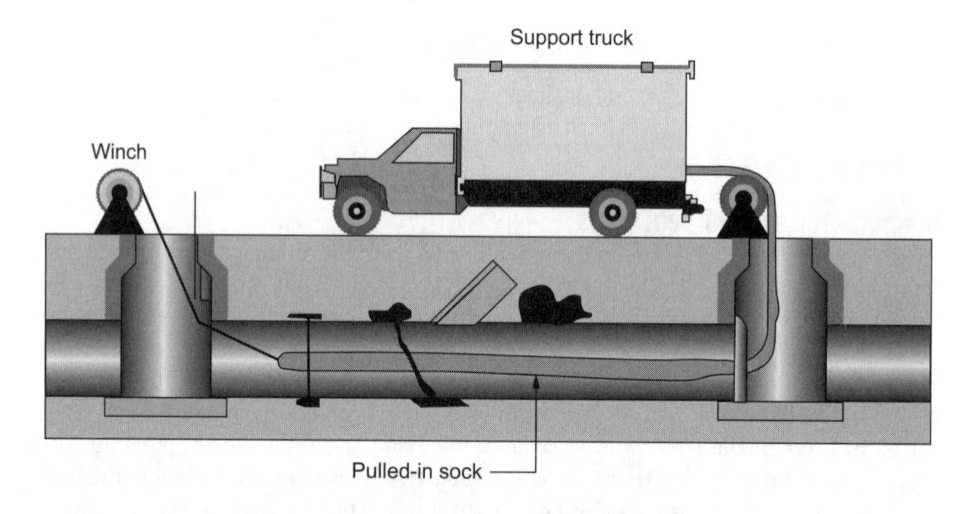

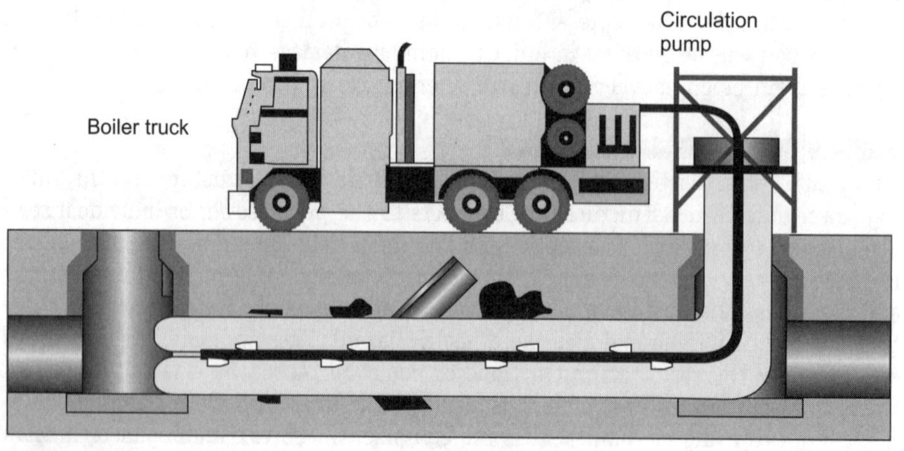

Figure 1–1
Sewer relining processes for malfunctioning sewers have been refined to reduce costs by eliminating excavation and pipe replacements. The installation details and products vary from one supplier to another; however, they all have the same concept.

This requirement makes sense when you consider that the discharge from incoming waste will ultimately discharge untreated to the ground and ground water table.

Problems with the systems are serious in nature in that leaks will damage the building and its goods. Obstructions may cause overloading, resulting in roof failure (costly and dangerous conditions). The indications of improper operations may not be as prevalent as in sanitary systems due to the lack of inside building fixtures being connected to the storm system. Several codes have very specific secondary emergency roof drainage requirements to ensure building safety. (See Figure 1–2 for an example of a building roof that has drains and terminals at several different roof levels).

However, the secondary systems are not required when the building storm load can be dissipated by simply directing it over the edge of the building. The solution for obstructions is cleaning roof drain strainers of leaves and twigs and utilizing drain cleaning equipment for problems in the piping. Storm system obstructions are of course far less common than those in sanitary systems due to the lack of solids and waste products that build up in sanitary drain lines.

In the Field

Several codes include regulation of outside roof collectors commonly called semicircular gutters and associated piping. It is very common for local jurisdictions to delete those portions of the code regulating this trade because they contradict the jurisdiction's licensing requirements. For example, eaves trough installers are not considered as qualified for a plumbing apprentice program leading to licensing.

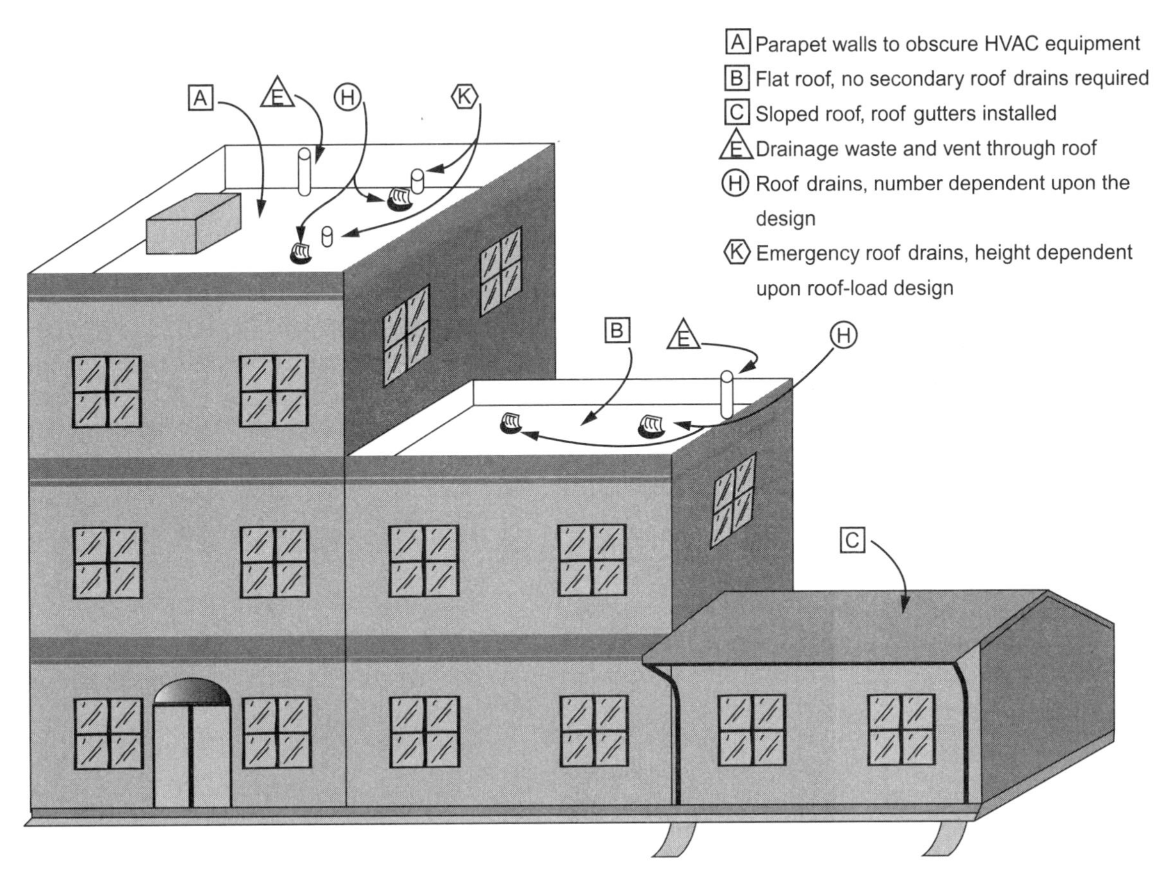

Figure 1–2
This roof illustration indicates the various roof drains, secondary roof drains, and vent terminals utilizing several different roof levels. Actual roofs may have HVAC equipment and many other products that must be considered in design and maintenance matters.

REVIEW QUESTIONS

1. A professional service technician must have good mechanical skills and a strong emphasis on ______ ______.
2. One of the first rules for a service technician is to treat the customer with ______.
3. A service technician should always properly ______ the drainage problem prior to using the first available line cleanout.
4. What business practice should best describe your actions during service calls?
5. ______ versus repair may be a very wise decision economically when performing service and repair work.

CHAPTER 2

Residential and Commercial Service and Repair

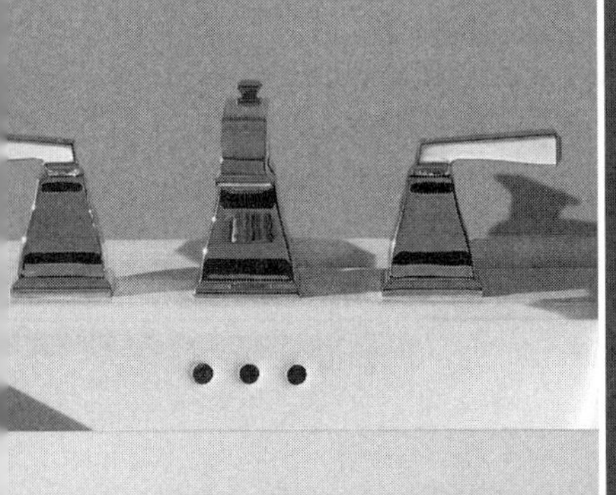

LEARNING OBJECTIVES

The student will:

- Describe the repair and servicing of various residential plumbing fixtures and fittings.
- Describe the repair and servicing of different commercial plumbing fixtures and fittings.
- Identify professional practice and code applications involved in service and repair procedures.

RESIDENTIAL SERVICE AND REPAIR INSIGHTS

This chapter will discuss the most common water controlling mechanisms that are used to supply or operate plumbing fixtures. Successful repair of residential and commercial fixtures and associated components is best accomplished with an understanding of the operation of each part in the assembly. Telltale symptoms of fixture problems or failures lead to the correct diagnosis for repair or replacement solutions. Also, it is extremely common for the success of a service call to be based upon good judgment regarding whether to replace or repair.

The majority of residential plumbing fixtures are reasonably economical items, and replacement should be presented to the customer as an opportunity to update and as an economical benefit. Customers must be informed of the economical benefit of replacement because it often will be the best choice for them in the long run. Most people will agree with this analysis if it is presented to them in a friendly, factual manner.

Always be tactful when suggesting to a customer that a replacement kitchen faucet is a better value than just replacing the most-worn parts of the faucet. For example, when a customer is involved in a kitchen sink faucet repair decision, she may decide that the features of a pull-out spray head model could be a better investment than a repair. Another consideration is whether you have the repair parts on your service vehicle. If not, cost and time considerations may also contribute to your customer explanation.

One of the most common difficulties with faucet repairs is in identifying and obtaining the correct replacement parts. Over the years, many manufacturers have made numerous models, so even a large repair company will be hard-pressed to have all the seats, stems, and washers on hand. Larger manufacturers will often refer you to major parts suppliers for repair components. In most cases, the same products are found within geographical areas. For example, housing projects that used the same major installation contractor create areas that have uniformity in product installations from one project to another. As a service technician, you will become aware of these service trends and stock your vehicle appropriately.

The next section focuses on various fixtures and their associated components.

In the Field

Good workmanship resulting in customer satisfaction is extremely important in residential service and repair work. Often, an unsatisfied residential customer has a greater potential of providing negative advertising than commercial customers. This negative effect may not always be immediately apparent to your employer, but will certainly affect your opportunities for advancement.

In the Field

Proper service and repair work must include considerations of code compliance. Original installations such as leg tub faucets did not have acceptable air gaps on the water supply to adequately address cross-connection issues. A faucet replacement with a gooseneck spout above the tub overflow rim is necessary in this case not only to meet the code but also to protect individuals in the structure and adjoining buildings with the same water supply against cross-connections.

RESIDENTIAL LAVATORY FAUCETS

Single Bibb—Compression and Gasket Seat Type

A single bibb faucet controls a single water flow only. Some have a replaceable seat and a rubber washer that is pressed into the seat when the closure mechanism is operated to stop the flow. This type can be easily repaired when it does not shut off tight as evidenced by leakage or a drip. If the faucet seat has been eroded or grooved by a long-term drip and the seat is not removable, the faucet will have to be replaced. Quality seat regrinding tools are available, and their use may be an option when the faucet replacement involves circumstances that complicate other service considerations. However, nonremovable seat faucets are generally cheaper and may have seat areas with insufficient metal thickness to make the repair worthwhile.

Another common failure is that of water leakage around the stem when the water is flowing. Packing or tightening the packing nut will usually correct this. Water service and meter valves with these types of packing nuts are common problem areas because of overtightening by the customer. In extreme cases, the stem will have been roughened and repacking will not be successful. Thus, a replacement stem or new faucet must be installed (see Figure 2–1). Note that if the stem threads are worn so that the stem is loose in the faucet and both the male and female threads are worn, a replacement stem will result in a very poor repair.

In the Field

Residential fixture and fitting repairs cover a broad range of devices. The following are general comments and cautions in addition to those covered in Chapter 1:

1. Use the proper tools. Smooth-jawed tools and protective pads will prevent scratching of finished surfaces.
2. Ensure that the device is operating satisfactorily when you complete your work. Often, repair work may contribute to the malfunction of adjoining fixtures or lines.

Continued

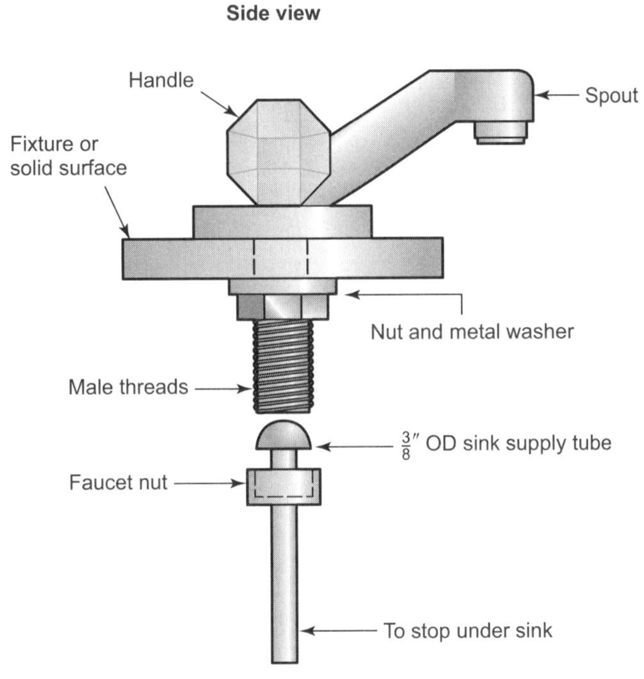

Figure 2–1
Conventional single-stem faucets compose the backbone of the faucet industry. The basic concept is used throughout plumbing installations from fixture water supplies to boiler drain valves.

3. Use proper repair parts. Several look-alike parts are available that are not made with proper tolerances to operate satisfactorily. Your service vehicle may be stocked with the appropriate parts, but the customer may insist on your using his previously purchased items.
4. Be sure that replacement components conform to code requirements. Manufacturers and suppliers are not prohibited from making products that do not conform to the code.
5. Check all operating parts of the fixture fitting for wear, not just the portion that is the problem of the moment. For example, replacing the ball and seat assembly in a traditional kitchen sink faucet would be the proper time to replace the O-rings in the spout assembly.

Combination Two-Handle

Combination faucets control two or more water flows (see Figure 2–2). Older models consist of stems, seats, and washers that are basically the same as single bibb models, and the repair processes are similar. Later models use cartridges that contain the seat, washer, and stem in a single assembly.

Several new products use a stem assembly with a polished face that sits on a gasket tensioned by a lower spring. The gasket may become brittle and adhere to the stem assembly face. The gasket can be replaced and the stem assembly shined. However, when considering the cost, the best repair is to replace the stem assembly because the new assembly ensures a longer service life by preventing an O-ring leak around the shaft and prevents an unhappy customer at a later date.

Single Lever

Single lever faucets control two water flows with a special cartridge inserted into a precisely machined housing (see Figure 2–3). Other single lever faucets may use a ball assembly. Repairs to these products must be completed according to the manufacturer's instructions. The orientation of the parts is important; therefore, check to be sure that hot corresponds to the left side of the faucet after making the repair as a cartridge pivoted around 180 degrees will result in hot water coming out when the user positions the handle to the cold side.

Faucet aerators cause air bubbles to be blended into the stream of water flowing from the faucet. These air bubbles eliminate splashing. Dirt, chips, solder, or other solids that collect on the inlet screen will upset the desired flow and pressure patterns. In extreme cases, flow will be totally stopped; a clogged aerator will create minimal flow when either the hot or the cold stem is activated. Usually, cleaning the aerator will correct the problem. If not, it must be replaced.

In the Field

When the customer is present and "looking over your shoulder," never be afraid to share a little knowledge with her, such as an explanation for cleaning an aerator to avoid minor problems later. Customers appreciate the advice, and their confidence in your communication abilities may assist in closing billing discussions.

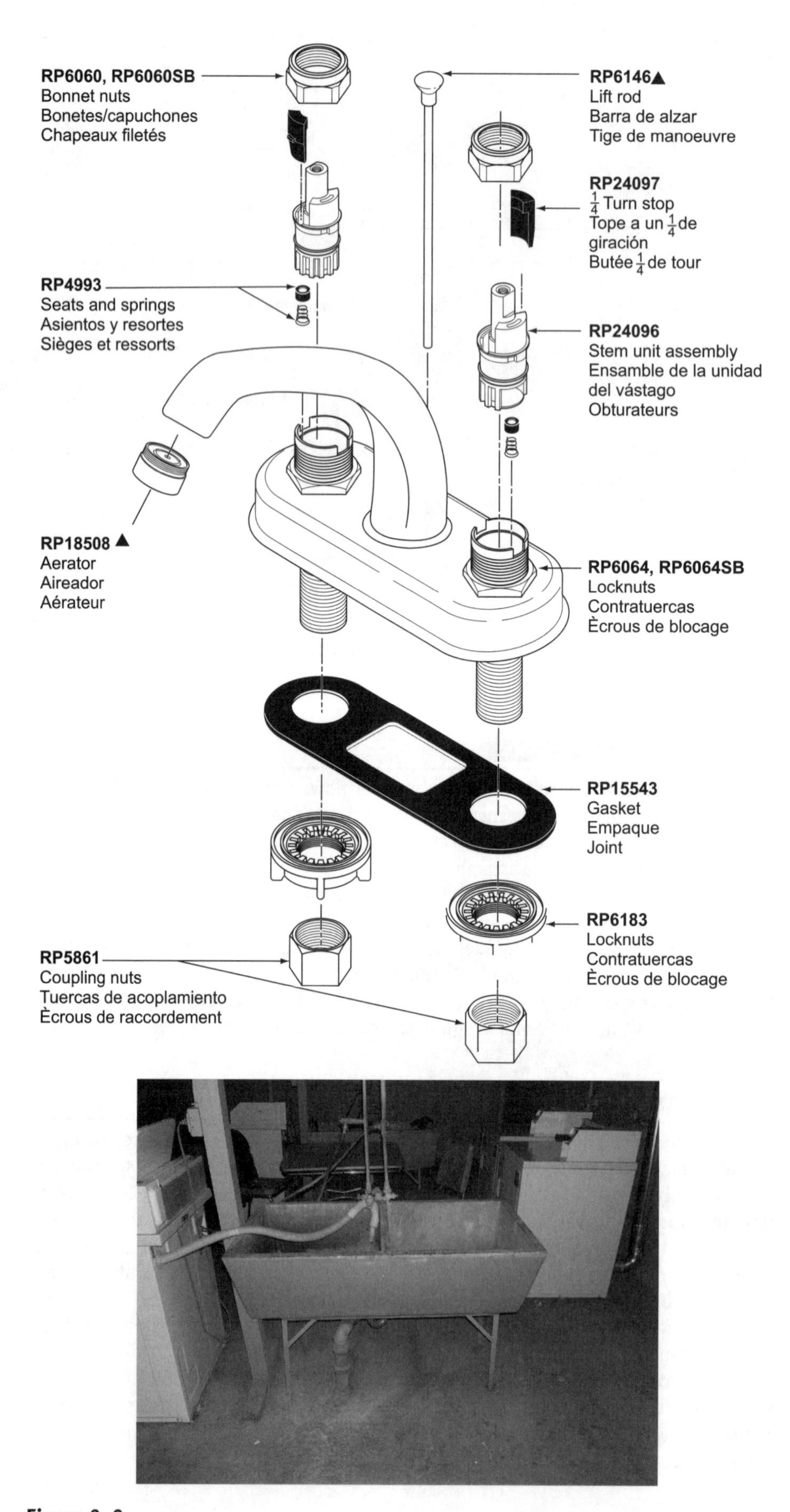

Figure 2–2
The valve concept of having a polished stem face sit on a flexible seal has been widely accepted and offers many advantages, one of which is ease of service.

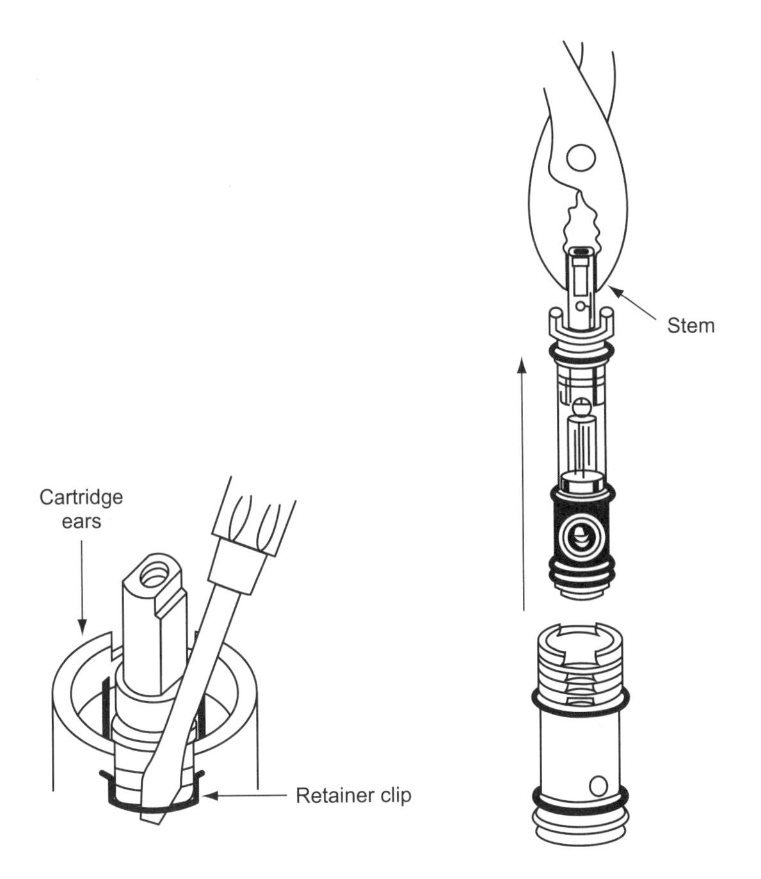

Figure 2–3
Cartridge-type faucets offer a long service life and have been embraced by technicians for many years.

RESIDENTIAL BATH FAUCETS

Most bath, shower, tub, and combination tub and shower faucets are designed similarly to and are therefore repaired by methods similar to those used for lavatory faucets. Commonly, tub or shower valves are not serviceable or accessible from the back side.

The valves are designed to be repaired from the front by removing the escutcheons or plates around the stem or cartridge assembly. In such cases, special deep-socket repair wrench sets and seat repair shaft kits are required to remove stem assemblies and replace seats. Temperature control types are serviced by replacing not only flow control mechanisms but also stems and the operating cartridge, either pressure balance or thermostatic types.

Service calls involving the repair of shower or tub and shower combination valves should involve considerations of plumbing code requirements. Codes now require the use of an American Society of Sanitary Engineering (ASSE) 1016 standard conformance valve to address temperature safety requirements for the user.

A service technician about to repair an older valve should discuss this matter with the customer, who may decide to pay for the added safety assurance of a new valve instead of a repair. Another important consideration is the permit or code requirements for the replacement of the valve. Some codes require a permit, resulting in required conformance to a current ASSE 1016 device when installing a new valve. Serious practical concerns are presented if the previous valve was a two- or three-handle valve. A three-handle valve has a built-in diverter assembly rather than a diverter on the tub spout. The norm in ASSE 1016 valves is a single wall opening, which may require a wide escutcheon cover plate to cover the former hot and cold openings. For appearance reasons, this may be objectionable to the customer.

An extremely important part of the repair is proper sealing of the escutcheon to the wall during the final repair stage. This prevents water splashing back up against the wall and finding its way behind the tile or wall surface. The vast majority of homeowners have seen this type of damage to walls or flooring and never understood

In the Field

Newer codes have required access panels behind tub enclosures to service the slip connections on the tub waste and overflow assembly. This panel in older structures traveled from the floor up to and including the valve for convenience's sake.

its cause. Water-soluble caulking, which comes in several colors, should be applied neatly around escutcheons and spouts. This is also a case where good salesmanship can be used to show the customer how to caulk and maintain the wall area over a period of time. Do not be afraid to educate the customer on how the tub and wall area expand and contract, causing small leakage areas. This can quickly be cared for in the caulking process and will improve the appearance of the overall work.

RESIDENTIAL SINK FAUCETS

Sink faucets are available in styles similar to lavatory faucets with a minimum of one main additional item that requires maintenance—namely, the swing spout. Swing spouts have sliding surfaces with a packing to prevent leakage. When the packing (usually O-rings) or the sliding surfaces become worn, repairs are necessary. In this case, repacking is seldom successful and faucet replacement is the best solution.

Many sink faucets come with a spray hose accessory, which is pressurized whenever the water is flowing. When a trigger faucet is released in the spray head, a diverter valve in the base of the swing spout stops the flow to the spout. Typical failures include failure of the diverter and deterioration of the hose. The diverter also acts as the backflow preventer for the hose assembly. When a back siphon condition occurs, the diverted assembly opens to allow air from the spout to break the vacuum. This diverter assembly is accessible when the spout is removed.

Pull-out spout models have become very popular in recent years for residential kitchen sink faucets. In these, the valve area assembly may be more complicated in its construction. However, manufacturers' literature and repair instructions have been radically improved and are readily available. Backflow prevention has become very sophisticated and is commonly built into the unit in much the same way as it is with diverter assemblies. Standards that govern the type of preventers may recognize several different methods such as nonmetallic duck bill checks, ball checks, or miniature backflow preventers. Caution should be exercised for these faucets as several manufacturers use checks and restrictors on the hot and cold supply lines.

RESIDENTIAL LAUNDRY TRAY

Laundry tray faucets are generally repaired by methods similar to those used on sink faucets (see Figure 2-4). However, the majority of older style, top-feed laundry tub faucets do not have replaceable seats, and on past service calls technicians may have

Figure 2-4
Cement laundry tubs were very popular prior to the use of durable, fiberglass laundry tubs. The top feed faucets for cement laundry tubs were unique and generally did not have replaceable faucet seats.

tried to get by with cone washer replacements. Cone washers are at best a very temporary repair effort. The top-feed laundry tub faucets were bolted with a compression screw to a laundry tub block or directly to the tub for the old cement tubs.

In the Field

You may consider selling the customer on the idea of a style update in the form of a replacement tub in order to update her faucet. Remember, you may be the individual who carries the extremely heavy tub out and ensures its proper disposal.

RESIDENTIAL WATER CLOSETS

Ballcock or Fill Valve

The most common service call related to malfunctioning residential water closets is attributed to ballcocks. It has been customary to replace a ballcock (also called a fill valve) that is not working properly. Some designs are arranged so that the complete working assembly is a cartridge component and the entire product does not have to be changed. Many of the more common quality manufacturers utilized a ballcock with a removable cap that is taken off with a $\frac{1}{2}''$ socket wrench. A replaceable seat and washer are under the cap assembly.

For code compliance, you should consistently install a model that has an antisiphon design for all replacements. Be certain that adjustable ballcocks are installed with the critical level marking 1 inch above the top of the overflow tube. Ballcocks in some older one-piece models and newer 1.6 gallon per flush water closets may be of a special design, which is necessary for optimum fixture performance or complete backflow protection. Be certain to select the proper replacement unit as recommended by the fixture manufacturer.

Diaphragm valves, which respond to the depth of the water in the flush tank, may not operate properly with water that contains significant solid debris materials. Float arm valves will perform better in such circumstances.

When you finish, be sure that the tank and closet bowl refills to the proper level after each flush. Also take the time to explain to the customer the extreme cost the malfunctioning unit may have caused, such as utility water expense, and that often water usage amounts generate projected sewer treatment costs on the customer's water bill.

In the Field

A commonsense repair of the seat and washer will be deeply appreciated by the customer when an explanation is provided. However, the greatest benefit is that you have reduced the number of potential callbacks by not becoming involved in changing the water closet supply tube with its different connections. Some of the old lines were unconventionally sized lines using cone washer compression seals.

Flush Valve

Located at the outlet of the gravity flush tank, this device controls the discharge of the tank contents into the bowl. At one time, the only flush valve design used a brass seat, into which was connected an overflow standpipe, and a rubber conical ball to close the opening. The ball is raised by lift wires or a light chain linked to the flush trip lever. More recent tanks contain a flush valve seat formed in the pottery. These seats very seldom become damaged. Repairing the flush valve assembly usually does not require changing the brass seat assembly, but occasionally the seating surface will become worn and rough. In those few cases, replacement will be needed. In other cases, an economical repair can be made by placing an adapter assembly held in place with epoxy over the original flush valve. This procedure eliminates the need to disassemble the tank, remove the flush valve seat, and reassemble it. Reassembly considerations are important as manufacturers have used differently sized gasket assemblies for the tank to bowl connection. *They are not universal seals.*

In the Field

A service vehicle that has been properly stocked with parts based on your area's most common fixtures is a tremendous asset. Repairs of flush valves with tip-away fill tubes and built-in gaskets can be accomplished only with like parts.

Flushometer Tank

Flushometer tank water closets use a small tank with very low-flow water consumption. It operates upon activation by using the system's water pressure captured in its own enclosed pressure tank, creating a type of turbocharged flush.

When improper functioning of a flushometer tank occurs, consult the manufacturer's recommendations for servicing. Use only replacement components that are recommended by the flushometer tank manufacturer.

In the Field

Winterizing flushometer tanks will involve additional time and attention to detail. Be prepared by obtaining detailed information from the manufacturer's website.

COMMERCIAL SERVICE AND REPAIR INSIGHTS

Commercial fixture and fitting repairs use a vast number of the same principles as residential service work. However, the product quality is usually heavier and more expensive and has a longer life expectancy than residential equipment. Another consideration is that commercial fixtures may be very specialized, such as medical facility equipment.

One of the greatest considerations in commercial service work is that mistakes and errors in judgment have far greater consequences than in residential work. This can be clarified when you consider that a poor repair affects the environment of the business you are working in. Leaks resulting from a poor repair may appear after you leave the site, resulting in far greater damage to costly equipment and products nearby. Further, any operation that has to be shut down may result in huge production losses by the customer. You might consider losses based on cost, and your boss will have in mind expensive litigation.

COMMERCIAL LAVATORIES

Commercial lavatory faucets have very similar functioning features to residential units. However, with increased emphasis on water conservation and new technologies, self-closing faucets have become far more popular. Self-closing, spring-actuated faucets have been around for some time. Restroom users have been known to move from one lavatory to another to gain increased time of operation during their hand washing when users perceive the operation is too brief. For that reason, your calibration and time settings are extremely important in the repair process. Faucets with electrical or battery-operated sensors have gained a great deal of popularity. These faucets are discussed later and can be repaired by following the manufacturer's directions using approved parts.

The codes are involved in hot water supply systems for consumer protection in several areas. The issue of lavatory supply for hand washing may also be addressed locally by requirements for physically disabled persons. Plumbing codes and other requirements define tempered water with temperatures lower than hot water. Those temperatures range from 110°F to 120°F. The codes may have ensured that the tempered water is controlled by devices conforming to ASSE 1016 or, more recently, by ASSE 1070 standards (see Figure 2-5). Do not remove these devices from the lavatory supply systems.

Commercial lavatories are supported by either a vanity top or are wall mounted by concealed hanger arms called carriers. Caution should be exercised if the lavatory is to be replaced. Anchoring, leveling, and security adjustments are accessible through punching holes on the underside of the lavatory.

COMMERCIAL CULINARY SINKS

Culinary sink faucets found in food preparation areas tend to be back mounted and serve more than one sink compartment. Faucets in these areas are specialized for different processing functions, such as a stainless steel clean-up countertop with a faucet spray tower connected to a flexible hose and positioned over a commercial garbage disposal that uses a water-supplied connection. Whatever their specific purposes, the faucets will have similar control mechanisms. Should a faucet replacement be necessary, be sure to obtain one with the proper spout length.

Sinks in food service areas are required to have grease traps or larger grease interceptors. It is extremely important that the facilities conduct cleaning operations on a timely basis to avoid plugged waste lines. When cleaning or servicing these devices, the cover bolts must be removed and the grease scooped out for proper dis-

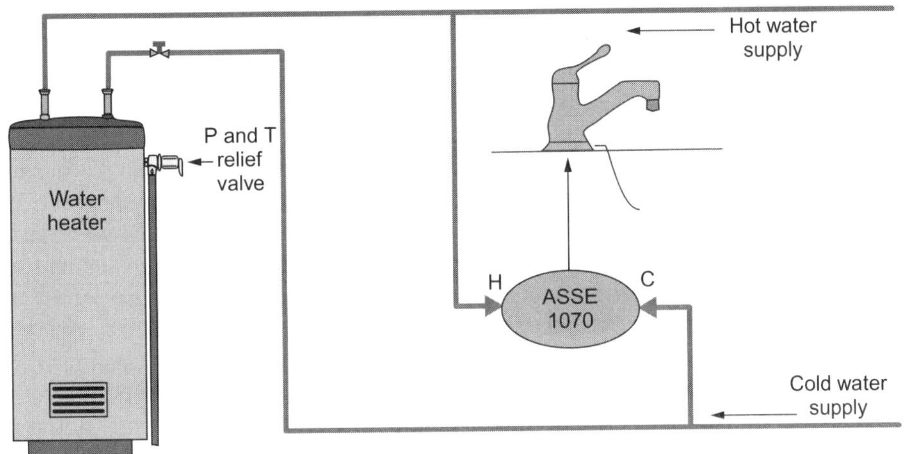

Figure 2–5
Several codes now recognize ASSE 1070 water-temperature-limiting devices on lavatories for burn protection. The devices do not provide thermo-shock protection as required by ASSE 1016 devices.

posal. Check the cover seal before reassembling the grease trap, and repair it as necessary to achieve an air-tight closure. Clean up any spilled grease for safety and sanitary considerations. Newer grease traps may be grease retention devices with sophisticated components. Make sure you have all the necessary instructional material and understand the system's operation prior to working on it.

Another device for consideration in kitchen areas is the commercial soap dispenser. They are very popular and aid customers in reducing costs and reducing employees' exposure to cleansing chemicals. It is very common for chemical companies to own, rent, and service these devices. Plumbing service technicians become involved because they are the water supply experts who may be licensed to install and service backflow preventers.

> **In the Field**
>
> Most grease traps have heavy air vent control fittings on the grease trap inlet side. These fittings mix air and the sink waste together to improve grease retention in the trap. They also have an orifice that controls the flow rate to improve retention. Make sure to reinstall the fitting if you replace the waste piping to the grease trap.

COMMERCIAL WATER CLOSETS

Commercial water closets, like their residential counterparts, are manufactured with several different bowl designs such as washdown, siphon jet, and others. An additional model in commercial applications is the blowout, which incorporates a greater flushing volume. A vast majority of closet bowls are floor mounted; however, it is very common to find wall-mounted bowls utilizing supporting systems in the back called chair carriers. Commercial settings for floor-mounted and wall-mounted bowls obtain their water discharge from flushometer valves. A good technician understands that setting a water closet involves much more than using a seal or gasket and bolting in the bowl. Care must be taken with floor-mounted bowls to not rock the bowl during the tightening-down procedure; the technician must shim the final position and then caulk the bowl to retain its position. Greater care must be taken in mounting bowls to carrier hangers. One of the most important things to understand in mounting a four-bolt wall-hung water closet is that three nuts are to be snugged up with a wrench, but the fourth nut must be only hand tight. Always exercise great caution when bolting closet bowls to the floor or to the wall and when bolting the tank to the bowl. As silly as it may sound, it is like playing catch with a raw egg or bolting two china dishes together.

> **In the Field**
>
> Plumbing codes will have conformance standards for soap dispensers, and you should become familiar with the requirements to ensure safe potable water supplies, such as which backflow preventer should be selected.

Most of the minor service calls for water closets, other than those involving the flushing supply, are a matter of reduced flushing capabilities. The $\frac{5}{8}''$ jet hole or rim holes installed to aid in rinsing the bowl may become obstructed. These openings can be cleaned out manually by holding a nail with pliers, moving from one hole to the next, and pushing the nail in while wiggling it around. You should have a visual indication of clearing the obstruction by seeing sediment fall into the bowl from beneath the bowl rim. For larger buildups, a chemical treatment of a cleaner designed to remove rust, calcium, and lime deposits may be necessary. Another common problem is an obstruction in the trap, which is not visible. Hold a mirror under the water with your pliers and illuminate with a flashlight to identify the obstruction. A pencil or comb in the waterway will pass liquid but hold waste discharges. Plunging is a temporary fix, and the use of a bowl auger for removing the obstruction can be very difficult. This problem occurs in school settings often to the point that custodians become removal experts.

In the Field

Your company's reputation is always at stake based upon your worst critic, your company's competitor.

Technicians should be aware of several code considerations when dealing with fixture replacements for commercial water closet service work. The applicable code may require a permit that will mandate using current code requirements. A water closet with 1.6 gallons per flush or less will have to be installed. Several other older code requirements such as elongated closet bowls and open front seats are also mandated. Licensed plumbers around the country tend to judge the quality and professionalism of a facility or business they may be visiting based upon such basic fixture placements like regular closet bowls in restrooms for public and employee use.

In the Field

Newer editions of the code now require privacy partitions between urinals to allow occupants greater privacy. One of the justifications given in the code update process was that individuals with "bashful kidneys" desired greater privacy and would use a water closet with its privacy partitions instead of provided urinals. Because commercial restrooms have fewer water closets than urinals, delay may result in the number of occupants moving through the restroom at key times such as intermissions.

COMMERCIAL FLUSHOMETERS

Flushometer valves have been manufactured for many years and have undergone a few improvements, such as the addition of water-saving features and automatic flush sensors. Most flushometer valves are provided with a stop to make servicing easier immediately prior to the flush valve. When servicing the valve, close the stop, release the pressure in the flushometer valve, and then disassemble. The orifice in the flushometer diaphragm or piston must be clear for the valve to operate properly. A vacuum breaker is provided in the vertical position at the outlet of the flushometer. A rubber diaphragm much like a duck bill check is the working part of the vacuum breaker. If the diaphragm is deteriorated, spillage may occur. Electronic flushometer valves should be repaired in accordance with the manufacturer's instructions.

COMMERCIAL URINALS

The performance operations of urinals, like most plumbing fixtures, are governed by international standards. One of the newer standards that might not have been updated in your area's code includes waterless urinals (see Figure 2-6).

In the Field

Caution should be used when replacing conventional urinals with some waterless urinals because the height of the fixture lip may be altered to a higher level. This occurs because of the lack of a trap in some fixtures, which changes the rough-in dimension of the urinal.

These fixtures require the same maintenance methods as water closets and are available in several different styles and models. Waterless urinals do not have a flushometer valve or any means for rinsing the interior portion of the urinal. Cleaning instructions for the facility are a consideration addressed in documents provided by the manufacturer. The early waterless urinals operated with a cartridge canister that contained a chemical liquid seal solution which allowed urine to pass into the drain while retaining the fluid above to maintain a trap seal protection. Part of the maintenance, then, is the projection of occupant usage to determine the schedule for washing and replenishment of the chemical solution. Portions of our industry have opposed the fixtures due to their fear that increased odors may occur from fluids not being "washed down" and that the lack of urine dilution by water may radically increase the amount of drain line buildup.

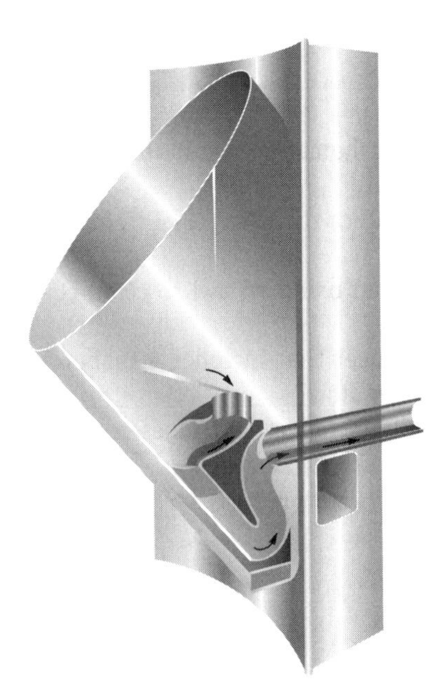

Figure 2–6
Waterless urinals have the advantage of addressing water conservation concerns and have been accepted by several code groups. The code-listed performance standards vary from china to plastic-based materials.

REVIEW QUESTIONS

1. What is the cause of a flushometer vacuum breaker leaking around the air ports?
2. What are urinals that protect their trap seal by using chemical liquid seals called?
3. What is accomplished when using an awl to clean clogged jet holes in a water closet?
4. What is the name of a faucet that includes a rubber washer that is pressed into the seat to stop the water flow?
5. Name one of the benefits of faucet cartridges containing the seat, stem, and washer in a single, replaceable assembly.

CHAPTER 3

Industrial and Institutional Service and Repair

LEARNING OBJECTIVES

The student will:

- Explain the use-group definitions and settings for industrial and institutional structures.
- Identify the various safety concerns involved in industrial and institutional plumbing fixture repair settings.

INDUSTRIAL SERVICE AND REPAIR INSIGHT

This chapter discusses the service and repair of industrial and institutional fixtures. In order to understand and apply proper repair and service of industrial fixtures, a clear understanding by way of a definition is necessary. The term *industrial applications* commonly refers to large-scale manufacturing business activity, such as a machinery production plant.

Despite their sturdy design and construction, industrial fixtures and fittings are subjected to heavy usage and are not immune to requiring service and repair. These heavy-duty fixtures should generally be repaired rather than replaced because of costs. In fact, many of the fixtures, fittings, appliances, and appurtenances that are installed in industrial buildings will require periodic maintenance to ensure their proper operation.

A thorough knowledge of industrial fixtures based on a strong installation background is necessary to understand the major concepts of industrial repairs. *Plumbing 301* addressed the fixture installations and gave detailed insights. The customer's consideration of whether to schedule service work is often based on the customer's continued operation, which is necessary to support his production operation, and translates to the industrial customer as a financial necessity.

Service work will now be considered for several fixtures in an industrial setting. The same fixtures for different settings, such as commercial installations, have been covered in earlier chapters and will be considered in this chapter with their slight variations due to their different occupancy uses.

In the Field

The plumbing code occupancy classifications for factories and industrial occupancies are F1 and F2. Buildings in these categories are designed for the assembling and processing of materials or products. The code, through use group designations and occupancy numbers, informs the user of what type of and how many fixtures are required in new and remodel construction.

In the Field

Communication with the facility manager is of the utmost importance because he will make the decision as to the degree of your service involvement. For example, the facility manager may opt for minimal repairs now with other work scheduled later because of production schedules, a consideration of after-hour repairs, or pending plant shutdowns.

INDUSTRIAL DRINKING FOUNTAINS

Drinking fountains in industrial settings are often heavy-duty, free-standing, self-contained, mechanically refrigerated, drinking-water coolers that conform to Air-Conditioning, Heating, and Refrigeration Institute (AHRI) standard 1010. These units are commonly in high-traffic areas, such as beside forklift material mover lanes. Plugged drains are common due to the fountains' convenient location for employees, who dump liquids and even coffee grounds into them. When servicing these drinking fountains, be sure that the outlet orifice is clear and undamaged. The water control mechanism can be repaired only with proper replacement parts.

A good service technician should always observe the area around him. For example, more experienced service technicians will tell the story of finding water being supplied to drinking fountains from fire suppression system lines and drains being connected to the base of a roof drain downspout.

In the Field

Contact the public sewer authority to check the sewer through its manholes, which are on public property such as streets. There are great safety concerns due to manhole cover weight, traffic issues, and entry into a confined space. The sewer authority will have the experience and can give authorization for you to check this area.

INDUSTRIAL DRAINAGE VALVES

Backwater valves are installed in larger drainage system lines within buildings. The valves are designed to separate interior piping from exterior systems that are subject to back-pressure from plugged or surcharged sewers, as shown in Figure 3–1. Upper floor levels commonly are not subject to these same back-pressures because the sewer system can be relieved by discharging through a street manhole, which is at a lower level than the sewer obstruction.

Backwater valves are commonly installed in basements when lower level fixtures such as floor drains would allow waste or sewage to backflow into a habitable floor area. They are maintained by disassembling and inspecting the interior for solids or nonworking flappers that obstruct the valve closure. Great care should be taken in servicing; before accessing and disassembling the valve, ensure that the system is not under a back-pressure condition. Common sense indicates that extensive flooding would leave contaminants in the habitable space.

In the Field

Never run a drain-cleaning auger through a backwater valve as the cable will break or become irretrievable. A flat sewer tape with a roller may also be irretrievable.

Figure 3–1
A building drain backwater valve prevents sewer obstructions from entering the structure. Courtesy of Josam.

INDUSTRIAL FLOOR DRAINS

Floor drains are commonly installed in structures but less often used in industrial applications. Equipment cooling will normally occur through a circulating system because this process saves the industrial manufacturer from purchasing water from a public utility when the water used in their applications may be recirculated instead. When water from manufacturing is discharged to floor drains, it will be run through a floor sink. A floor sink with its lip above the floor level will keep unwanted materials out of the drain. Technicians often scoff at the consideration of floor drain problems and have been heard to say, "No moving parts. How difficult can that be?"

Stoppages are usually caused by an accumulation of dirt and debris in the trap. Well-washed drains seldom become stopped. A hose stream will usually clear the drain; otherwise, conventional rodding or a special plunger can be used.

Many industrial models include strainer baskets, which only need to be lifted out to be cleaned. Other designs are cleaned in similar fashion to floor drains in commercial applications.

The trap seals on floor drain traps must be maintained. Earlier textbooks in this series explained the purpose of trap seals, which is to eliminate foul odors with noxious fumes that could cause health risks to building occupants. A trap seal primer is required by many codes to replenish floor drain traps and prevent trap seal loss from evaporation. The primers fall into one of two categories. First, the replenishing water comes from potable water distribution systems that use differential water supply valves or solenoid timer valves; or second, the replenishing water comes from waste discharges connected just above a trap inlet such as at a lavatory. These supply lines must be maintained for proper functioning of the primers. You may have to service the solenoids and flush out the smaller, commonly $\frac{1}{2}''$ supply lines to the traps.

In the Field

Always wear personal protective equipment (PPE) when working with drainage system cleaning devices. Be aware that the building user may have poured caustic or toxic solutions down the drain in order to clear the obstruction. Have proper ventilation circulating around the work area.

INDUSTRIAL INTERCEPTORS

Interceptors come in a multitude of sizes and shapes, and most importantly they serve many functions in industrial settings. Interceptors are devices designed to separate out undesirable material and retain it for removal or to reclaim products such as precious metals from the drainage system. The average plumber encounters grease interceptors primarily in industrial settings and smaller grease traps in food service areas. All these devices require regular removal and cleaning, and you will perform periodic inspection and service functions. Be sure that you observe all regulations for disposal of materials removed from any interceptor system. Keep complete records of all such operations, including date, type of material, and method of disposal. Always wear appropriate personal protective equipment (PPE)! After servicing any type of unit, be sure that you wash thoroughly. Remember, any interceptor can be a breeding place for bacteria. In addition, clean up any spills, especially oil or grease, because these materials on the floor can increase the chance of falls.

Several types of interceptors and insights for their servicing will be covered next.

Precious Metal Recovery

Precious metals recovered in drain interceptors may include gold or silver. The owner or manager of the facility or laboratory should witness or supervise any service performed to ensure that all valuables are turned over. These interceptors are commonly considered strainer-type units; however, they may be very sophisticated recovery devices with solutions and electronics.

Grease Interceptors and Grease Recovery Devices

Grease interceptors and recovery devices are commonly present in food service areas; however, they may also be found in industrial applications and, for the sake of study, are included here with other interceptors (See Figure 3–2). Always inspect and clean the flow restrictors on the inlet side of the device. A clean-out is usually installed on the lower side of the fitting. Carefully check the baffles, interior partitions, and gasket of the interceptor. Be sure that the cover seal is intact before reassembly and operation.

Grease Recovery

Inspect the blade, gears, and moving parts for wear or damage and replace as necessary. Check the recovery container for excessive water and adjust the timer according to manufacturer's installation instructions to provide the best oil and grease recovery.

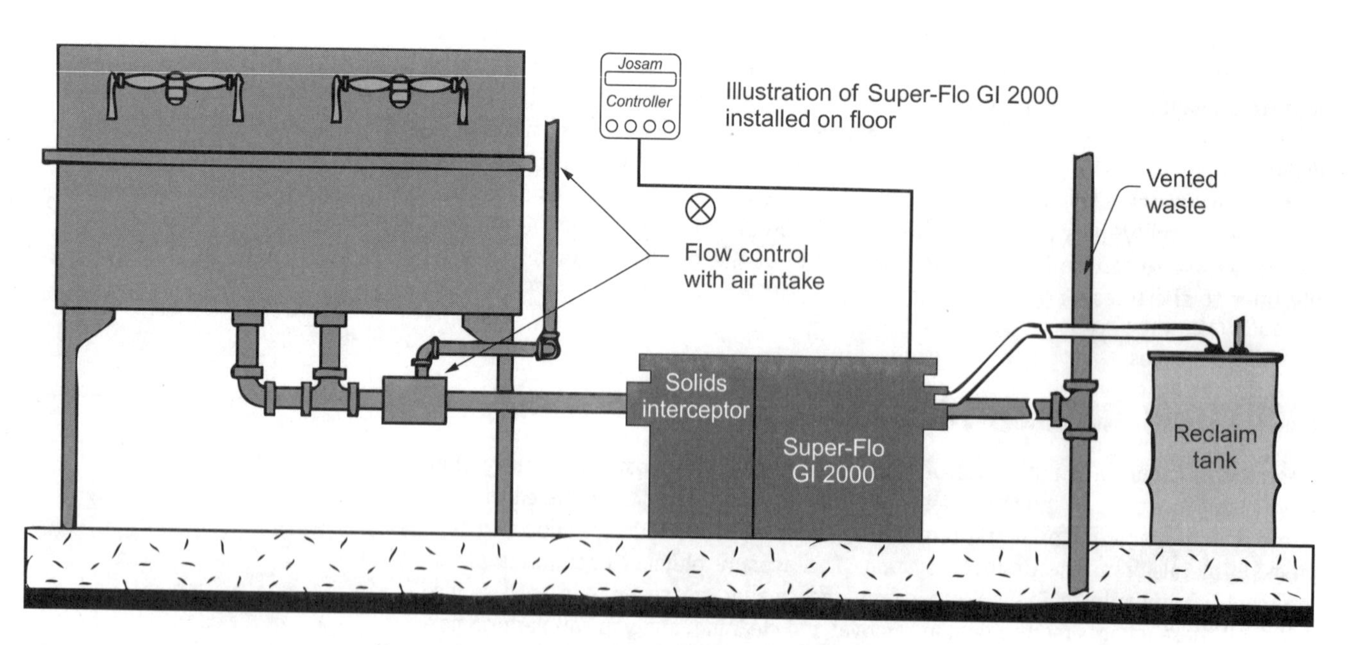

Figure 3–2
A grease recovery device used to accumulate the separated grease in container for ease of disposal.

Oil Interceptor

Oil separators are similar to grease interceptors but are much larger scale and are drained to an oil recovery tank. Thus, servicing consists of pumping out the recovery tank, usually done by a pump-and-haul contractor. The units are commonly vented and may contain flammable material, so exercise caution.

Sand Interceptor

Sand interceptors are cleaned using a pump device to remove the accumulated material. Verify that the inlet and outlet pipes are clear and that the floor inlet grates are intact.

In the Field

Familiarize yourself with the receiving portion of the system. The floor drains or sumps may have trap seal problems, which allow fumes and flammable vapors back into the work area. Production workers in these settings may have dulled senses to the odors and may not be protected from the dangers.

INDUSTRIAL MOP SINKS

Industrial mop sink service considerations are very similar to those in other use-group settings. However, this fixture is commonly abused and used as a receiver of other liquids such as those from the water heater pan, relief valve discharges, and condensate drains. Your service sales expertise may be employed here to inform the building owner of better piping solutions. This is also an area where the faucet supplies are misused, such as vacuum breaker removals of chemical solution take-offs. Plumbing codes, standards, and backflow prevention device manufacturers have excellent literature addressing these situations.

In the Field

Attention: This Information Could Save Your Life

Never enter a pit, interceptor, or manhole for service without following proper safety procedures. The atmosphere you breathe in a confined space may not support life. Do not work alone in these situations. Your employer has confined space entry procedures that you must follow. Plumbers are often in areas of concern such as container storage closets during medical gas installations when nitrogen brazing or purging procedures are being conducted.

INDUSTRIAL MULTIPLE-USE WASH SINKS

These fixtures are usually equipped with foot-pedal or sensor-activated faucets that control the water flow to a spray ring. The foot-pedal or knee activated faucets are spring-loaded to close when the pedals are released. As with most other special products, only approved parts are suitable for repair. The water temperature is controlled by preset mixing valves. Check the temperature of the spray at normal flow volume after repairing the pedal valve. Be sure the total flow is satisfactory for the application.

INDUSTRIAL WALL HYDRANTS

Nonfreeze (often called frost proof) wall hydrants are made with different body lengths. When replacing these, it is easier if you replace the faucet with the same length unit; otherwise, the piping to the faucet will have to be modified. Most nonfreeze faucets will break as a result of freezing if they are not installed to drain or if a hose is left connected to the faucet during the winter. In either case, the water cannot drain out usually because a vacuum is present in the hose, and the remaining water freezes (expansion), causing a rupture. The newest models will drain even if a hose is attached to the outlet. Some wall hydrants contain two control elements so that they can be connected to both hot and cold water supplies. This allows the user to select the water temperature desired when in use.

In the Field

Building code occupancy classifications for institutional occupancies are commonly I-1 through I-4. These buildings are designed for the care or supervised living environments, such as medical treatment and correctional purposes, where people are restricted. As stated earlier, the codes, through occupancy use-group descriptions and subcategories, clarify what types of and how many fixtures are required in new and remodel installations.

INSTITUTIONAL SERVICE AND REPAIR INSIGHTS

A definition of institutional use and occupancies is necessary in order to understand and perform proper repair and service of fixtures used in these settings. Institutional applications include structures that have a public setting in which people are cared for medically or where people are detained for correctional purposes, for example, a hospital or prison.

Our study of institutional equipment includes those fixtures, fittings, and appurtenances that are found in schools, hospitals, and penal institutions. These institutional fixtures are similar to industrial fixtures in their sturdy design and construction and are subject to added requirements such as those related to higher sanitary and security concerns. Your service role in ensuring proper institutional operations is just as critical as the pressure from owners dealing with manufacturing financial pressures in industrial settings. The critical aspect comes from a basis of limited public funds used for operations of institutional settings.

Institutional fixtures and fittings are similar to the types discussed in earlier chapters. Variations that may be encountered in institutional applications include the following.

In the Field

Licensed professional plumbers generally feel compelled to explain their profession to the general public as the uninformed public does not understand the intricacies of the profession. However, the medical field often deals with similar unglamorous responsibilities and has the respect of the general populace.

INSTITUTIONAL BEDPAN WASHERS

Bedpan washers are special fixtures very similar in use to water closets, and in fact water closets are often fitted with bedpan flushometer valves or spray hose cleaning accessories. The flushometer device controls water flow to a tailpiece assembly. Bedpan washers must be equipped with vacuum breakers or other approved backflow preventers. These must be repaired using approved parts and carefully inspected after repairs are completed. If the assembly leaks, O-ring seals must be replaced. The spray hose model is serviced with methods described for compression faucets. Also, spray hose assemblies may have foot-controlled valves in order to reduce the attendant's exposure to bodily fluids.

Flushometer valves with swing-down bedpan washer arms are common to conventional water closets in patient rooms. The combination toilet and pan cleaning feature reduces construction costs by having one fixture rather than two and improves medical staff services by reducing time involved and distance to a central fixture separate from patient rooms.

In the Field

Your training as a plumber has centered on the plumbing code in order to ensure an understanding of proper installation practices and aid in licensing. Unfamiliar requirements from "the authority having jurisdiction," such as a health department, are very common. Unfortunately, you may not have been exposed to the requirements. They may vary from one geographical area to another. Ask questions of your supervisor, obtain the published requirements, and keep notes for later use to aid in your service work.

INSTITUTIONAL FAUCETS

Frequently, faucets will be equipped with knee- or foot-control arms. Often, these special arms are spring-loaded to close. It is also common for hospitals to prohibit the use of faucet aerators as some health departments feel that unwanted vapors or air can be intermixed with the flow of water.

Showers

Institutional settings often have column showers for multiple users at one time. These column showers are supplied with tempered water from a central valve, and the user may be able to control the volume of flow. The water flow control is a compression, spring-closed, or metering type faucet, and the temperature controller is either a pressure balance or thermostatic type. These devices must be repaired using only manufacturer-approved parts and methods and should be adjusted for maximum hot water settings per local codes, manufacturer's instructions, or jobsite specifications. Outlet temperatures are commonly limited to a maximum of 120°F (49°C).

In the Field

Local code requirements have changed over the years, and the devices are commonly described by plumbing standards, such as those by the American Society of Sanitary Engineering, referred to as the ASSE.

Institutional Interceptors

Interceptors in institutional settings are often limited to smaller, individual-fixture applications such as plaster traps. These interceptors are serviced like other interceptors discussed in earlier chapters. When servicing the devices, do not allow any of the intercepted material to fall into the body of the interceptor. Be sure the cover seal is in good order when reassembling the cover.

Penal Institutions

Penal institution fixtures require similar operation procedures as other fixtures and fittings, except for their ability to provide added security and the manner of obtaining access to the fixtures. It is very common the find these correctional fixtures manufactured from stainless steel, as shown in Figure 3–3.

Stainless steel, while providing greater strength, also eliminates shards (weapons) caused by broken china, which of course is a safety matter. Some penal institutions use different plumbing appliances and appurtenances designed to reduce inmate misuse. These devices include antiflood valve controllers and hoops installed in the water closet discharge line to capture linens flushed into the system. Refer to the manufacturer's installation instructions for service on flood control products. These items are serviced in chases behind the cells. When in the chase, check all details to be certain that there are no other problems.

Several penal and similar high-security areas are fitted with vacuum waste systems rather than conventional drainage and venting systems. Such systems are preferred by prison security agencies, as they offer the following advantages:

- Prevent the transfer of contraband between cells
- Allow easy recovery of contraband placed in the system
- Can be maintained easily
- Use less water

Institutional Schools

Schools and other educational facilities may have special water temperature considerations, but the actual repair of fixtures or fittings will follow the descriptions in earlier chapters. Local plumbing codes often limit the maximum temperature for all fixtures in a lower elementary school setting. The devices to control those temperatures vary and are commonly dependent upon ASSE standards. Special problems with school fixtures involve heavy, frequent use and vandalism. Vandalism-resisting methods must be applied as conditions allow. Thus, conceal as much pipe and equipment as

Figure 3–3
A combination stainless steel water closet and lavatory for penal institutions designed for durability and to prevent broken vitreous china pieces from becoming weapons.

possible. Use heavyweight products, and clamp down or anchor accessories with Allen wrenches or other special screws. If you are servicing this equipment, be sure you release such protections before dismantling the equipment.

FINAL PRECAUTIONS

After repairs are completed, test the operation of the device. Verify the performance of any thermostatic control with a thermometer. Be sure that the temperature differential is sufficiently close so that thermal shock will not be imposed on the users. Consider that incoming water temperatures may vary throughout the year and affect your final setting. For example, the difference between winter and summer temperatures in reservoirs providing water that is later treated may affect your final adjusted outlets.

Be especially careful of your exposure to waste products in hospitals, clinics, medical centers, and other institutions. Use full personal protective equipment, wash thoroughly and carefully, and consider sterilizing your tools. Hospital personnel may be able to offer advice on how to accomplish these precautions.

REVIEW QUESTIONS

1. Name the two categories of trap seal primers.
2. Vandal-resistant fixtures in correctional facilities are essential for what reasons?
3. Why are faucets with foot-controlled valves advantageous in institutional settings?
4. What is a major concern to address when opening an interceptor pit cover to service a broken baffle?
5. Is good personal cleanliness practiced during institutional service work primarily to aid your employer's service contract?

CHAPTER 4

Water Piping Service and Repair

LEARNING OBJECTIVES

The student will:

- Identify the safety concerns and proper procedures for underground water service repair excavations.
- Implement procedures to locate both above- and below-ground leaks to reduce repair costs.
- Describe a service format that ensures that the appropriate repair methods are employed in above- and below-ground repairs.
- Recognize hazards related to fire safety and health-compromising materials.

IN-GROUND SERVICE AND REPAIR OF THE WATER SERVICE

Chapter 4 contains service and repair information for above- and below-ground water lines serving structures. The first consideration is the outside installation, the water service, which is commonly installed first. The installation of the material outside a structure is in a much more hostile environment than inside due to its underground location, which includes rocks and soil-trenching conditions. A water service by definition is the line from the public source, commonly at the property line or well, to the structure. The water meter for public water supplies is used for tracking and charging for water usage. The water meter may be located at the edge of the public right-of-way or in the structure. The accuracy of this plumbing appurtenance is important to building owners as they may be charged a public sewer rate based on the water usage.

Whenever water piping is located below ground, there is the chance that a leak may develop along its surface or in a pipe joint. These leaks may be detected by observing rates of water usage or by wet or moist soil in the leakage area. It can happen, however, that the water leaking from such a break will be absorbed by the ground and not show up on the surface. Sometimes these leaks can be heard. Both visual and auditory observations may be difficult in northern regions because the services are installed at much lower elevations to avoid freezing conditions, which can be "pushed down," so to speak, by heavy vehicle traffic. Large problems such as broken mains, which occur in northern climates due to freezing and thawing conditions causing soil movement, are very obvious and always result in shutdowns.

The common service procedure for smaller water service problems is to excavate in some small measure and analyze the degree of failure. This effort will enable the building owner to make an informed decision related to repairs or replacement of the line based upon cost and longevity. Should the decision be made to replace the line, you should consider making the replacement with a new route. The present location was more than likely dictated by the sewer excavation installation location, which is far less flexible. The following discussions present the concepts needed to repair buried piping and make replacements.

1. Be aware of other utility lines. Some areas are heavily congested with buried lines and piping. Contact local authorities or utility-location services to mark the utilities. A national 811 phone number is available, and operators at this number will connect you to your local or regional utility-locator service. A national utilty-locator website is also available at http://www.call811.com. In some cases, it is a violation of the law to excavate without first locating other utilities, and if it is not a violation, there is still a significant likelihood that any problems encountered will result in extended litigation. Many localities have a single contact number that coordinates and notifies all utilities, which then physically mark the utility locations. These utility-marking services have various names around the country.
2. Establish a written plan in order to facilitate other team members' involvement. These team members would include the excavator, building employees, and the owner. The need to keep the occupants informed that you are working on the water line and advising them that the water will be turned off is very important.
3. Locate and deal with any equipment that must have water available continuously (medical needs, cooling, washing processes). It may be necessary to find an alternative source of supply before turning off the suspect water line.
4. Obtain all necessary safety equipment such as trenching boxes and shoring components. As always, safety should be your first concern. Wet soil will slide more readily than dry soil. Do not take risks to finish the job quickly.
5. After opening up the break or exposing the joint failure, dig down below the pipe so that water may accumulate below it, allowing the line to stay dry. If necessary, pump out this low collection area to keep the line dry. Remember to follow the appropriate safety rules when using electrified pumps in ditches.

In the Field

Exercise common sense when dealing with water meters and associated valves such as the curb cock, a valve located at the end of the public service. Plumbers are familiar with the codes and laws addressing what can be done and who can service plumbing systems. However, public utilities may be very conservative regarding who can perform which functions and especially what material can be used. The public utility has great responsibilities, such as operating cross-connection control programs as a result of the Safe Drinking Water Act. Contact the public utility prior to major water service repairs to familiarize yourself with its local ordinances and rules.

In the Field

The plumbing codes list all the accepted products for water services, and often state or local legislation establishes that products cannot be restricted based upon cost or discriminated against. It is very common for the public utility to have a conservative acceptance of the highest grade metallic products. These positions are often the result of experience based upon success of past materials or joint failures. Your customer may be pressuring you because of cost considerations based upon the service distance, which translates to costs. Contact the public utility early in the discussion stages for a resolution on the choice of materials.

In the Field

Some plumbing codes require that hospitals have two separate water services when constructed.

6. Should brazing procedures be used on a copper system and the supply or building service valves are leaking, a dry system must be obtained. Freeze equipment and kits are available to create a plug in the pipe by freezing a short section.
7. The actual repair should be made by a method appropriate to the water service pipe material and the exact nature of the break. If the broken section is the result of deterioration, remove the bad material, support the piping through the break, and install proper material for a permanent repair. The connections and material must conform with the manufacturer's installation recommendations. After the leak is repaired, pressure test the line to be sure that additional leaks are not present.
8. Return the building operations to normal following water sample testing acceptance. Three major areas of concern are explained in greater detail here for consideration and to encourage you to identify and remedy other pending problems. First, relieve the building system pressures at sufficient intervals in the structure. The compressed air in a refilled system is dangerous and has been known to shatter china fixtures when a flushometer valve is used. Second, the previously discussed problem is the beginning of the classic cross-connection covered in many cross-connection horror stories. Your vigilance and actions to address this problem are of the utmost importance. Third, disturbing a water main may free sediment, impurities, and other harmful products into the system. Recently, a case of Legionnaires' disease was reported and attributed to construction being conducted on a large line serving a hospital. Using technical support in testing water samples is a wise choice.

In the Field

For prevention and record keeping purposes, a report should be created to identify the exact location and expected cause of failure. The cause of failure could be attributed to material defects, improperly applied materials, aggressive water or soil, faulty installation, improper backfill methods, or freezing, among others. Similar conditions may result later and cause another failure, and your report will allow more effective completion of the repair.

TRENCHING SAFETY

Those who have been buried in plumbing excavations and survived can attest to the necessity of trenching safety. Your employer will have operation procedures, and you should have had safety training for excavations. Always apply those safety rules and regulations when involved in excavations. Trenching safety will be specific to many areas of this book such as water and waste installations, service, and repair. The equipment operator is understandably one of the most important people on site due to his or her ability to view the trench, side walls, and entire site. Safety regulations require an OSHA-certified "competent person" on site to address safety procedures. The hazards most often associated with trench work are cave-ins, materials (tools, pipe, or debris) falling into the trench, machinery falling into the trench, debris falling from digging machinery, power line hazards, and unprotected jobsites that jeopardize the safety of all individuals including passersby.

In the Field

There is considerable danger at all times when working in a trench. Three simple facts must be realized about safety when doing trench work.

1. No job is so important that safety can be ignored.
2. A person can live only minutes without breathing. If breathing is hindered for only a few minutes, the person may suffer brain damage.
3. A person does not have to be totally buried to suffocate. One must only be buried to chest level to suffocate; if the chest and lungs cannot expand, breathing will stop.

Cave-Ins

The degree of cave-in potential relates to the soil conditions, ditch depth, maximum allowable ditch slope, weather conditions, and shoring design. In our study overview, we will consider a ditch, a trench, and an excavation as being similar or the same. Soil is very heavy; 1 cubic foot may weigh 100 pounds. One cubic yard could weigh as much as 2,700 pounds, equal to the weight of a small pickup truck and the size of a small drafting table. These comparisons illustrate the seriousness of cave-ins. Prior to excavations, the soil supports itself by spreading its weight-bearing capabilities against itself. When excavations occur, the weight becomes force and tends to spread outward due to compression. The trench, which causes a lack of soil stability, becomes subject to failure, a cave-in. A potential cave-in is often referred to as soil distress and at times provides a visible warning condition. These conditions are named tension crack, sliding, toppling, bulging, and boiling (ground water coming up at the trench bottom). Figures 4–1 through 4–4 illustrate these conditions.

The type of soil surrounding the trench is of great importance because it affects the stability of the trench and its ability to withstand a cave-in. For example, solid rock could have nearly vertical sides and not be subject to cave-ins. Sand, however, is very unstable and would have to be sloped back at an angle to prevent the soil from sliding or caving in. Individuals who have experience with moist beach sand can attest that the moisture content also has a great impact on the trench's ability to withstand movement. This explains why sewer excavations move ahead at the pace of the pipe installation rather than outpacing the installation. Keep in mind that normal conditions often present several layers of different soil, such as gravel, then sand, and finally clay, each having different stability values.

Considering the soil stability value before excavating can prevent cave-ins, as referenced earlier, by adhering to "maximum allowable slope" values. The *maximum allowable slope* is a more current term for the angle of repose. By definition, the maximum allowable slope is the greatest angle above the horizontal plane at which a material will lie without sliding. These angles were determined through engineering principals and testing. Remember, this varies depending on the type of soil. One method to avoid cave-ins is to slope trench walls equal to or less than the maximum allowable slope. An accurate method of determining the unconfined

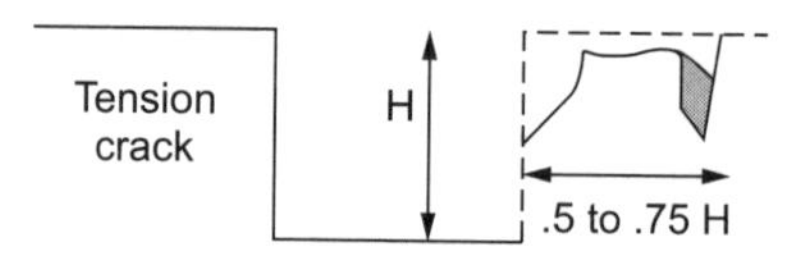

Figure 4–1
A drawing illustrating a tension-crack excavation danger.

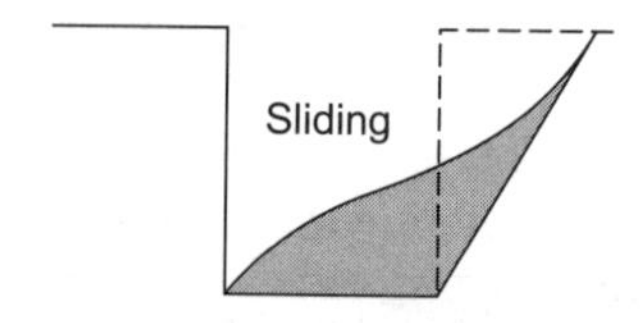

Figure 4–2
A drawing illustrating a sliding excavation danger.

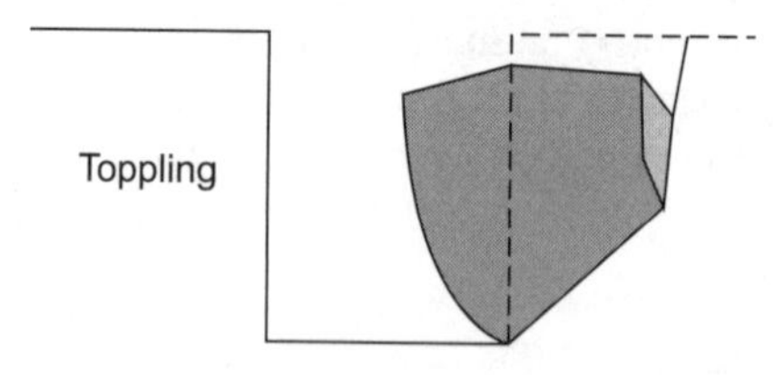

Figure 4–3
A drawing illustrating a toppling excavation danger.

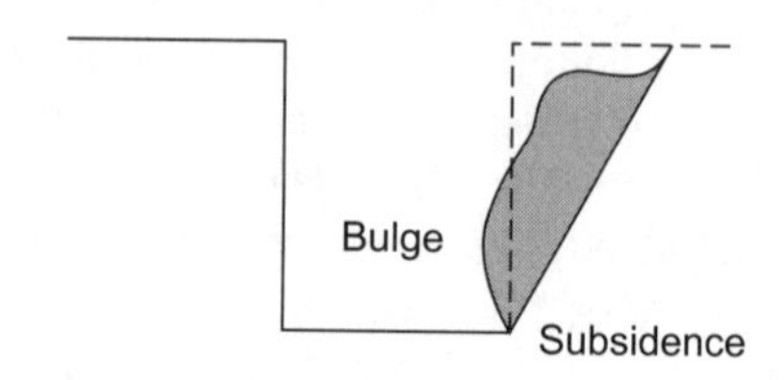

Figure 4–4
A drawing illustrating a bulging excavation danger.

strength of soil is with a handheld penetrometer. You would excavate to a level of 5 feet in each type of soil and push a penetrometer in the side of each bank, noting the reading of tons of pressure per square foot (TSF) on the penetrometer. Apply your reading to Figure 4–5 to find the correct angle.

For example, a reading of 1.5 TSF in clay would require a slope angle of 56°. Safety requirements generally specify that all trenches 5 feet or deeper must be cut back to the proper angle, which is the maximum allowable slope, calculated using the tangent of a triangle. You can check your angle by using a mathematical tangent table and dividing the trench depth by the distance of the slope back from the trench bottom.

Due to extended depths, you may not have the room to slope each side out the necessary distance. For example, a 7-foot-deep trench with 2 feet for the pipe and working room (trench bottom) in stiff clay (60° angle) needs a 4-foot slope on each side, for a total of 10 feet of width. But considering a 10-foot trench in sand with a far greater angle (increased setback) indicates an impractical installation in built-up areas where space is restricted. Your solution is proper shoring or a trench box. There are several methods of shoring, and the industry seems to be very receptive to trench boxes, which reduce time needed for setting up as you move along the trench. The bottom of the trench box can be no higher than 2 feet above the trench bottom and must extend to 2 feet above the maximum allowable slope level. You can use boxes or shoring in addition to digging the maximum allowable slope above, and the boxes can be stacked. Employees cannot be in the box while it is being moved, and access to the trench is provided with a ladder. Ladders must extend at least 3 feet above the box and provided every 25 feet. The excavated soil or spoil material must be deposited at least 2 feet from the edge of the ditch. Shoring involves the temporary placement of supporting members that are used to hold the trench walls in place. Shoring retains the walls of earth that would otherwise collapse during trenching operations.

Excavated Debris Falling into the Trench

All materials and debris should be located at least 2 feet away from the trench edge. This policy will prevent debris from falling into the trench and will also deter overloading the trench wall with extra weight. Boxes or shoring materials, if

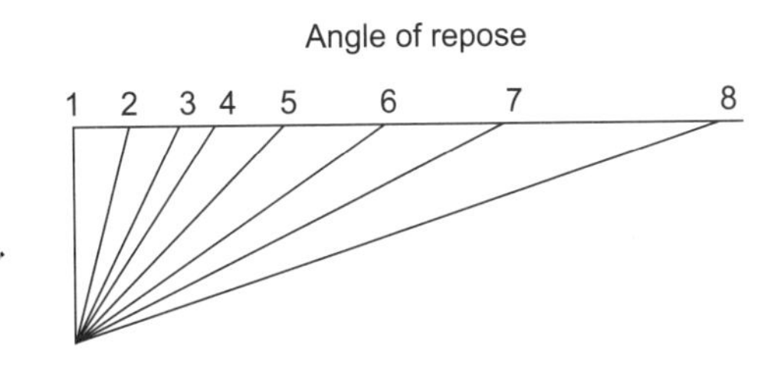

1. Solid rock formation (90°)
2. Fractured rock formation (75°) $\frac{1}{4}$:1
3. Stiff clay (63°) $\frac{1}{2}$:1; 2.5 TSF minimum
4. Firm clay (56°) $\frac{2}{3}$:1; 1.5 TSF minimum
5. Granular soil—dry (45°) 1:1; 1.0 TSF minimum
6. Granular soil—wet (34°) 1$\frac{1}{2}$:1; <1.0 TSF
7. Saturated granular soil (26°) 2:1
8. Running soil (18°) 3:1

Figure 4–5
A drawing illustrating a trenching angle of response to ensure excavation safety.

used, will have higher sides than the trench, which adds additional safety margins for debris.

Debris Falling from Working Machinery

Your position in the trench is of critical concern. Common sense dictates the need to be out of the way of any machinery being used, and the machine operator must be able to see you. Another good reason for keeping a good distance from machinery, especially if in the trench, is the possibility of debris falling into the trench from working machinery. Hard hats and goggles provide protection against some falling debris.

Materials and Tools Falling into the Ditch

Any pipe or tool used on the job should be placed at least 2 feet from the edge of the ditch for the same reason mentioned regarding debris. When pipe is transported into the trench, it should not be dropped or rolled into the trench, but rather it should be lifted and properly supported so that the spoil material will not topple into the excavation. Do not stack material by the edge of the ditch, and only transport pipe when it is required. Hard hats are required when working on a ditch.

Machinery Falling into the Excavation

Machinery falling into the excavation is not an infrequent occurrence. The primary concern for the plumber is to avoid injury by staying out of the reach of the equipment at all times. Remember that the trench sides are either shored or made at the maximum allowable slope, but the equipment is not able to have shoring near its base or be at an extended distance from the trench bottom due to the extended slope explained earlier. Another good reason for maintaining a safe distance is that machinery may vibrate significantly. This vibration is transmitted to the soil and may induce a cave-in.

In the Field

In the event that a power line is struck, do not touch the machine, the operator, or any wiring until it has been determined that the line is no longer energized. The wise technician will have all utility numbers immediately accessible in case of emergency.

Machinery Hitting Power Lines and Other Utilities

Underground utilities include all water, gas, telephone, electric, steam, storm, sprinkler, and other equipment piping. Overhead utilities include power, lighting, and telephone cables, as well as trees and similar hazards. Underground utilities must be located prior to beginning any excavation. Often, it is mandated that the utilities be located by hand digging. The systems, when found, must be identified. In some cases, large municipal high-pressure water lines have been misidentified and sewer installers have started the process of tapping, thereby jeopardizing the system.

Passersby Falling into an Open Trench

Open ditches must always be secured, as an open trench is a magnet to the curious. Observe all rules and regulations, providing warning lights and barricades to alert passersby that the excavation is there. It is the responsibility of the technician to properly secure the jobsite at all times, especially at night, to protect the public against accidents and potential falls.

ABOVE-GROUND SERVICE AND REPAIR OF THE WATER DISTRIBUTION SYSTEM

Leaks in above-ground water piping are usually easier to locate and repair than below-ground leaks. Above-ground leaks are made apparent by the appearance of water. Leaks in water piping are continuous, whereas leaks in waste piping are present only when water is discharged from a fixture. The water may appear on the wall or floor quite near the leak, but not always. Sometimes, the water will show up many floors below the actual location of the leak. Occasionally, a leak may make a sound that can be detected by careful listening, but do not rely on this method to the exclusion of other, more scientific methods.

Evaluate the original service problem prior to performing any work. The basic considerations are what caused the failure, what affect it will have on the building services, and how future problems can be avoided. Attempt to determine the cause of the leak. Table 4–1 may be used as a guide.

Leaks caused by excessive stress due to expansion or contraction can better be understood by reviewing Table 4–2. This valuable information may be used in new construction planning as well.

In the Field

The following procedure can be used to locate a leak in a concealed riser. Isolate the riser from the system and install a pressure gauge at the riser base. After filling the riser, the pressure will drop until the water level reaches the location of the leak. Observe the pressure drop caused by the leak and record the reading. This stabilized reading will indicate the distance above the gauge you should find the leak. For example, if the final gauge reading is 11.5 psi, the leak must be –11.5 psi divided by 0.434 psi/foot = 26.5 feet above gauge elevation. Recall from earlier training that water develops a pressure of 0.434 psi for every foot of elevation.

Repair Procedures

Repair or replacement decisions were previously discussed in Chapter 1. If the piping is generally eroded or corroded, a complete repiping installation should be considered. This possibility should be discussed with your supervisor or with the owner. An alternative material that is less likely to corrode should be considered.

In the case that a replacement is not feasible, the following discussion presents the concepts needed to repair above-ground piping. Once the leak area is determined, continue by shutting off the water as close to the leak as possible. Usually, the piping will have to be drained to make the repair. Opening a faucet at a high fixture will permit air to enter the pipe and allow the water to drain out through an open faucet below the leak. Be sure that all affected persons are notified that the water must be turned off. Again, confirm that there is no special area or need for water before cutting off the line. Check all the options you can before opening walls or ceilings to observe the leak.

If piping appears to be leaking in a floor-ceiling assembly, it is almost always less expensive to open the ceiling system rather than the floor. After it is located,

Table 4–1 **Leaks in Water Distribution Piping**

Expansion and/or contraction	Hangers should be secure but should allow for movement. Do not install piping where it is bound between joists or similar obstructions. Thermal expansion will make creaking noises in bound piping and will ultimately result in pipe failures. The lack of proper thermal expansion devices or piping loops in hot water distribution or recirculation systems may also result in leaky piping systems. Consult piping handbooks for methods to compensate for expansion.
Joint failure due to poor installation techniques	Review joint assembly procedures and recommendations by the manufacturers. Avoid joining dissimilar metals or dissimilar plastics without approved adapters.
Water hammer	Install an air chamber or shock arrestor as close to the point of use as possible and be sure that it is accessible for service and air replenishment.
Erosion/corrosion	Calculate water flow velocities. In some cases, the erosion/corrosion problem is attributable to excessive velocities. Take a sample of the water if velocity is suspected to be aggressive. Do not overflux copper pipe and tubing. Use noncorrosive fluxes recommended by the manufacturer or the Copper Development Association.
Freezing	After repairs have been made, provide the necessary heat and/or insulation to the pipe to prevent further mishaps. Advise the building owner of the proper methods to freeze-proof the vulnerable piping.
Stress	Relieve the stress. If necessary, use additional fittings and pipe to do so. Use hangers and supports to relieve stresses in piping.
Vibration	Do not allow two piping materials to touch each other where they cross. Do not allow a pipe to rest directly on a beam, duct, or similar surface in which vibration, expansion, or contraction could wear a hole in the pipe.

Table 4–2 Linear Movement Caused by Thermal Expansion and Contraction

Material	Coefficient in/in°F	Change in Length per 100 Feet in Inches				
		Δ10F	Δ50F	Δ100F	Δ150F	Δ200F
Metallic						
Brass (red)	9.2×10^{-6}	0.11	0.55	1.10	1.66	2.21
Brass (yellow)	1.0×10^{-5}	0.12	0.60	1.20	1.80	2.40
Cast iron	5.6×10^{-6}	0.07	0.37	0.67	1.00	1.34
Copper	9.4×10^{-6}	0.11	0.54	1.08	1.62	2.16
Steel	5.6×10^{-6}	0.07	0.37	0.67	1.00	1.34
Wrought iron	6.6×10^{-6}	0.08	0.40	0.79	1.19	1.58
Non-metallic						
ABS	5.5×10^{-5}	0.66	3.30	6.60	9.90	13.20
CPVC	2.5×10^{-5}	0.42	2.10	4.20	6.30	8.40
PB	7.5×10^{-5}	0.90	4.50	9.00	13.50	18.00
PE	8.0×10^{-5}	0.96	4.80	9.60	14.40	19.20
PVC	3.0×10^{-5}	0.36	1.80	3.60	5.40	7.20

Δ means change in temperature. For example, an increase from 30°F to 40°F is Δ10°F.

In the Field

Some companies specialize in leak detection and can isolate a leak to a 1-square-foot spot in an inspected area that is the size of a football field. The services of such companies should be sought out when it is determined that the leak may be too difficult or too time consuming to locate with conventional methods.

In the Field

Be particularly careful when removing old pipe insulation. Frequently these materials will contain asbestos, which is white or gray in color and flakes off when disturbed. Consult your supervisor when asbestos-containing insulation is suspected. By law, only authorized persons using approved methods may remove, transport, or dispose of asbestos material. There are severe penalties for the violation of this law. Do not take chances with your health or the health of others. ASBESTOS IS DEADLY.

the leaking section of pipe can be repaired by conventional methods. Modify the piping as required to minimize the chance of recurrence of the leak by the previously identified problems, such as those in Table 4–1.

Different materials, of course, use different joining methods. New technology now offers tremendous choices in joining methods for water distributions system and may eliminate the need for open flames in the repair process. Should you be involved in soldering or brazing, remember that some building materials will smolder for a long time before bursting into flame. Use flame-retardant shields when soldering around combustibles. Wet down the surrounding dry combustible surfaces with a water spray bottle before soldering. Fire extinguishing equipment should be on the site, and be sure that you inspect the repair area very carefully for fire before you leave the job.

After the repairs are made, restore system pressure and check for leaks. Be sure all the air is vented out by opening the faucets at the highest elevations. Check all flushometer valves for proper operation. After restoring a line to working pressure, these valves frequently are fouled by small pieces of scale or solder that are dislodged by the surges and transients of the process.

REVIEW QUESTIONS

1. Why is it important to identify what contributed to a water leak?
2. Why are there specific rules for handling old pipe covering that may contain asbestos?
3. Explain why not all water line leaks in buried piping result in water appearing at the surface.
4. What is the reason to keep an accurate record of where and how a water service repair is made?
5. Why is it important to be informed of the water department's involvement in water service utilities when making repairs?

CHAPTER 5

Drain, Waste, and Vent Piping Service and Repair

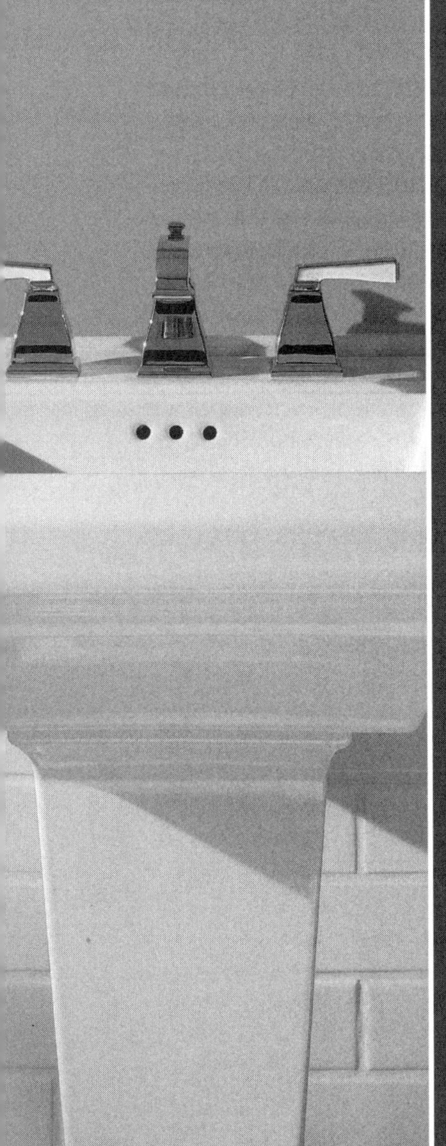

LEARNING OBJECTIVES

The student will:

- Identify the various installations that contribute to drainage pipe failures.
- Identify which applications require specific material to be used in the repair service.
- Explain the code concerns related to subsoil piping repair methods.
- Explain when drainage pipe should not be relocated.

IN-GROUND SERVICE AND REPAIR OF DRAIN, WASTE, AND VENT PIPING

Chapter 5 will address the repair and service of the drain, waste, and vent piping in a structure, referred to as the DWV sanitary system and the storm system. This chapter will include the subsoil storm piping as well as conventional storm systems. Both DWV and storm systems include the mains, branches, and risers called stacks and conductors. The first portion of this chapter will address the most difficult area in which to diagnose occurring problems, the underground systems. We will consider the recommended procedures to correct the obstacles posed by these systems and alternative solutions for difficult situations.

The problems that usually occur in underground sanitary and storm drainage systems include broken lines or fitting failures, causing leaks, settling, or stoppages. These problems, except for stoppages, will require replacement to some extent. Also, broken lines or fitting failures will generally cause stoppages in the system over time. Removing the stoppages requires determining the location of the problem based on evidence provided by cleaning tools and cameras, which send a signal at the location and video to the operator. Video camera equipment will pinpoint the problem and assist you in making a decision for the best solution. When these problems occur below grade within a building, the repair can be difficult and expensive. In these cases, consider rerouting the line as an alternative.

All of the malfunctions mentioned above may be a direct result of poor installation procedures such as the settling, improper pitch or grade, or the lack of solid, tamped earth for support of buried piping. Your replacement must consider and correct these errors in order to be successful. Plumbing codes and installation procedures require pipe to be laid on a firm, solid, and continuous trench bed. Clean backfill should be placed in 6-inch layers and tamped in place until the pipe is covered by 12 inches of tamped earth. Different types of pipe have specific instructions for backfill and compacting. If the trench is dug too deep or if the trench at the pipe depth is rocky or unstable, the trench should be overexcavated by a minimum of 6 inches and sand or fine gravel used to fill the trench bottom to the proper elevation. If material removed from the trench contains rubble, rocks, broken concrete, or frozen ground, install the pipe and then backfill the trench with sand or fine gravel in 6-inch layers, compacting each layer. Only after the pipe is covered with 2 feet of this type of material may the original material be returned to the trench.

The recommended slope of lines, both sanitary and sewer, is critical as lines with backfall or level lines will be a problem. Sanitary lines are designed with a slope of $\frac{1}{16}''$ per foot for 8-inch or larger lines, $\frac{1}{8}''$ per foot for lines 3 inches to 6 inches, and $\frac{1}{4}''$ per foot for lines $2\frac{1}{2}$ inches or less. The minimum slope for horizontal storm branches is $\frac{1}{8}''$ per foot. The important consideration is to maintain a slope that will give approximately 2 feet per second (fps) velocity for sanitary lines. At this velocity, water and waste flow together and will not separate prematurely. Earlier plumbing codes sought to maintain a 4 fps velocity for sanitary branches that contained greasy waste.

In the Field

Plumbers who are well versed in plumbing codes are often unfamiliar with other engineering practices such as civil engineering, where the slope is extremely different and considers greater distances. These differences are usually applied to multiple-dwelling sewers and municipal locations such as streets or public easements.

System Failures

System failures are basically damaged systems that do not operate as designed and may cause stoppages. Broken lines may become root-clogged line stoppages. In order to be considered a system failure, either the pipe's or fittings' integrity must be compromised. In layman's terms, a hole or crack exists in the pipe or fittings, and in extreme cases the pipe may have collapsed. Deteriorated pipes and fittings have many causes, such as internal and external corrosion (rust), chemical waste contributing to corrosion, and poor material composition. Chemical waste deterioration must be discussed with the building owner, and a suitable resolution must be implemented to eliminate future problems.

System failures will result in exfiltration (fluids leaking out) and infiltration (fluids leaking in). Waste (sewage) leaking out is a health hazard and cannot be tolerated. The first concern is for the building occupants, followed by the surrounding ground water table and the surrounding earth. Infiltration of ground water surrounding a failed pipe or fitting will impair a treatment system's proper operation. This includes both a septic tank with its drain field and municipal treatment systems. Storm drainage lines are subject to extreme failures due to infiltration. High volumes of storm water flowing through a system will escape to the surrounding area, elevating pressures surrounding the failed pipe or fittings. When the flow is diminished, infiltration will occur at elevated levels, carrying surrounding earth back into the lines. The lines may be able to operate over an extended period of time by carrying the sediment away in a gradual process. However, the absence of earth may develop into huge voids that can contribute to cave-ins and surrounding structural failure. This is commonly observed in parking lot area drain cave-ins.

Pipe failures related to broken lines or deteriorated pipe may justify replacing the entire line with different material. You will have to determine whether the pipe contents or the soil conditions contributed to the failure, but a change has to be made. In some instances, the material composition was not adequate, such as with bituminized fiber pipe. The product has been proven to be inferior over time and no longer has code acceptance. Many years ago, material shortages during national emergencies resulted in changes to several major products such as cast iron, copper, and steel. New products with appropriate standards were developed and used for a short period of time. If you encounter problems and these materials are present, replacement is the best option, but the owner must be consulted first.

Fitting failures are generally caused by random material failures or improper installation. Poor installations will place stress on the fitting, causing it to fracture. Also, today's increased technology provides many new products and joining methods, which increases the likelihood that installers may not assemble the seals properly or may mix fittings that were not intended for one particular type of piping material.

Stoppages and grade problems are closely related, and both cause repeated problems. Obstructions in drainage systems with larger pipe sizes should heighten your level of investigation. Smaller lines will naturally have a buildup of materials, especially grease in food service areas, but buildups in larger lines are not as frequent. Improper grade of the pipe or settling will result in pockets of water being formed in the line. These grade errors affect the foot per second discharge rate of the waste, which causes slower flow and allows obstructions to form in the line.

The grade errors will show up when a line is being cleaned and a video camera is used. Do not assume that removing the obstruction means the conclusion of your service call. You should run a camera through the line after flushing the system. When greater grade problems are present, the video will show nonsloping lines, lines with back fall, or lines with pockets, indicated by standing water in the pipe. The increased use of plastic materials that are not as resistant to deflection movement results in pockets due to poor installation methods or unstable ground beds. Identifying these problems allows the technician to discuss them with the owner and determine the proper, long-lasting remedy.

In the Field

A wise service technician will have the owner or his representative view the system failure upon excavating in order to explain the system problems. If that is impractical, taking pictures or video for the owner's review at a later date is very sensible.

Materials

Cast iron is a generic term for a series of alloys primarily composed of iron, carbon, and silicon, with small amounts of other elements. It is commonly called soil pipe because of its common acceptance and larger size that conveys soil discharges, which are sewage containing fecal matter. Traditional plumbing systems utilize galvanized pipe and cast fittings for waste, while the soil lines are cast iron. Joining methods for cast iron soil pipe systems are shielded hubless couplings (commonly called hubless bands), compression joints (utilizing rubber gaskets compressed in the joining process), and caulked joints (using lead and oakum). Hubless couplings permit an easier soil pipe repair.

In the Field

You should consult the plumbing code in your area for acceptable products listed with their required standards applying to underground sewer and building drainage.

Defective sections of plastic pipe and copper tubing have to be removed and replaced with new material. Copper tubing can be joined to existing material with slip couplings and soldering or brazing. Soldering is generally used for waste piping, while brazing is generally used for underground water systems and medical gas installations. Copper waste piping is not installed underground as frequently as other materials, but plastic piping has gained wide acceptance in residential and light commercial areas. These plastic materials include many different types and compositions of the generic term *plastic*, such as acrylonitrile butadiene styrene (ABS), coextruded composite plastics, and polyvinyl chloride (PVC). The plastic systems are joined by the use of cleaners and solvent cement and can be incorporated into other material systems with relative ease.

Other accepted materials are available and use appropriate joining methods (usually some version of the hubless coupling) to make repairs. Those other materials include special acid waste piping, stainless steel, and vitrified clay pipe.

Subsoil Drains

In the Field

When performing underground repairs, always be sure that you observe all required safety rules in your excavation and trenching methods. Chapter 4 discussed several of these.

Subsoil storm drains deserve consideration in addition to the previous discussions of soil, waste, vent, and storm lines. These systems are specialized for receiving ground water from surrounding earth and may be considered in some code areas as falling under the building trades' jurisdiction, not plumbing. Subsoil drains are commonly 4-inch perforated pipe or tile systems extending around building foundations and may provide a piping retention grid under the lowest floor level. They are not required to have clean-outs and may become infiltrated with sediment or sand. Replacement of problem systems is expensive; however, there are several retrofit methods for consideration. Use caution in a selection as building codes do not allow basement wall structural conditions to be jeopardized by removing load-bearing floor thicknesses around the floor to wall area.

Above-Ground Service and Repair of Drainage Waste and Vent Piping

The repair and service of above-ground sanitary drain, waste, vent, and storm piping problems occurring in a building are far more frequent than those occurring for in-ground piping. Further, above-ground DWV piping problems make themselves evident more quickly than in-ground difficulties because they are more noticeable and have the potential to cause greater damage if left unattended.

Most problems in this area are stoppages, leaks, or noises. Stoppages result from improper installations, abuse of the fixture, or breakage of the drain line. Also, stoppages occur if there are defective joints or backward slopes in the piping combined with frequent, heavy use of the fixture or with improper waste being placed in the fixture. "Problem" wastes are almost any solids placed in sinks or lavatories, as well as some sanitary napkin products placed in water closets. Even the so-called disposable products will occasionally be a problem, especially in low-consumption water closets. Liquid leaks can relate to broken pipes or joints, inadequate flashing, or condensation. Leaks may release liquids or sewer gases, and sewer gas leaks may be the result of dry traps or faulty gaskets around sewage lifts. Noises originate either from flow inside the pipe, expansion or contraction of the pipe, or vibrations in stacks during discharges when they are not bound in position.

In the Field

Plastic pipe threaded jam nut assemblies have a tendency to loosen due to nut expansion and thread contraction after installation. These assemblies are used in plastic traps, sink continuous wastes, and tub waste and overflows. A service call for what could be termed as "plastic creep" may be necessary. Strangely, these problems seem to occur only once after installation.

Locating Problems

Above-ground problems with leaks are usually easier to find than those below ground because the defect can be observed directly and the leak can be traced back to the pipe defect. Possible problems include leaking tub drains, water closet connections, roof flashings, and all types of joints. Consider whether the leak occurs only after rains, after fixture usage, only in certain types of weather, at a certain time of day, or only after a certain person uses the fixture.

For sewer gas leaks, use similar queries to locate the source of problems. With what circumstance does the odor occur—weather or wind direction—and does the trap leak water or lose its seal because of other causes? One problem that occurs occasionally is the exhaust fan pulling sewer gas into the bathroom when the putty or wax ring seal of the water closet to the closet collar is faulty. Replacement of the water closet seal is then necessary. Another common sewer gas problem may occur when the wind, under certain circumstances such as blowing from a specific direction and with specific force, may push sewer gas from roof vent terminals. The sewer gas then travels down into the roof overhang or air vents, which then conduct the sewer gas into the occupied space.

Noises can be identified through careful checking of the installation. Bubbling and gulping noises may be an indication of stoppages caused by poor septic and drain field operations or plugged vents where the drainage system is seeking air. Internal noises are made more noticeable by copper or plastic pipe and less so by cast iron or steel pipe. For this reason, cast iron soil pipe has been advertised as the "quiet pipe" at various trade show exhibits. External noises can be reduced by making sure that the pipe is not restrained or bound in the installation. Riser clamps should be used to support vertical piping, but wooden wedges should not be used to support the pipe in floor penetrations.

In the Field

Plastic waste piping in wood framing construction is famous for making dripping noises due to expansion and contraction. This can be avoided by using a separation method such as wrapping roofing felt (tar paper) around a pipe or fitting where it comes into contact with a wood framing member such as a drilled stud for a lavatory waste line.

Repair Concerns

Pipe line or joint leaks are repaired by methods appropriate to the type of material encountered.

Conduct all possible tests and investigations before cutting into finished surfaces (ceilings, floors, or walls). Several areas of instruction have emphasized that replacement lines may be economically relocated during repairs. Caution should be exercised here as local health or plumbing codes will prohibit the installation of drainage lines over food service areas to prevent contaminants from dripping or leaking onto sanitary products. This is easily understood for service technicians

In the Field

Never conduct repairs with any type of unapproved method. The plumbing code specifically prohibits obstructions and connections that would retard or obstruct flow in the drainage system. See Figure 5–1.

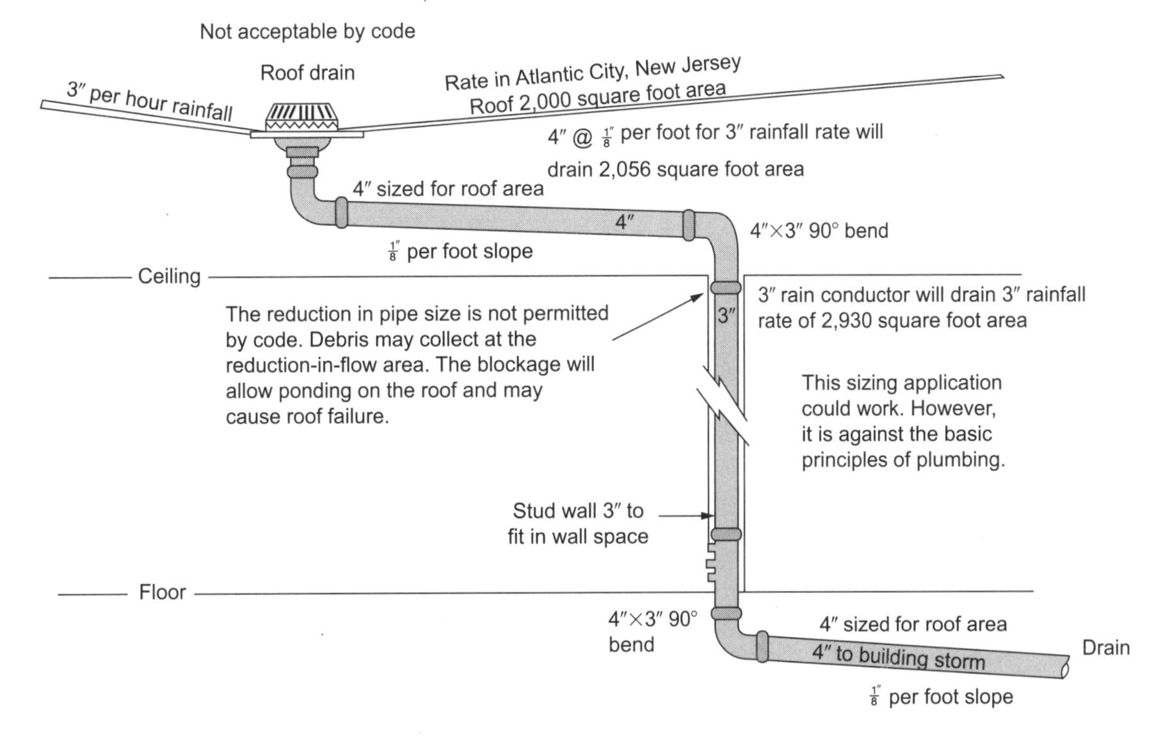

Figure 5–1
A roof conductor illustrating a reduction in flow from the horizontal run. Plumbing codes prohibit these changes in piping because of increased likelihood to create drainage obstructions.

who have worked in food storage areas located under bar sinks areas with their many indirect wastes, which will often overflow.

Check to be sure that any leakage is from piping rather than from around the fixture, especially leaks associated with showers, bathtubs, or water closets. Careless use of shower curtains or shower doors may be the cause of the leak. Improper sealing around the tub-shower valve escutcheons may cause leaks. However, your service work should include sealing the areas of escutcheons and fixture-to-floor contacts with a high-quality sealing caulk. Males, especially small boys or incapacitated adults, may be careless when urinating. This is a serious matter and not the subject for joking; you cannot repair a problem that is not caused by the plumbing. Discuss this very carefully with your customer if you believe that this is the cause of the problem.

Condensation from cold surfaces can be eliminated by insulation and vapor seals. It is very common to find storm drainage lines with insulation to reduce condensation. Be advised that after replacing pipes or fittings, the owner's insulation provision must be returned to its former operational design. Insulation will also reduce the level of noise from piping.

In the Field

Past practices for food service areas such as those in grocery stores have been to plan the stores' fixture layout and pipe the waste system accordingly. Recently, store owners have seen the need to remodel and relocate fixtures accordingly. Rather than perform costly below-concrete changes, they have elected to have portable service sinks with drain-pumping devices installed. These pumps raise the waste to overhead pipes and discharge it to a remote location. Some of the systems may operate with an assisted vacuum waste system and have secondary troughs under the waste piping. Code officials have very serious concerns with these systems.

REVIEW QUESTIONS

1. Name a pipe material that is subject to collapse due to material composition.
2. Identify the methods that may reduce the drain pipe noises in the above-ground drainage system.
3. Explain when replacement by rerouting should not be considered for waste piping in specific use groups.
4. Explain which pipe materials may be more prone to ground settlement and develop a pocket in the line that contributes to repeated stoppages.

CHAPTER 6

Fuel Gas Piping and Lead Product Service and Repair

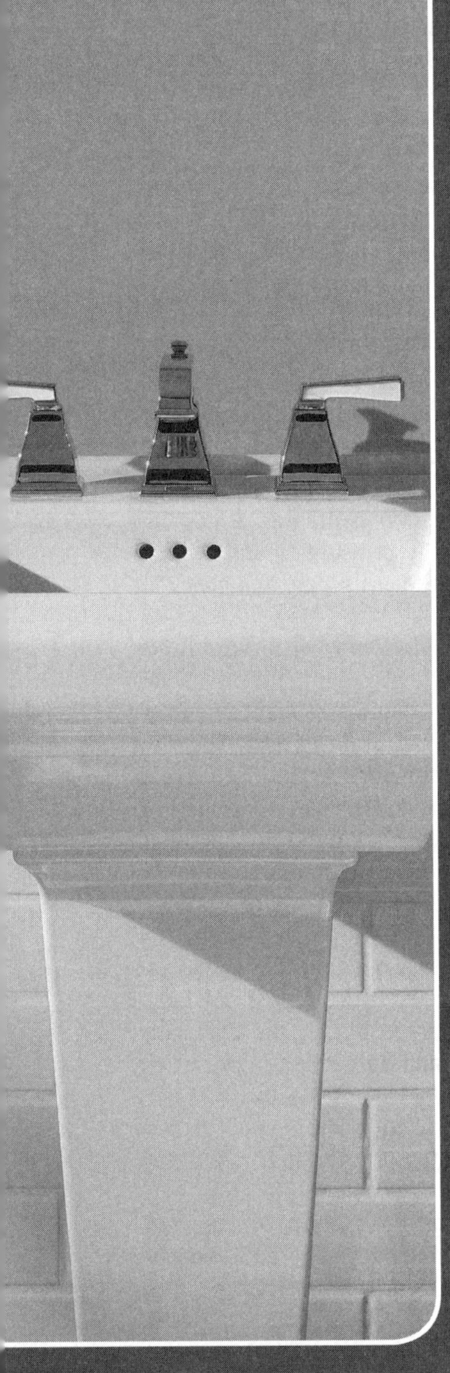

LEARNING OBJECTIVES

The student will:

- Compare the different gas system leak detection methods for proper service decisions.
- Describe how to perform testing for locating gas leaks.
- Discuss the process of lead shower pan replacements.

THE REPAIR AND SERVICE OF FUEL GAS PIPING

This chapter presents the safe procedures for repair and service of fuel gas piping above and below ground. Leaks in fuel gas piping are a serious hazard. There are many tragic stories related to explosions caused by gas leaks. In some cases, the plumber is called upon to detect, locate, and service these leaks. For your conscience's sake and that of your employer, you need to treat any gas leak as a very serious matter until the precise nature of the leak can be determined.

Generally, building occupants will smell a gas leak due to the gas odorizer (which is added to the gas for easier leak detection), and the leak will be minor. But until otherwise determined, any evidence of a leak must be considered to be a severe hazard. Gas leaking into a closed space will leave a telltale odor. The gas odorant is detected by most individuals at very low concentration, so the first indication of a leak must be reported and serviced as soon as possible. Always check such reports with urgency and great care because any gas leak represents a hazard.

Most leaks, especially in above-ground piping, will be at joints. Pipe shifting or building settling can put stress on joints that can cause leaks. In some cases, the pipe joint sealant will dry out or otherwise change so that the joint will begin to leak. It is possible for the odorant to enter a building from an underground leak outside the building, so if inside piping is in good order, you will have to expand the investigation to outside the building.

If a gas leak is suspected to be serious enough that an explosion is possible, evacuate the building immediately. Avoid all risks of injuries if possible. Do not operate any electrical switches (on or off) if gas is suspected to be in the atmosphere. Turn off the gas and electric services at the building mains for safety.

In the Field

A thrown electric switch such as a light switch is basically an ignition source, much like throwing a lighted match into a combustion source.

In the Field

Rural areas generally have trucked-in liquefied petroleum (LP) and the necessary gas storage for LP gas systems, while metropolitan areas have natural gas systems. LP gas has double the heating value and a higher specific gravity than air. For that reason, leaking LP gas will settle to the floor, trenches, or floor drains, unlike natural gas, which disperses into the air. LP gas leaks are very dangerous as the pockets of concentrated gas are highly combustible.

Testing

The following procedures are applied for testing modified or newly installed gas piping or for searching for leaks in existing gas piping. New gas pipe installations (or extensions to existing systems) are required to be tested at 3 pounds per square inch gauge (psig) or 150% of operating pressure, whichever is greater. A gas line that has been in service need only be tested at operating pressure when checking for leaks. This means that all you have to do is turn on the gas supply and check all pipe, joints, and appliance connections for leaks.

To determine if the system is leaking, the easiest procedure (with natural gas when a meter with a test dial is present) is to use the meter hand/needle check. Shut off all pilot burners and gas cocks to appliances, or cap off appliance connections. Do not depend on automatic gas valve closure to perform this test. The flow of automatic gas valve leakage (permitted by industry standards) when the valve is closed may blur the accuracy of this test. Mark the test hand/needle—one method is to wet a small piece of paper and stick it to the meter glass with an edge lined up with the test hand/needle—and observe it for a period of time, according to Table 6-1. If there is no movement of the test hand/needle at the end of the indicated time, the system does not have a leak. In this case, purge the piping and light a small burner. Observe the meter to be sure the test hand/needle moves (as it should). If there is no movement of the test hand/needle, the meter is not functioning correctly. Turn off the gas supply and notify the utility. If the test hand moves, the piping system must be checked for leaks using bubble testing or other methods.

Do not use open flames to test for leaks. If you decide to pressure test the system, be sure all devices are isolated that could be damaged by high pressure and use air, nitrogen, or carbon dioxide as the test medium. NEVER USE OXYGEN.

If leaks are detected, mark them for repair at the completion of the test. After all leaks are repaired, retest the piping and continue until all leaks are eliminated.

In the Field

The bubble test is conducted by working a liquid soap solution (dish soap diluted with warm water or a commercially made leak detection tester product) with a small brush or dabber around the joints and connections. System leaks will appear as small bubbles at the leak point. This is caused by the interior gas, which is under pressure, escaping from the piping system.

Table 6–1 Test Observation Times for Various Meter Dials

Test Hand/Needle—Volume/Revolution in Cubic Feet	Time to Observe in Minutes
$\frac{1}{4}$	5
$\frac{1}{2}$	5
1	7
2	10
5	20
10	30

In-Ground Leaks

When the source of the problem is not immediately located, enlist the aid of the gas supplier. All of our discussion to this point has assumed that the piping is the responsibility of the building owner. Use caution here as fuel gas services may be the property of the gas supplier, a utility company.

For in-ground leaks under the building's ownership, small and definitely local leaks can be temporarily repaired with an emergency clamp as a safety measure. If the piping is generally deteriorated, the entire line should be replaced. For a system corrosion problem, obtain recommendations from a corrosion specialist before proceeding. See *Plumbing 301* for a discussion of these actions and preventive methods. Further, if metallic pipe has failed, consider replacing it with modern plastic material or coated metallic material. Below-ground piping may be attacked by erosion or corrosion processes.

Good Customer Relations

Always take a few minutes to instruct the building owner or her representative in the proper safety methods when a gas leak is suspected. This is particularly critical for use group classifications with multiple occupants (hotels, motels, apartment facilities, places of assembly, or other public use spaces). Management and maintenance staff should know the smell of the gas odorant, the most expedient methods to evacuate the building, and the methods to shut off the gas with accessible gas valve shut-off tools.

REPAIR AND SERVICE—LEAD PRODUCTS

Plumbing, based on the Latin word for *lead,* means "lead working." This portion of the chapter will discuss the servicing of leaks that develop in lead plumbing materials, although many plumbers will never have to work with lead. The typical repair is to replace lead products, but if the defect is small and does not indicate general deterioration, actual lead repairs can be made as described. Leaks in these lead products are usually a result of deterioration of the material because of chemical attack. If the material is generally in poor condition, complete replacement is advised.

The increased concern over contamination of the water supply system and the use of plastics and synthetic materials has resulted in a vast portion of lead work becoming history in the plumbing industry. The greatest reason for the decline of lead shower pans was economics as the option to select less costly, prefabricated base and side walls constructed of plastic materials became available. These units are, of course, much easier to install. Previously, all roof flashings, shower pan liners, and safe waste drains were made of lead. Although present industry trends point to synthetic materials, some areas still require sheet lead for flashings or

In the Field

The advent of modern indoor plumbing allowed lead to become a primary material used in the plumbing profession. Sanitary pipe, for many years, meant lead was used as the material for water service supply, smaller drainage pipes, and drain pans. Today's current plumbing includes the increased consciousness of health concerns based upon using proper, sanitary piping and codes enforced through licensing programs. Every plumbing professional who researches the history of lead will have a far greater sense of appreciation for the trade and direction in the industry.

pans under showers. Flashings and pans are now made of plastic, composition rubber, or similar newer materials, and in some jurisdictions liquid coating with reinforced corners have been accepted for pans.

Repair Methods

Whether repairing or replacing, joints are made in lead by using a torch or soldering iron, solder flux, and solder on the cleaned joint areas in the lead fabrication. A small, clean flame is used to heat both sides of the joint, and the solder is worked onto the joint. A patch piece is used to cover any hole in the lead. Solder (usually 40/60 or 50/50 tin-lead) has a lower melting point than either of the parent metals. This characteristic permits getting the lead hot enough to melt the solder and still not melt the parent lead sheet or pipe.

Shower Pans

When a leak is associated with a field fabricated shower installation, check the tile grout and the wall and floor area in the vicinity of the shower entrance and door. Also, check the grouting around the faucet handles, the shower control, and the shower walls and base. It is very common for leaks in these areas to be misdiagnosed as drain or pan problems.

If all indications suggest that the shower pan is leaking, install a test plug in the drain at a level below the floor. Fill the shower base to about a 3-inch depth, and allow the water to stand for several hours. Check for leakage on the ceiling below and for a change in the water level. If no water appears below and the level does not drop in the shower base, then the drain line must be leaking. If the water level does drop, the pan has failed.

Follow these steps if the pan has been diagnosed as the problem and replacement is necessary. First, remove the strainer and then put a test plug in the drain opening to keep debris from falling into the pipe. Next, remove the tile from the base to a height of 6 inches up the wall. Be sure to wear gloves and safety glasses when removing the tile.

Remove the grout from the pan so it can be observed for failure points or regions. If the failure is caused by nail heads or other pressure points on the subfloor, these should be corrected before the replacement pan is placed into the shower area. Remove the nails and replace with countersunk galvanized screws.

The replacement pan should be made to be $\frac{1}{2}''$ smaller than the original pan total base dimension plus the two side dimensions. This lead dimension is generally 6 inches larger than the base on each side as the sides are folded up to form a point, which is then folded against the side (see Figure 6-1).

Measure the drain opening location accurately so that the pan can be cut after it is in place. Make one final check of the floor surface to inspect for raised or protruding areas.

Fold up three corners somewhat to permit placing the pan through any door opening. After it is in place, flatten out the bottom, and nail the sides in place, placing the nails near the top edge. After cutting the drain hole and setting the secondary collar, coat the lead with tar or asphaltum to prevent attack by the mortar material. Make sure the weep holes around the clamping collar are free to allow water, which may seep through the tile to drain out properly.

The pan should then be tested by putting a test plug in the drain opening and filling the new pan with water.

In the Field

Plumbing codes recognize the following materials for shower pan liners: plasticized polyvinyl (PVC) sheets, nonplasticized chlorinated polyethylene (CPE) sheets, sheet lead, and sheet copper.

Other Materials

Other pan materials introduced in recent years are said to have superior characteristics compared with lead.

Remember that lead was used because of its imperviousness to moisture. Your shop's or area's practice will dictate the favored material. The techniques of installing

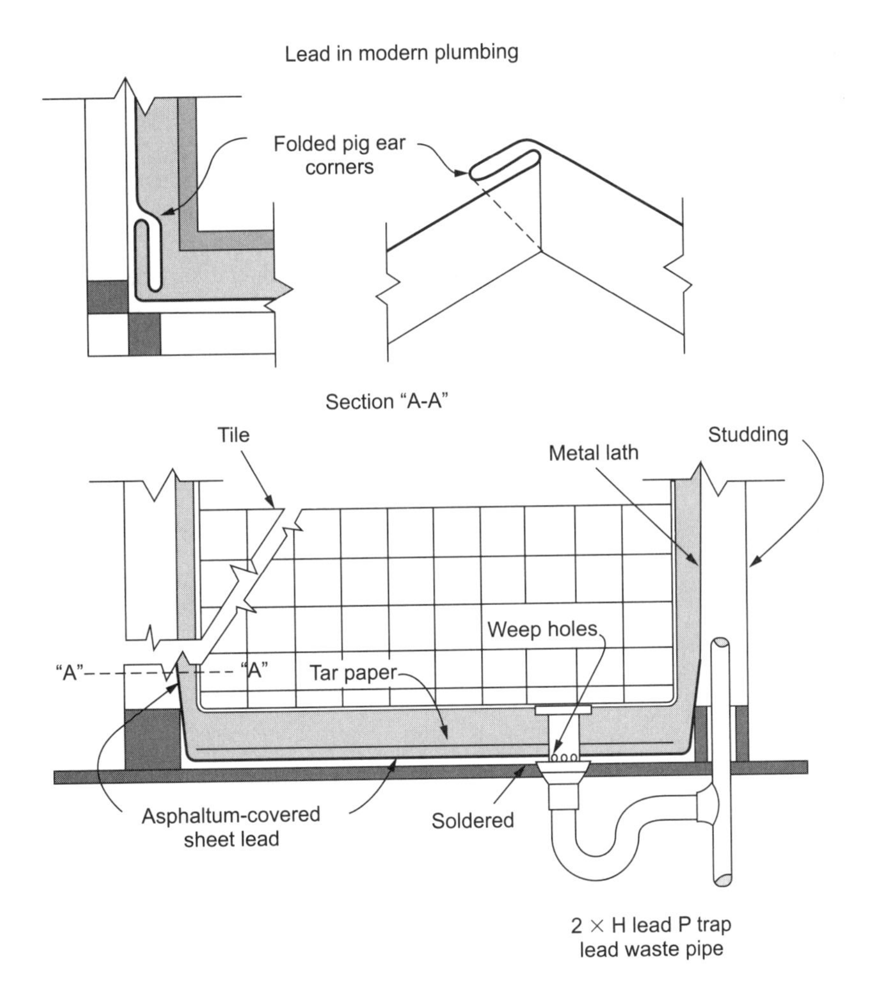

Figure 6–1
This drawing of a lead shower pan liner illustrates a method to ensure waterproofing of the structure.

pans made of other materials are similar to those required for working with lead. Do not coat these alternative pan materials with asphaltum or tar. Asphaltum and tar paper were formally used to isolate the lead from the lime, which adversely affected lead during the curing of cement for tile setting. Certain plastics will be adversely affected by the oils in asphalt.

In the Field

Many plumbing professionals have enjoyed reading and collecting older books related to plumbing, such as the *Questions and Answers on the Practice and Theory of Sanitary Plumbing* series written by R. M. Starbuck. Several readers have commented that you almost believe early plumbers were great chemists because of the early products' mechanical properties.

REVIEW QUESTIONS

1. Why is bubble testing the best way to locate a gas leak?
2. Why are materials other than sheet lead used as shower pan base liners?
3. Can the easy gas leak test with the test hand on the gas meter be used in every case?
4. Are suspected gas leak situations to be treated as general service calls or serious hazards until actual conditions are determined?

CHAPTER 7

Water Heater Service and Repair

LEARNING OBJECTIVES

The student will:

- Describe the two tubes located inside the water heater and explain their functions.
- Describe the necessary troubleshooting procedures to service the heat sensing controls in fuel-fired and electric water heaters.
- Interpret code concerns when installing water heater drain pans.
- Discuss water quality issues related to heaters and heat exchangers.

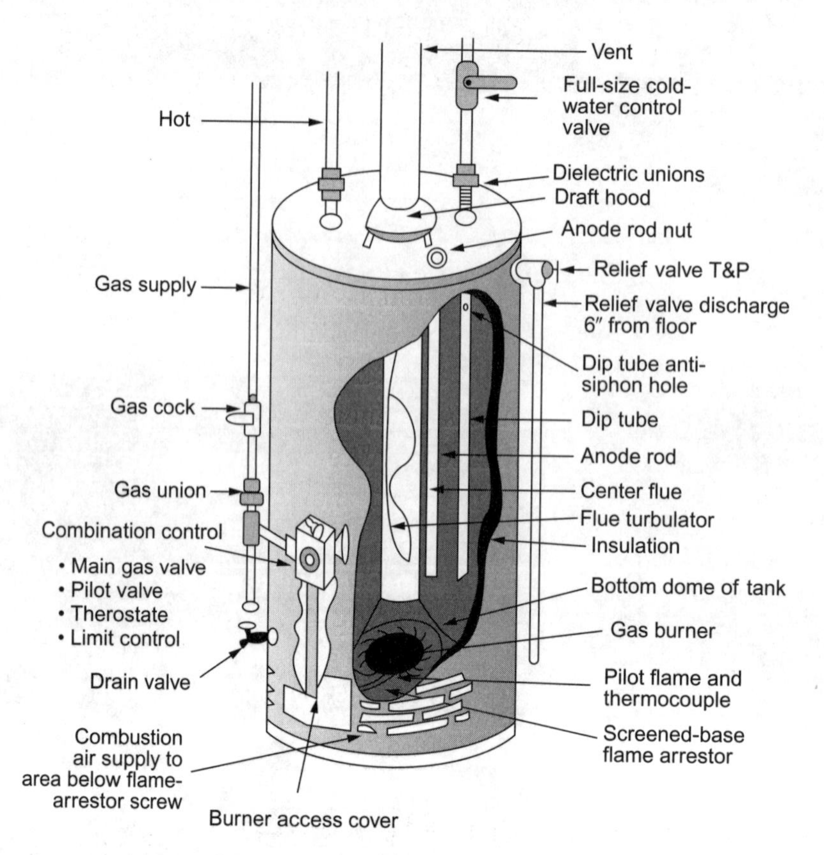

Figure 7–1
A gas water heater cutaway illustrating the various components of the unit.

SERVICE AND REPAIR—WATER HEATERS

This chapter will study the service and repair of water heaters by analyzing special service and repair techniques for different types of heaters and their basic components (See Figure 7–1). We will then discuss service and repair techniques for different types of heaters, including new products on the market. The installation, replacement, and troubleshooting of water heaters was discussed in basic terms in *Plumbing 201* and *Plumbing 301*.

Certain problems may occur with water heaters that either reduce their capacity or make them inoperative. Such problems include the following.

DIP TUBE PROBLEMS

If the water heater dip tube becomes deteriorated, cold water will mix with the tank contents above the lowest level. The effect will be to reduce the apparent capacity of the heater because the heated water is significantly cooled by cold water dilution at the top of the water heater. The dip tube must be replaced with a pipe that will convey incoming cold water near the bottom of the storage tank. The typical dip tube is a manufacturer-recommended high-temperature plastic replacement dip tube flared at the top with a small gasket or the original, which is now built into the extension inlet nipple. The tube has an antisiphon hole drilled in the top 6 inches of the tube that is generally $\frac{1}{8}''$ in size.

Within the past few years, a unique problem has occurred in that the tube has deteriorated over a period of time. This failure is usually indicated by the customer noticing that he has a reduced supply of hot water and his faucet aerators have become plugged by very small granular particles or flakes of a white substance. Some customers have been able to determine that the items appear only on the hot supply lines, for example, not in their water closet supplies. Manufacturers initiated massive service replacement procedures over a period of time. However, some of the problems still appear from time to time in the heaters that were not identified. Should you be called upon to service one of these units, the most difficult aspect of the service call will be to remove portions of the dip tube left in the heater and then flush the heater and system. A less common dip tube failure is the collapse of the tube to a nearly closed position. In this event, very little water can enter the heater.

To replace the tube, turn off the heat input and the water supply to the tank and remove the cold water inlet piping. The old tube can then be removed. Place the new tube in the inlet nipple and reinstall the inlet piping. Restore water pressure to the tank and check for leaks.

Occasionally, you may encounter a heater in which the hot and cold connections have been reversed in replacement situations. In this circumstance, the cold water is automatically diluting the hot water in the tank. Switch the piping to fix the problem but only if you ascertain that there is a problem. An installer may have exchanged the cold and hot supply nipples on a heater replacement in order to accommodate the required access to heater components. You may have expected that the customer would have noticed and had the installer make the correction. However, a customer may believe that the operation is what he should expect, perhaps due to new heater conservation methods.

One final consideration of dip tubes is similar to all other product servicing—that is, always use the manufacturer's original replacements or materials with the same specifications. For example, different water heater manufacturers or different models have unique dip tubes that are designed to reduce stratification in a heater or increase agitation of the incoming cold supply to reduce sediment in the tank bottom. A conventional, straight dip tube replacement may not be the correct length and may not perform as intended by the manufacturer.

FAULTY RELIEF VALVE

Water heater relief valves are one of those plumbing appurtenances taken for granted but supply the highest degree of safety in a plumbing system. All plumbers should be well versed in their operation to ensure proper plumbing safety. Every time you check a water heater, use the relief valve test lever to be sure that the valve mechanism is free-working. Manufacturers recommend operating the valve at least once a year and examining by removal every three years. If it is stuck or if it leaks water, it must be replaced.

T&P relief valves provide protection from explosions caused by faulty controls by discharging the overheated water and allowing cooler water to enter the tank to safely reduce the temperature. This is accomplished by the valve's thermostat expanding, which lifts the valve disk from the seat and allows the hot water to escape. A reduction of the 210°F water by less than 10°F allows the thermostat to contract and the loading spring to reseat, stopping the discharge. The greater the temperature or pressure is, the greater the force and the more the valve opens.

Water heater T&P valves must be set to open at 210°F and must be able to remove energy at a rate equal to or greater than the energy input rate of the water heater. Manufacturers of water heaters are required to mark the heaters with the maximum working pressure. The T&P valve, in accordance with code, must have a maximum opening pressure of 150 psi or be less than the working pressure rating printed on the heater.

Ensuring that the heater has the properly rated valve is accomplished by referring to Figure 7-2 and this commentary.

- Item A - The name of the manufacturer is provided.
- Item B - The serial number date code is provided. The first two digits indicate the year, and the last two digits represent the week of the year it was manufactured.

In the Field

Plumbing codes require all storage water heaters to have temperature and pressure relief valves or a combination (T&P) thereof conforming to ANSI Z21.22 and that they be self-closing. Exercise great caution here because several authorities having jurisdiction view any water amount, however small, within a pipe as storage due to their understanding of water heater explosion potential. This indicates to them that T&P valves are required on tankless water heaters, not just pressure relief valves.

In the Field

Water heated above 212°F will boil, expand, and turn into steam, thus dissipating its energy. Water can be heated far above 212°F in a closed, pressurized environment without flashing into steam and giving up its latent heat energy. However, if the water is heated and pressurized and then exposed to atmospheric pressure (14.7 psi, normal habitable pressure) by a rupture, a much higher level of energy is released—an explosion.

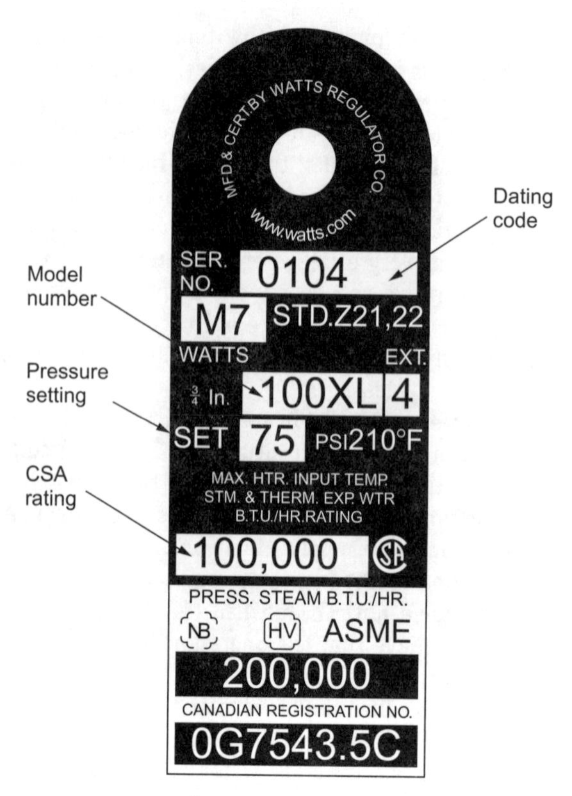

Figure 7–2
A residential water heater relief valve name tag that provides installation details addressing safety matters. Courtesy of Watts Regulator Co.

- Item C – The size and model number is provided.
- Item D – The pressure setting of the valve is provided, which must be less than the working pressure of the tank posted by the manufacturer on the heater.
- Item E – The temperature setting is provided to ensure that the valve is a T&P device.
- Item F – The AGA valve's BTU rating must be greater than the BTU input rating of the heater.

The T&P relief valve location in the heater is critical so that it can sense the tank water temperature properly. Plumbing codes identify that the valve must be located in the top 6 inches of the heater. This provides accuracy for the sensing element, ensuring that temperatures greater than 210°F will be moderated. Temperature stacking information inside the heater was provided in *Plumbing 301* but we should also consider thermal temperature lag. This term basically means that in moving the sensor away from the heater's highest temperature (the top 6 inches), a lag or loss in sensing occurs. A rule of thumb is that for every inch a $\frac{3}{4}''$ valve is moved away from the tank, 8°F of sensing are lost. For example, a $\frac{3}{4}''$ relief valve mounted in the hot discharge line 4 inches away from the heater may sense a temperature of 212°F when the temperature inside the heater top is 244°F ($8 \times 4 = 32$, $32 + 212 = 244$).

The relief valve discharge piping is also a critical safety consideration. The line must have an adequate temperature rating to prevent collapsing and cannot be closed off by a valve or threads that would allow a pipe cap. It must be 6 inches from the floor to prevent splashing of a nearby occupant, yet not be so close to the floor or a floor drain to be a violation of a water supply air gapping, which would violate cross-connection rules.

Table 7–1 lists specific relief valve service/repair approaches based on the amount of water discharged through the valve.

In the Field

Most plumbing codes that have a large geographical consideration (north to south and east to west) will mandate that the discharge line terminate where the heater is located or continue through to the outside of a building after having an air gap in the room. This is a commonsense issue when you consider that a dripping relief valve drain line in cold climates would freeze shut. In effect, the valve would be inoperable and subject to a heater explosion. For that reason, the air gap is required in the room.

Table 7–1 Troubleshooting Water Heater Relief Valves

Problem	Cause	Solution
Relief valve drips (Thermal expansion)	1. Sediment on relief valve disk 2. Pressure surge or thermal expansion due to closed loop, e.g., backflow preventer on water service	1. Replace relief valve 2. Install an expansion tank at water heater
Relief valve discharges large amounts of hot water (Excessive temperatures)	1. Incorrect thermostat operation 2. Excessive water pressure	1. Replace thermostat 2. Install an expansion tank at water heater

The replacement of a combination temperature and pressure relief valve is relatively simple. The tank pressure is relieved, the discharge line is removed, and then the old relief valve is removed. The new relief valve is installed in the opening and the water pressure restored to the heater. Check the completed job for leaks. Before you leave, check to be sure that the test lever opens the new valve.

MAGNESIUM ANODE

The magnesium rods in most storage water heaters act as a sacrificial anode to protect the steel in the tank construction. The interior of water heater tank is coated with a thin layer of porcelain enamel to resist corrosion. If a tank flaw occurs, the magnesium dissolves, creating electrons that move to the tank flaw to form a type of patch and extend the tank life. The rod naturally deteriorates and may have to be replaced.

The life of the rod is largely determined by the water quality. In some water, a chemical reaction with the magnesium gives rise to a very unpleasant odor, commonly described by customers as a rotten egg smell. The odor will always be associated with the hot water. The solution is to remove the rod and thoroughly flush the heater. Follow the manufacturer's recommendations, such as using an aluminum replacement from the manufacturer, in order to not void the warranty.

The vast majority of the rods can be very difficult to remove. It is common for technicians to remove the rod using a large pipe wrench and a cheater extension. Use caution here, however; that amount of torque can disturb other piping and create leaks. The rods can be replaced using the same steps as those involved in replacing the relief valve.

LEAKY OR PLUGGED DRAIN FAUCET

Probably the most common water heater problem (except for a leaking tank) is the drain faucet becoming packed with debris. The typical boiler drain's globe valve design is easily stopped up with the solids that frequently collect in a water heater. Also, for a period of time, several manufacturers used a plastic valve with a larger, round, hand-turning mechanism surrounding a hose-threaded connection in the center shaft with its O-ring. Normal high heating temperatures warp the plastic and cause the O-ring to leak. The best repair is to remove the valve and replace it with a nipple and ball valve.

To replace a failed drain valve on a water heater, shut off the water to the tank and be sure all faucets on the plumbing system are closed. The water heater will then be air locked. The drain valve can be exchanged for a new unit without significant flow from the heater. The new valve should be prepared with a nipple (if necessary) and thread-seal to make it ready to install. We recommend that the standard drain valves furnished on heaters be discarded and ball valves installed in their place. The ball valve provides a straight-through waterway that can be rodded out if scale or dirt blocks the passageway.

FAILED GAS CONTROL UNIT

Gas controls on the typical residential water heater have immersion rod sensors placed in the water. The controls fail either by not turning on when the water gets cold or not turning off when the water gets hot. The valves are expensive, and your troubleshooting technique should not be the "replace and try it" method. You should first research the thermocouple operation with its ignition source, fuel supply to the valve, and setting. The thermocouple sensing probe lets the regulator know that a pilot flame is available so gas can be sent to the burner assembly and ignited by the pilot flame.

Air lock the heater as described above and shut off the appliance gas cock. Remove the gas control unit and replace it with a similar model. Then restore the water pressure and check for leaks. If the threaded connection into the tank is not leaking, reassemble the gas piping and relight the burner.

FAILED ELECTRIC HEATING ELEMENTS

Electric water heaters are fairly common in residential applications and will require service work, which most often involves the replacement of heating elements or their controls. Both are accessed through panels on the front of the heater. The most common residential electric water heater has two thermostats and two elements. The thermostats are mounted on the outside of the heater storage tank underneath the access panels and insulation.

There are several different wattage ratings of heating elements. The higher the wattage rating is, the faster it can heat the water. The length of time it takes any water heater to heat its contents is called the recovery rate, and of course the capacity to the heater is also considered. Nonsimultaneous electric heaters will have two elements, called an upper and a lower element because of their locations, immersed into the water through the tank wall. Only one of the elements at a time will be energized. The top element heats the upper 30% of the tank and is the primary element as the hottest water will be at the top and used first (initial demand). The lower element is responsible, of course, for the remaining 70% and will take a greater amount of time to complete its heating. When the upper thermostat is satisfied, the lower element will be activated.

A standard residential electric water heater is classified as a 240 volt, 4,500 watt, nonsimultaneous water heater. Electric heaters are measured according to the wattage of the heating elements rather than the BTUs, as for combustion heaters. One watt equals 3.412 BTUs; therefore, 1,000 watts (1 kilowatt) equals 3,412 BTUs. In order to compare the heating value of electric to gas, you would consider the 4.5 (4,500 watt nonsimultaneous element) multiplied by 3,412, which equals 15,354. For that reason, a 4,500 watt element would be a 15,354 BTU element. Remember that the nonsimultaneously operating (one at a time) element in an electric heater with the equivalent of 15,354 BTUs can be compared to a standard 35,000 BTU gas water heater. You can now understand why electric heaters have a lower recovery rate and why a plumbing contractor will select a larger storage capacity in electric heaters.

Three wires feed these nonsimultaneous water heaters, one grounded and the other two energized by 120 volts each. The energized two are referred to as line voltage L1 and L2.

In the Field

Many years ago two sets of wires were supplied to heaters with their respective grounds (two separate lines) in order to have a timer remotely located at the meter. That concept has passed, but this knowledge will explain why you might see an extra line in the heater vicinity in older installations.

A path of current using both line voltages occurs when the thermostat calls for activation. Either one of the elements, when operating correctly and calling for power by the thermostat, will have a combination of the lines working together to provide 240 volts.

Two methods may be used when servicing/repairing the high-limit thermostats and elements in an electric heater. The options are testing with the heater energized and testing with the heater deenergized. The option you choose may be dependent on the type of test equipment you have and your comfort level with the conditions around the heater. You would be wise to use the deenergized method when your surroundings are confined, dimly lit, or damp.

Testing with the heater deenergized is accomplished by using Figure 7–3 and an ohmmeter. First, remove the power to the heater by toggling the circuit breaker to the off position at the circuit breaker panel. Each of the controls and elements will be tested with an ohmmeter. Further, the parts being tested should be disconnected from the system in order to obtain accurate readings. For example, when testing an element with the wires attached, the continuity of a burned out element (open, infinite resistance) may show very low resistance if read through other components and around to the other side, thereby providing a false reading. Another important consideration is that good elements will show a minimum resistance and

In the Field

Do not assume that one of the elements will be safe (power free). When power is supplied to the heater, *both* elements will have 120 volts supplied to them at the same time.

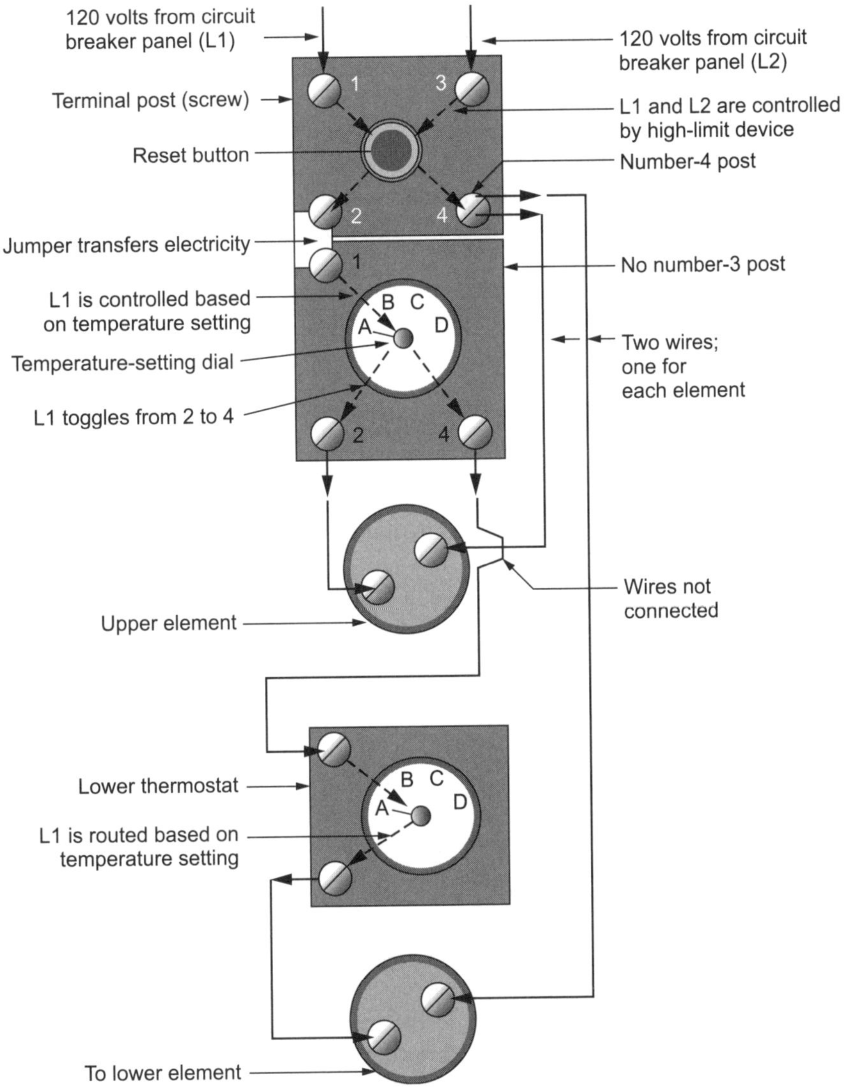

Figure 7–3
Wiring connections for a nonsimultaneous, dual-element, residential water heater.

that switches will indicate little or no resistance when closed and infinite resistance when in the open position. The following test procedure provides instructions for verifying the proper operation or necessary component replacement for electric water heaters:

1. Check for continuity between terminals 1 and 2 on the high-limit device. If you find near-zero resistance, check between terminals 3 and 4; both sets should read about the same. If not, push the red reset button and check again. If the device will not reset, it should be replaced. There should not be continuity between 1 and 4 or between 2 and 3.
2. If the high-limit device is good, then check for continuity between terminals 1 and 2 or between terminals 1 and 4 on the top thermostat. There should be continuity between one set of terminals or the other. If not, the thermostat is defective and should be replaced.
3. If both the high-limit and the top thermostat are good, then the top element should be checked. The element is a load device, so there should be more resistance when compared to a switch. A normal element will have approximately 13 ohms across it. If it reads zero ohms, it is shorted closed, and if it has infinite ohms resistance, then the element is shorted open. In either case, the element is defective and should be replaced.
4. If the heater is producing some hot water but runs out quickly, it should have passed all of the above tests, meaning the problem is with the bottom circuit. If the water in the lower portion is cooler than the lower thermostat's temperature setting, then the contacts should be closed and there should be a near-zero resistance reading between the two screw terminals on the thermostat. Usually, turning the thermostat to its maximum setting will cause the thermostat to close, which would make the resistance move to zero. If not, the thermostat is bad and should be replaced.
5. If the lower thermostat is good, then the lower element is the last to be checked. It should be tested in the same manner as the upper heating element (item 3).

Another test for calculating the rated current flow checks the elements with the power on. First, verify that the proper voltage is being supplied to the heater element. Next, use an ammeter to check current to the element. The current should be 10% of the calculated value. If the ammeter does not show the calculated current between the value of plus or minus 10%, the element should be replaced.

Example 1

$$\text{Rated Amperes} = \frac{\text{Rated Watts}}{\text{Rated Voltage}}$$

$$21.6\ \text{A} = \frac{4{,}500\ \text{W}}{208\ \text{V}}$$

21.6 A plus or minus 10% = 23.76 A to 19.44 A

In the Field

Elements must always be in contact with water when energized or they will overheat and burn out.

This testing information, along with additional information, is also presented in *Plumbing 301*.

When an electric element must be replaced, be sure that you turn off the electric power and air lock the heater. Remove and quickly replace the heating element. Restore water pressure and be sure the installation is watertight. Reassemble wiring and then restore power to the heater. Check the current flow to the element with a clamp-on ammeter, and check the operation of both the upper and lower thermostats.

NEED TO REPLACE CIRCULATING FLUID IN SOLAR SYSTEM

Solar systems and others use the principle of heat transfer from another source to the heater, such as a solar collection unit or a miniboiler located outside the structure (a heat transfer module). In colder climates, the fluid will contain some type of antifreeze chemicals, which is a very general term and may be equated with toxicity levels, much like the medical term *Gosselin rating*. In any case, extreme caution should be used in the selection of fluids that by code could determine the type of heat exchanger used, such as single-wall or double-wall exchangers with a leak path open to the atmosphere.

Replacement of solar fluid is performed by turning off the supplemental power to the heater and covering the solar collectors. Drain the heat transfer fluid into a drum or other receiver to later discard it or save it for future use. Be sure to follow all Environmental Protection Agency (EPA) and local environmental rules regarding the fluid disposal. Flush the collector with water at high velocity, and then use a scale-removing chemical if local water is scale-forming. After flushing, refill the system with heat transfer fluid. Remove the collector covers, restore power to the auxiliary heater, and check operation of the system.

In the Field

Consult the plumbing code for your geographical area and the authority having jurisdiction, such as the local health department, when making decisions related to the transfer fluid selection.

WATER TREATMENT

In locations where the water contains problem chemicals, a common condition involves the formation of scale on heated surfaces. At first, this scaling only reduces the efficiency of the heat transfer process, but eventually scale will build up enough to permit heat transfer surfaces to burn through, thus destroying the heat exchanger.

If the water in your area poses problems such as excessive hardness, high iron content, or chemicals, you should become familiar with the general principles of treatment. Numerous plumbing contactors have focused on water treatment often out of necessity due to local water conditions. Become familiar with specialty organizations—for example, your local water quality association—that provide services to minimize the problems in the plumbing system and meet owners' expectations.

WATER HEATER PANS AND THEIR DRAINS

Plumbing codes now require the use of drain pans beneath water heaters to reduce damage to the structure. Your service call or a heater replacement may involve the installation of these pans. Check your plumbing code when considering the installation requirements, which may vary from one jurisdiction to another.

Two major considerations should be pointed out when considering water heater pans. First, the relief valve by code might not be allowed to discharge into the pan (an air gap is required). Several code authorities believe the high relief-valve discharge rate will overcome the pan drain. Second, the pan drainage piping by code may require the drain material to be acceptable water distribution material. This is significant because many technicians will install larger piping, which would be PVC. The referenced code requirement would force the installing technician to use other materials, which in larger sizes may be far more expensive. The larger size is based upon a consideration of apartments' having some type of pan stack header accumulation for multiple floors.

REVIEW QUESTIONS

1. What is the purpose of the magnesium rod placed in most water heaters?
2. Explain when the term *Gosselin rating* should be considered in the plumbing system.
3. Explain why the usual dip tube failure reduces the volume of hot water available upon request.
4. What type of a water heater drain valve is best for purposes of cleaning out sediment at the valve location?
5. What is the primary method for observing operations of a relief valve and the recommended time frame for the observation?

CHAPTER 8

Waste Stoppage Service and Repair

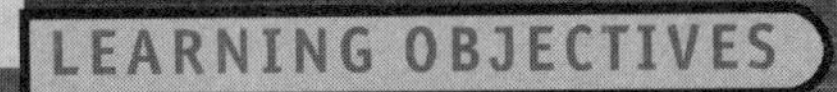

LEARNING OBJECTIVES

The student will:

- Recognize the system in its entirety and better identify how to correct the problem in the most efficient manner.
- Interpret where to locate the cleanouts in accordance with the code requirements during original construction.
- Identify drain-cleaning hazards for protection purposes and to reduce clean-up costs.
- Identify the different drain-cleaning equipment in order to select the appropriate device to use.

WASTE STOPPAGE SERVICE AND REPAIR

This chapter will address the service and repair of stoppages in fixtures, drain lines, and sewer lines. One of the most common service calls in both residential and commercial systems is for the removal of stoppages in drainage systems. Waste line stoppages can occur in any size drain line. The problems with these inoperable drains can vary, including a homeowner plugging a kitchen sink with excessive carrot tops to an industrial food preparation plant's sanitary sewer grease obstruction. Smaller diameter drain lines, serving sinks or lavatories, are cleared with small cable machines (See Figure 8–1) or hand-operated snakes. If the problem is in the building drain, large sewer rods or drain-cleaning machines will be needed to open the line.

Special Problems

The consideration of special problems will be beneficial in determining where to begin in drain cleaning, as pointed out earlier in the commercial storm sewer problem. Do not be misled by the symptoms of a stoppage so that you work on the wrong thing. Many times, a water closet will not flush properly because the drain or vent is stopped up, not because the bowl is obstructed. Check carefully for the fixtures that are draining properly and those that are not to determine just where the stoppage is located. You should also determine whether the stoppage is sudden or has been gradually developing. The former suggests a solid object in the drain; the latter suggests a gradual accumulation of soap, hair, scum, etc.

Back-to-back fixtures may be a problem, especially if improper drainage fittings were used in the rough-in. The drain cable will go from one fixture to the other without going down the drain where the stoppage is. It may be necessary to remove the piping right up to the wall opening so that you can control the cable direction going into the drain line.

There are also special opportunities in service calls related to waste stoppage service work. Those opportunities may be in the form of meeting special challenges, which will require service and repair technicians to be creative. A classic opportunity was presented when a plumber was called to a dental office several years ago by his dentist who was having problems with one of his exam room's $\frac{1}{2}''$ copper vacuum lines not pulling the normal aspiration vacuum. The dentist thought, "OK, a copper line.

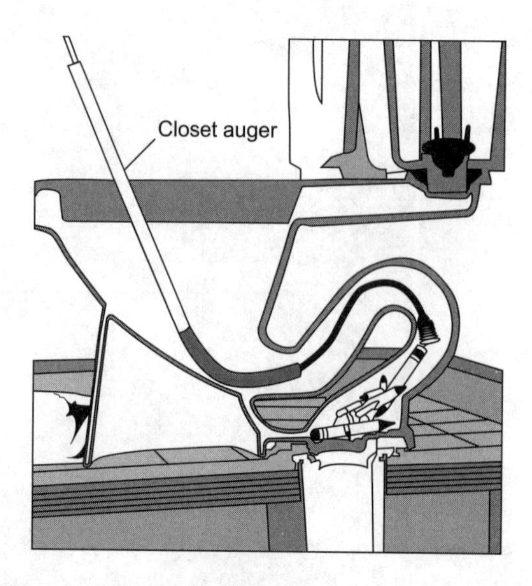

Figure 8–1
A closet auger being used to remove a blockage from a U-bend. The auger mimics the shape of the drain and enables navigation of tight bends.

Call a plumber." The plumber wisely considered the possibility that amalgam from fillings was hung up in a copper joint that may have not been reamed during installation. The service representative quickly purchased a very small and cheap do-it-yourself hand auger from a local hardware store. The coil head was cut off, and the cable was pulled out to make a smaller head that fit down the $\frac{1}{2}''$ line near the exam chair. Success—or as the plumber described his creative work, "just routine."

Residential

Drainage problems in residential applications are generally due to one of two issues. The first issue concerns the residents and how they may overtax or misuse the system. The second problem is the age of the structure, which of course has a bearing on the materials that would be found in the system.

Residential occupants' living habits vary greatly and have a bearing on the drainage system's performance. A brief list is provided to challenge your thinking in the following items:

1. Bathing oils and pet bathing may contribute to slower tub drains.
2. Occupants with longer hair may contribute to lavatory drain pop-up assemblies plugging.
3. Small children are prone to placing toys or toothbrushes in the water closet.
4. Guests may flush sanitary napkins or cotton swabs in the water closet.
5. Guests or owners unfamiliar with kitchen sink disposals may overload the kitchen sink continuous waste lines.

In the kitchen sink example, gaining proper drainage deserves more consideration than grabbing a plunger. Traditionally, the kitchen sink continuous waste line is held together by jam nuts and washers, which may be pushed apart by plunging. With waste water in both basins, that sink could cause a mess on your customer's new hardwood floor.

The age of the structure must always be considered during service calls for drainage problems. Extreme care should be used with handheld electric cleaning units in systems having old lead waste lines such as tubs. Obviously, you would not want to move through and out a bend in the line and damage the system. The quality of older galvanized systems may be marginal in the threaded joint area because of past corrosion or misuse due to caustic drain-cleaning solutions. Once the obstruction has been removed, the customer may still be left with the old, poor-performing, corroded galvanized waste piping in the system. When the waste piping is accessible, practice good communication and sales skills, which may benefit the customer through your suggestion of replacing the piping with a material that lends itself to much quicker installation, such as PVC waste piping.

In the Field

Some technicians cleaning a kitchen sink drain line in a typical single-story ranch home will rod the drain through the vent on the roof. If the obstruction is downstream from the sink, the clean-up time is tremendously reduced through the use of a garden hose for washing the roof and cable outside, which prevents having to clean the drain within the structure, which of course reduces clean-up time and accidents.

Commercial

Commercial and industrial drain line obstructions are more likely to involve larger cleaning equipment due to larger lines and more extensive systems than residential structures.

The greater likelihood of encountering safety concerns such as toxic substances requires increased vigilance on the part of the service technician when dealing with inoperable lines.

Diagnosing the problem correctly prior to taking action is extremely important. A classic case occurred when a contractor was called to address a small leakage around an older-style PVC cleanout's O-ring seal in the winter. The cleanout at the base of a 6-inch roof conductor was located in a grocery store stock area. The technician's first thought was to remove the cleanout, redope, and tighten to stop the leak. Before taking action, though, the technician speculated as to why any water would be in a roof conductor in the winter. The technician created a $\frac{1}{8}''$ hole in the cleanout, which was later used as an access cover screw hole, and observed leakage with a great deal of pressure. The technician understood that the building's storm line had a major obstruction in the system.

The blockage caused the storm lines to be filled with melted snow and surcharged the entire system. The roof was also surcharged with water, slush, and snow. The small cleanout leaks were a sign of potentially greater problems, to the extent of a roof collapse due to the extremely increased weight load. Consider that the slush and snow prevented run-off on the flat roof. Further investigation clarified that an area drain in the parking lot through which the storm sewer flowed had become frozen and plugged. A specialty drain-cleaning contractor was called who removed the catch basin cover and steamed the line in reverse to thaw out the frozen obstruction. The most critical factor in repair work is for you to determine what the problem is in order to provide a solution.

Kitchen equipment in restaurants and food processing areas where large amounts of grease are present provides unique drain-cleaning challenges. Obtain as much information as possible prior to using your equipment. This is necessary to keep from rodding through grease traps or interceptors, which of course could destroy the interceptor's baffles or break your equipment. Retrieving broken cables is a very expensive and difficult operation.

Commercial cleaning and the use of camera equipment will often identify greater problems such as a blockage that resulted from pipe settling or collapse or deteriorated joints that are permitting root penetration. These problems can be only temporarily resolved and obstructions will recur, so the sewer line should be excavated and replaced. Refer to earlier chapters on sewer repairs for solutions.

PLUMBING CODES

The plumbing codes do not discuss pipe cleaning equipment or their operations; however, many areas of the code address the need for equipment access to the drain, waste, and vent system (DWV). DWV systems' location, direction, size, and piping construction are deeply related to stoppage concerns. Recently, codes have changed to accommodate improved drain-cleaning equipment and technology; for example, the distances between cleanouts have been increased.

While codes are always changing as they are in the business of being updated and improved, the updates apply only to new construction and remodeling projects. However, a few basic rules will provide an understanding of where to look for clean-outs and what to expect.

A. **What constitutes a cleanout?** A cleanout is an opening in the DWV (includes the storm) system for the removal of obstructions. Cleanouts include removable caps or plugs. Some codes will extend this concept to removable traps and fixtures. Be cautious, however, as several authorities having jurisdiction view this as a licensing issue because drain-cleaning contractors, in their opinion, may not have the expertise to remove and replace fixtures.

B. **Where are they located?** Cleanouts are generally required at the base of a stack or conductor, at the building drain and sewer junction, every 100 feet, and at changes over 45°.

C. **What size should they be?** The codes vary somewhat here, but they basically require the same size as the pipe up to 4-inch lines, except for cast iron test tees and P-trap connections.

D. **What about access?** Access must be ensured, which requires the cleanout to be brought up to grade.

E. **What about clearances?** Clearances are identified around the cleanouts to ensure adequate space for operation.

F. **What about direction?** The basic premise is that cleanouts should be installed in the direction of flow to aid with rodding.

The code venting section also becomes involved with drainage pipe obstruction prevention by requiring every dry vent to rise above the flood level rim of a fixture by at least 6 inches before connecting to other vents or running horizontally, as in Figure 8-2 below. This prevents waste from accumulating in a horizontal vent.

Service Hazards

A particular hazard, especially in smaller lines, is the likelihood that a resident has put a drain solvent or chemical cleaner in the stopped-up line. This chemical will still be there when you arrive to open the line. Be extremely careful about exposing your eyes and skin when opening a cleanout or removing a trap on a flooded line. Do not allow any standing liquid to come in contact with your skin or eyes or to soak your clothing because, if the fluid is aggressive, the soaked clothing will provide extended exposure to your skin. Many of these drain-cleaning products are serious hazards to you, especially your eyes!

Another hazard to be considered is the possibility of mechanical injury from the drain-cleaning machine. The rotating equipment can catch hair or loose clothing. Rings can harm your fingers if crushed by a twisted cable or caught in the cable. Remove rings and loose jewelry, and wear heavy, close-fitting gloves or snaking mitts as protection from the mechanical and chemical exposures.

Other risks that need to be recognized include electrical shock from a faulty motor or power cord and hazardous contents of the pipe line. Be sure to use a ground-fault circuit interrupter (GFI) outlet on your power cord for the supply to the power rodding machine.

Medical labs present new hazards. Infections are always a risk when working with drain lines. Be sure to wash carefully and thoroughly after each cleaning job. Some medical chemicals (azides) will form explosive mixtures in the presence of metals. Azides are common in blood platelet–counting solutions. Some drains have exploded when the azide compounds were agitated by the electric snake.

In the Field

Your employer should have written safety procedures for such things as medical gas installations, confined space entry, and trench safety issues. Refresh your memory from time to time by reviewing those and other procedures.

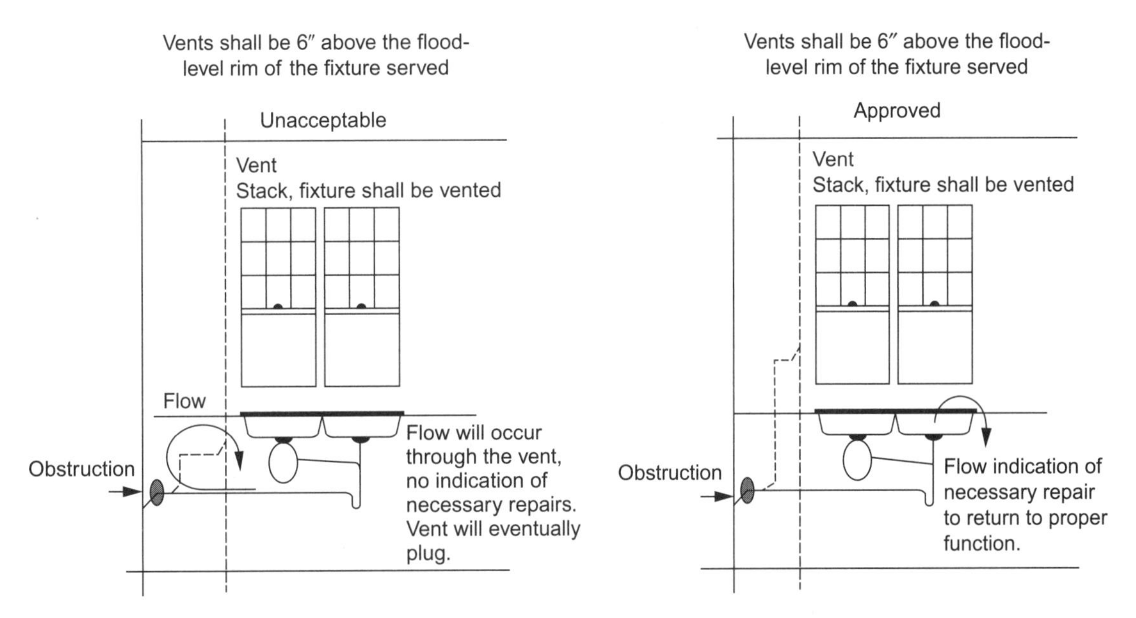

Figure 8–2
A kitchen sink vent above the fixture that illustrates vent connections to ensure proper operation of the venting system. When vent connections do not meet code requirements, line obstructions may continue without providing visual indications such as overflow conditions.

In the Field

Do not open up walls or other surfaces without approval of your supervisor and the building owner.

When cleaning drains, consider the following precautions:

1. Use eye protection. Chemical drain cleaners are frequently present and may be splashed into your eyes.
2. Plug drain-cleaning equipment into GFI outlets.
3. Check for the presence of backwater valves, grease interceptors, traps, or other appurtenances before snaking.
4. Know the torque limitations of your equipment. An electric snake can turn a specific size cable only so far in length. A lost or broken cable often means digging, and the negotiated retrieval costs may cost your employer future work.
5. Read the snaking equipment operational manual carefully and completely. Consult local tool representatives for specific insights into the limitations of the equipment.

Equipment

New equipment related to cleaning drains, video taping, and relining sewer systems has radically changed drainage service work to a perfected art. This section will describe the construction and use of the various drain-cleaning devices used to clear stoppages.

Plunger

A plunger is a cup-shaped rubber device with a short handle. When the rubber device is placed over the drain opening in a fixture, pushing down on the handle will apply a considerable pressure on the trap inlet. Note that if the stoppage is beyond the vent connection at the end of the trap arm, the plunger will not be effective because the pressure wave developed by the plunger will simply be relieved at the vent connection. Remember that water could be forced out of the overflow in fixtures so equipped. Do not use a plunger on a stopped lavatory or sink trap where an overflow may exist without first covering the overflow. Also, consider that if hazardous drain-solvent chemicals have been poured down the drain, they could splash back through the overflows if left uncovered.

Paize Plunger

A plunger made with a flat rubber disk attached to a handle is a Paize plunger. It is used to clear floor drain stoppages and is made in sizes specific to each size floor drain. The disk is pushed down the drain a few inches, water is put on top, and then the opening is sealed with rags. A quick motion with the handle will produce a pressure rise and clear the drain trap.

Closet Auger

A closet auger consists of approximately 4 feet of flexible cable encased in a sleeve that has a 90° bend and protective sleeve on one end. One end of the flexible cable has an enlargement for catching or dislodging stoppages; the other end is connected to a 3-foot straight rod and crank handle. Some units have an integral extension feature that will allow you to snake well into the closet bend.

The sleeve end with the bend is inserted into the bowl. By pushing down and turning the handle, the cable will proceed into the trapway. As the pushing and turning continues, the cable will pass completely through the bowl.

If the stoppage is fibrous (sanitary napkin, diaper, etc.), it probably will be caught and retrieved by the auger. Solid objects will be pushed ahead and into the drain pipe. However, some objects will not go either way.

Be sure to flush the water closet several times with plenty of toilet paper after using the auger. In some cases, a comb, pencil, or toothbrush was caught in the trapway. The obstacle would catch paper and solid waste, blocking the bowl, but the auger did not remove the original cause of the stoppage. In such cases, the water closet must be removed and checked for the problem. Often a compact mirror can be held down below the water level with channel lock pliers and a flashlight shined on the mirror to reveal the object on the other side of the closet trap partition. Understanding what is in the trapway will enable the technician to remove the obstruction.

All closet augers are not the same in their performance. Obtain a good-quality item and experiment with others from the shop. You will be surprised by the differences in various augers' performance.

Cables

Either manually or power-operated and available in various diameters and lengths, cables are introduced into the drain pipe and rotated continually.

They are easy to insert until the obstruction is encountered, when the going gets difficult. Once the obstruction is dislodged, the cable will move ahead again. Work the cable back and forth through the stoppage area several times to be sure it breaks up the stoppage. Be sure to also run plenty of water down the drain to flush dislodged contents ahead and away.

Check manufacturers' literature for cutting head options, cable diameters, and cable lengths to use for a particular problem. Do not overload the equipment. The motors on the equipment will provide only so much torque on the cable.

Be sure to control the cable at all times. If you punch through the obstruction at about the same time that the cable end comes off the drum, the cable could be drawn into the drain. There are retriever heads available to catch a lost cable, but they do not always work, in which case excavation would be required to extract the cable. Kinked or damaged cables will lead to lost cable portions. Check your equipment frequently and replace cables when necessary as part of preventive maintenance.

Sink Rods (Hand Snakes)

A sink rod is a small-size, relatively short (about 25 feet) cable that is used to clean lavatory or sink drains. The benefit to these units is that they are less intrusive to the system and will not create as great a clean-up mess as larger units.

Fish Tapes

A fish tape, also called a flat sewer rod coil, is a flat steel ribbon, available in various sizes, that is pushed into the drain pipe. This device is the predecessor of cables, which are rotated while being worked in the drain. The fish tape is satisfactory for the contractor who wants a low-cost device for occasional use. The units generally have a pinch-type handle that allows the operator to force the rod down the line. The oiled rods also commonly have revolving roller heads on the end.

REVIEW QUESTIONS

1. Why would a plunger be used in a water closet before a water closet auger?
2. Explain why a service technician should exercise caution with the contents of a plugged drain line.
3. Should the nearest clean-out be opened promptly to remove obstructions when found in a system?
4. Why is conducting a careful review of the pattern of the stoppage's development important prior to beginning your removal work?

CHAPTER 9

Water Hammer Service and Repair

LEARNING OBJECTIVES

The student will:

- Explain how water hammer is developed in water piping systems.
- Demonstrate how water hammer can be reduced to aid in a water system's performance.
- Explain the different methods of providing water hammer arrestors.
- Explain how to obtain the correct size device and place the device in the correct location.

WATER HAMMER SERVICE AND REPAIR

The study of water hammer and the methods of minimizing it are extremely important to ensure the proper operation of a water distribution system in a structure. It is a result of having full-flow to no-flow conditions. Water hammer, besides being very noisy and distracting, may cause a great deal of damage to the plumbing system. It has been known to rupture pipe, damage valves (especially the solenoid type), open relief valves, and break fitting joints. This chapter will study causes of water hammer and discuss methods used to prevent it, such as fabricated and mechanical water hammer arrestors.

Source of Water Hammer

A column of water moving in a piping system at a moderate velocity cannot be stopped instantly. If we attempt to do so by rapidly closing a valve, either by manual motion or by releasing a spring-loaded plunger, the moving water impacting the closure mechanism will produce a large pressure increase due to changes in velocity, causing the water to reflect off the closure and travel back down the water line. This pressure rise and reverse travel will produce a sound and impact the pipe as though it were struck with a hammer, hence the term *water hammer*. Water hammer can also occur without an audible sound.

Typical valves that generally create this problem are dishwasher, laundry machine, and softener solenoid valves. Single lever sink or lavatory valves can be closed rapidly enough to develop the problem, but this rapid closure must be done intentionally. For that reason, we would consider such hammer noises as results of vandalism or accidental misuse.

Conditions that increase the severity of water hammer include higher initial pressure, higher flow velocity, and faster closure of the valve. Loosely supported piping will increase the noise level, but poor methods of hanging pipe will not increase the water hammer phenomenon. Because a higher velocity makes the problem worse, a larger size line will reduce the velocity (for a specific flow rate) and therefore reduce or eliminate water hammer. Additionally, the pressure surge shock waves are smaller in plastic piping systems because the plastic material will absorb some of the energy to further dampen the force.

Prevention of Water Hammer

The usual method of controlling water hammer is to place a compressible gas container as close as possible to the quick-closing valve because the shock wave attenuates rapidly as it travels back from the closure source. A compressible gas at the site of the quick-closing valve allows the momentum of the moving water to be absorbed by reducing the volume of the gas with only a modest pressure rise.

The obvious method of providing this gas volume is to install a tee and capped riser adjacent to the valve, called a field fabricated absorber or air chamber. The air trapped in the riser provides the shock absorption desired. Unfortunately, experience shows that the air trapped in this piping arrangement is soon absorbed by the water as the water surges up and down into this side branch. This absorption of the air into the water eventually results in all air being lost, which causes the air chamber to become waterlogged. Also, fabricated air chambers, which are increased in size and length to hold more air, are filled by air under gravity force, 14.7 psi. Unfortunately, that atmospheric air will be compressed to $\frac{1}{5}$ or $\frac{1}{6}$ of its original volume by the line water pressure at the system's reactivation. To eliminate this problem, manufactured devices called shock absorbers or shock arrestors are used. The shock absorber contains an air volume sealed into a bellows or piston device that prevents the air from being absorbed into the water.

Other means of reducing the severity of water hammer include increasing pipe size to the quick-closing valve or changing the valve to one with a slower closing rate. Water hammer can produce a pressure rise in pounds per square inch (psi) of up to 60 times the water flow rate (in feet per second, or fps). Thus, if the flow is at 10 fps, sudden closure can produce a pressure rise of 600 psi. If the valve closure rate can be reduced or the pipe size to the valve increased, the pressure rise will be greatly reduced. The most effective way to prevent water hammer is to have a nearby faucet open and flowing when the fast-acting valve closes, but this is impractical. If another flow path exists, the pressure build-up cannot occur. Some ballcocks are available with pressure relief mechanisms, and these devices safely relieve pressures that exceed 80 psi by bleeding to the closet tank.

Figure 9–1 shows an air chamber design with typical dimensions for these devices. Such field-fabricated arrangements must be installed to be accessible. If the chamber becomes full of water, it can be recharged with air by closing the isolating valve and opening the air inlet and drain valves. After all the water has drained out, close the drain and air vent valves and open the isolating valve.

Figure 9–2 shows typical mechanical shock arrestors. These devices cannot lose the air charge unless the bellows or internal O-rings fail. Because these failures can occur, the devices must also be accessible. Smaller devices are usually used in residential use groups and are installed at the fixture, for example, at the automatic clothes washer box. Larger devices may be installed at a large commercial fixture or in an accessible pipe chase.

Tables 9–1 and 9–2 provide information for selecting mechanical shock absorbers. Note that there are standard sizes A through F, each of which is rated for a certain number of fixture units. These tables present data furnished by the Plumbing and Drainage Institute (PDI), an industry association of persons interested in these matters. The information is in standard PDI-WH201-06, available from PDI, which can be accessed at http://www.pdionline.org. The standard was developed to provide a uniform measure of performance and has a great deal of information for the plumber. This and other standards from PDI (such as G101 for grease interceptors) have been incorporated into many plumbing codes and integrated into manufacturers' installation instructions. Most of the referenced standards groups only sell their standards, which of course helps them recover their operational expenses.

In the Field

Several plumbing codes state that water hammer arrestors must conform to ASSE 1010, Performance Requirements for Water Hammer Arrestors. This standard is the criteria used by manufacturers to measure their products' performance. Code-required conformance to PDI-WH201, good designer practices, and installation instructions will address the installation methods.

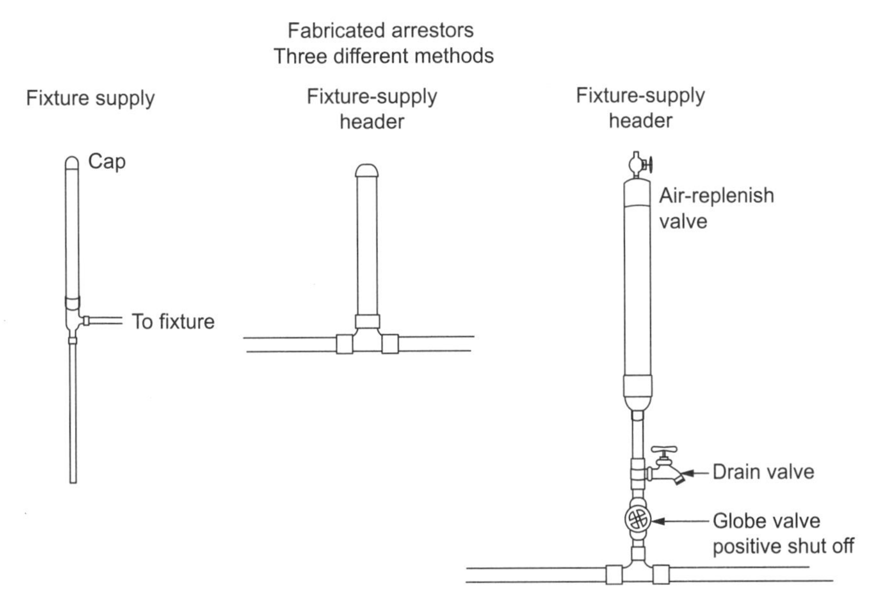

Figure 9–1
Field-fabricated pipe arrestors have been used for many years. Several codes do not accept them because of a prevailing opinion that the air becomes absorbed at a rate that allows them to become ineffective. The systems must be drained to replenish the air.

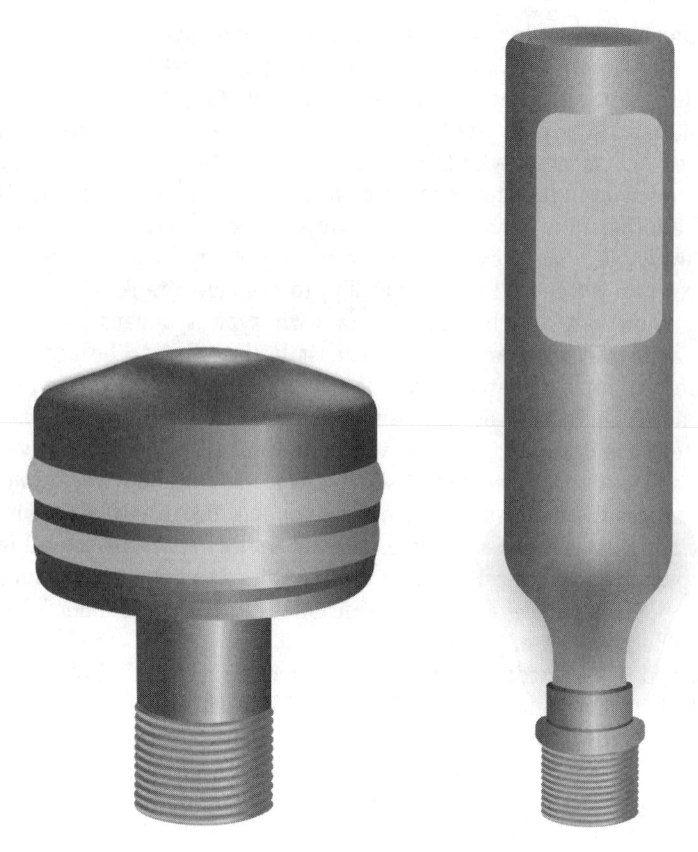

Figure 9–2
Manufactured shock arrestors reduce the water hammer within the piping system.

Table 9–1 Sizing and Selection—Water Hammer Arrestors

PDI Unit	A	B	C	D	E	F
Fixture Units	1–11	12–32	33–60	61–113	114–154	155–330

Table 9–2 Water Hammer Arrestor Selection for Water Flow Pressures up to 65 Psig

	PDI Water Hammer Arrestors					
Length of Pipe	Nominal Pipe Diameter					
	$\frac{1}{2}''$	$\frac{3}{4}''$	1″	$1\frac{1}{4}''$	$1\frac{1}{2}''$	2″
25	A	A	B	C	D	E
50	A	B	C	D	E	F
75	B	C	D	AE	F	EF
100	C	D	E	F	CF	FF
125	C	D	F	AF	EF	EFF
150	D	E	F	DF	IF	FFF

Table 9–1 shows the correlation between the PDI size schedule and the typical number of fixture units associated with that size. Fixture unit information is identified in PDI-WH201.

Table 9–2 shows the recommended mechanical arrestor size as a function of pipe size and pipe length for a maximum flow pressure (65 psig) and a maximum flow rate (10 fps).

There is seldom a need for the larger units because only the fast-acting valves create the water hammer problem, and the more fixtures that are on a line, the more likely that something will be flowing when the fast-acting valves close. Further, if no other device is flowing, the flow rate to only one device on a large line will be very low, thus limiting the energy available to develop water hammer. When long runs of piping are employed to serve a remote item of equipment, the water hammer arrestor should be located as close as possible to the point of quick closure. At this location, the water hammer arrestor will control the developed energy and prevent the shock wave from surging through the piping system.

The length of pipe to be used in Table 9–2 is the developed length of pipe size of the fast-acting valve and one size larger to the point where it connects to a line two sizes larger. For example, a $\frac{3}{8}''$ dishwasher valve would require that you measure the developed length of the supply to this valve back to where it connects to a $\frac{3}{4}''$ branch.

PDI-WH201 contains application examples and discussion for multiple fixtures, including those that are equipped with slow-acting valves, such as flushometers and lavatories. Such systems do not give rise to water hammer, and for reasons cited above, multiple fixture installations are very unlikely to generate significant problems.

REVIEW QUESTIONS

1. What is the cause of water hammer?
2. Explain why increasing the line sizing affects water hammer.
3. Explain the effect that solenoid valves have in a water supply system.
4. What is the challenge of having loose or inadequate hanger systems in a water distribution system?
5. Explain why field-fabricated air chambers are not as effective for water hammer protection as mechanical arrestors.

CHAPTER 10

Controls Troubleshooting

LEARNING OBJECTIVES

The student will:

- Identify the importance of pressure-reducing valves in the water system.
- Summarize where vacuum relief valves are used and how they benefit the plumbing water distribution system.
- Compare locations of the various temperature controls in the water distribution system and describe the various temperatures used in the system.
- Describe the troubleshooting procedures used to repair faucet sensors in the plumbing system.

PLUMBING CONTROLS TROUBLESHOOTING OVERVIEW

This chapter's consideration of plumbing control troubleshooting will focus on the piping and fixture controls dealing with **pressure and temperature.** Controls by definition are those devices that regulate the system or its outlets such as pressure, temperature, and vacuum mechanisms. The plumbing water and waste systems are generally simple stand-alone operations. Plumbing appurtenances and appliances are involved in electrical considerations much the same as heating, ventilation, and air conditioning systems (HVAC). The chapter will close with considerations of faucet sensors and float sensors. Backflow preventers, the most important plumbing controls, will be considered in a later chapter.

Water Pressure–Reducing Valve

In the Field

Most plumbing codes limit the maximum pressure to a building to 80 psi or less. They will generally go on to state that if the supplied pressure is over 80 psi, an approved pressure-reducing valve conforming to ASSE standard 1003 must control it. Always check your local code for requirements in your area.

Excessive water pressure in a water distribution system is detrimental and will adversely affect or damage system components. The most susceptible component in a plumbing system recognized by a service technician is a water closet ballcock. Technicians would next consider system solenoid valves for washers and then joint failure. Excessive pressures from the water supply are generally associated with a public water supply purveyor. City pressures are considered to be stable; however, they may vary due to high usage. A private well supply to a structure has its own pressure controls associated with the well and its pump controls.

In order to reduce system pressures to a specified maximum level, a pressure-reducing valve is installed, as in Figure 10–1. Pressure-reducing valves are also referred to as pressure-regulating valves or pressure regulators. They are designed to limit the system's output to a certain maximum level at no-flow conditions, called static pressure, and to maintain water pressure at predetermined rates during flow conditions. *Residual pressure*, also called flowing pressure, is a term describing the pressure at the point of discharge considering the pressure drop from devices such as backflow preventers and piping friction loss.

Pressure-reducing valves come in several different designs because of various system flow requirements in the valves, which would have been considered prior to installation and your service call. The two main designs with different characteristics are direct operated and pilot operated. The direct-operated valve has a lower cost and is less accurate than the pilot-operated unit. The pilot-operated valve has a greater degree of accuracy during a larger range of flow and pressure conditions. All valve types should have system strainers either built into the device or separately installed before the unit.

Figure 10–1
A pressure-reducing valve used to reduce the water supply to a desired water pressure. Courtesy of Watts Water Technologies, Inc.

A very important consideration when selecting a valve or its replacement is the bypass provision of the valve. This bypass affects the system tremendously to relieve excessive expansion problems by dissipating thermal expansion caused by the water heater. Pressure-reducing valves without a bypass act like a system check valve or system backflow preventer. This action basically seals the structure's system under static conditions and does not allow expanded water to migrate back out to the water service, which may cause the water heater relief valve to discharge. Plumbing codes generally recognize this potential and call for a pressure-reducing valve with a bypass or a device to compensate for the thermal expansion. Another device would be an expansion tank properly sized and mounted at the water heater.

Troubleshooting

Troubleshooting begins with a customer reporting excessive noise in the system, various system component failures, or relief valve discharges. The following items will outline your course of action to identify and resolve the problem.

1. Excessive noise – Review the system and check for loose hangers and inoperable water hammer arrestors, which should be replenished or replaced. Move to step 4.
2. Various system component failures – Review the components and determine whether a pattern is present. For example, one ballcock diaphragm failure would not establish a concern for excessive pressure. Several ballcock or solenoid failures are good indications of a pressure problem. Check to see if inoperable water hammer arrestors should be replenished or replaced, and repair components as necessary. Move to step 4.
3. Relief valve discharges – Review the system and check the relief valve as detailed in Chapter 7. Determine whether a pressure-reducing valve is present. Next, attach a pressure gauge to the system and note the operating pressure. Replace the relief valve if it is faulty. If the relief valve is functioning correctly, proceed to step 4.
4. Service the water hammer arrestors if necessary. If they appear to be operating correctly, determine whether a pressure-reducing valve is present. Next, attach a pressure gauge to the system and note the operating pressure. If the system pressure is far above normal (80 psi), proceed to step 5.
5. Several things may occur here to analyze and resolve the problem:
   A. The municipal water supply system may have been upgraded, which would require the selection and installation of a new pressure-reducing valve. It is expected that the water purveyor would have notified the building owner if that occurred.
   B. The municipal water purveyor may have upgraded its water meters to have built-in backflow preventers, which causes a thermal expansion problem with the relief valve. An expansion device would have to be incorporated into the system to provide resolution.
   C. The existing pressure-reducing valve needs to be serviced. Note that a plugged strainer would not cause excessive pressure at the valve outlet; however, it should be serviced.

In the Field

The cost of a relief valve in residential applications is minimal. You should consider valve replacement as a matter of preventive maintenance and explain this to the customer.

In the Field

Most codes that limit the system pressure as stated earlier will allow outside sillcocks with higher pressure levels and will also allow upper floor height considerations. For that reason, be careful that the outlet you select for your gauge pressure reading reflects the system's adjusted pressure.

Vacuum Relief Valves

A vacuum relief valve, shown in Figure 10–2, is a small device that is often overlooked in the plumbing system. Its importance and necessity are addressed by some plumbing codes in two different locations. First, all bottom-fed water heaters and bottom-fed tanks must have a vacuum relief valve. Second, pressure tanks, except those with internal diaphragm means, must have vacuum relief valves installed.

The codes generally require that the vacuum relief valve conform to the performance requirements of ANSI Z21.22, IV. The valve size and its connection are

Figure 10–2
A vacuum relief valve is required to avoid unit damage when bottom supplied tanks are subject to a vacuum. Courtesy of Watts Water Technologies, Inc.

relatively small: $\frac{1}{2}''$ and $\frac{3}{4}''$. A $\frac{3}{4}''$ valve can serve a much larger line size. The valve is mounted on top of pressure tanks and on the hot side of water heaters; remember that the water heater's cold feed is on the bottom of the tank. These bottom-fed heaters are only storage heaters and, generally, electric heaters with heating elements.

The purpose of the device is to activate and stop a vacuum when a tank is drained and siphoned. A vacuum relief valve allows air to enter and even out with atmospheric pressure, while a drained tank under a partial vacuum would allow external atmospheric pressure to damage the drained tank or burn out the elements (now dry).

Troubleshooting

For example, troubleshooting would begin with a customer informing you on a service call that, when the building maintenance person changed the office building's photo-developing water filters in the past, other problems seemed to occur. The customer would call an electrician to come out and replace the element in the water heater located above the men's restroom.

Diagnosis

You suspect that when the water is shut off to change the filters, the maintenance person shuts off the main water supply and drains the system pressure at the employee lavatory. Further, this employee does not shut off the power supply to the heater or shut off the heater gate valve. The electric heater is a small unit with a cold-fed bottom supply mounted above the restroom.

Solution

You explain the problem of the maintenance staff's poor practice of not following a proper shutdown procedure. Further, you provide an estimate for installing a vacuum relief valve and offer the customer a scheduled maintenance contract. The contract will address the filter changes and will reduce costs associated with an unnecessary electrician service call in the future.

Temperature Controls Overview

Having an adequate supply of hot water to a structure is not only a plumbing code requirement, it is essential for a building to operate and to ensure that building occupants have adequate facilities for proper sanitation. Higher temperatures of hot water are necessary in occupancies dealing with food preparation and processing. Hot water is defined by most codes as being 110°F or greater and is not limited in its maximum temperature by definition. Tempered water is generally defined as water between 85°F to 110°F. Many codes require that tempered water be supplied to hand washing facilities rather than hot water to protect the user from burns. Users who are specified in these considerations are elementary schoolchildren and accessible/barrier-free users. The latter group would of course include individuals who may have lost the sense of temperature in their hands, subjecting them to greater burn risks.

> **In the Field**
>
> Water over 125°F can cause severe burns.

Burns are classified according to their depth and the degree of tissue damage. First-degree burns are limited to the outer layer of skin and result in very tender, red skin but no blisters. Second-degree burns damage the skin (epidermis) and underlying tissue (dermis) and result in significant blistering, which is susceptible to infection. Third-degree burns are very serious and involve destruction of the full epidermis, extend to the dermis, and may destroy nerves. The tissue is very susceptible to infections. The amount of area involved in the burn is of great importance in the medical treatment.

The amount of time an individual is exposed to hot water as well as the temperature must be considered when discussing these relationships to burns. Table 10–1 provides a general overview of the water temperature and time necessary to produce serious burns.

Table 10–1 Serious Burn Temperature and Time Connections

Water Temperature	Time for Third-Degree Burns to Occur	
	Children	Adults
100°F	Safe for bathing	Safe for bathing
120°F	2.5 minutes	5 minutes
140°F	1 second	5 seconds

Over the past several years, plumbing codes have advanced in addressing burn prevention safety. Residential fixtures such as showers have been mandated to have safety devices built into the valves that conform to performance standards and are designed to address water temperatures. In addition, early temperature protection has been provided to protect an individual who is showering from thermal shock. For example, when another fixture is used, such as flushing the water closet next to the shower, the shower valve could have lost its cold-water volume. This thermal shock could startle the bather, who could then pull back and slip, resulting in an injury. The valve has also been required to have a high-temperature limit control. The control was built into the valve as a mechanical adjustment to limit the motion of the valve handle, thereby regulating the temperature. Municipal sources from a reservoir have been a problem in this area from winter to summer with widely varying temperatures supplied on the cold side. The different temperatures remove the so-called calibration from the mechanical setting.

Several codes now have expanded temperature safety from accessible lavatories and showers to include bathing tubs and bidets.

The expansions in the codes to address temperature safety issues are largely a result of manufacturers' developing more sophisticated and safer products and designers' seeking greater code uniformity. Members of both these groups travel to the various code agency update conferences in an effort to improve the codes. Table 10–2 provides a list of some to the code-recognized fixture applications, referenced standards, and temperature ranges. Exercise caution here as this information varies from code to code.

In the Field

There are presently four major plumbing codes in North America, along with numerous state and local codes. Often, local codes are composed of the national codes with local jurisdiction amendments. The four codes, in alphabetical order, are as follows:

- *International Plumbing Code*
- *National Plumbing Code of Canada*
- *National Standard Plumbing Code*
- *Uniform Plumbing Code*

Troubleshooting

Troubleshooting is accomplished by noting the device's deficiencies as reported by the building owner and evaluating the device's ability to provide temperature regulation. Temperature readings with an accurate device are critical to your diagnosis. Assistance from a helper is valuable because the helper can shut down the cold or hot supply, which enables your evaluation. Finally, consulting the manufacturer's information through their website is also useful.

In the Field

The code community considers limiting the temperature in a bathtub to 120°F for safety. Many homeowners have their water heater temperature turned up for several reasons, one of which may be that an owner runs extremely hot water into her cast iron tub in the winter prior to bathing and later adds appropriately mixed temperature water. Why? She does not want to have a cold bottom while sitting in a warm tub.

Faucet Sensors

Electronic flushing and faucet use is one of the most innovative concepts in the plumbing field in recent years. The commercial area of plumbing has embraced the electronic flush devices, primarily in flushometer valves and lavatory faucets, to address conservation issues. The residential side has appreciated the convenience of hands-free operation and the appearance of leading technology. The electronic flow mechanisms are practical and hygienic and increase water conservation. These hands-free electronic devices require the service technician to have more knowledge to install and maintain them. They also may require more detailed service because of their smaller port size and increased number of components, which include the typical faucet, sensors, transformers or battery packs, and a solenoid valve.

Whereas plumbing has always been fairly basic, these sensor devices offer many variances. Manufacturers have been very innovative, and therefore the products differ widely. For example, the power supply can vary from a 12- or 24-volt trans-

Table 10–2 Fixtures and Applications Referenced in Code

Fixture/Application	Referenced Standard	Maximum Temperature Discharge
Bathtub/whirlpool tub	ASSE 1070 – Performance Requirements for Water-Temperature Limiting Devices	Maximum of 120°F
Bidet	ASSE 1070 – Performance Requirements for Water-Temperature Limiting Devices	Maximum of 110°F
Combination tub and shower	ASSE 1016 – Performance Requirements for Individual Thermostatic, Pressure Balancing, and Combination Control Valves for Individual Fixtures; or CSA B125 – Plumbing Fittings	Maximum of 120°F
Public hand-washing facilities	ASSE 1070 – Performance Requirements for Water-Temperature Limiting Devices	Minimum of 85°F Maximum of 110°F
Showers, individual	ASSE 1016 – Performance Requirements for Individual Thermostatic, Pressure Balancing, and Combination Control Valves for Individual Fixtures; or CSA B125 – Plumbing Fittings	Maximum of 120°F
Showers, gang	ASSE 1069 – Performance Requirements for Automatic Temeratura-Control Mixing Valves; or CSA B125 – Plumbing Fittings	Maximum of 120°F
Systems, such as dual temp. systems in a restaurant	ASSE 1017 – Performance Requirements for Temperature Actuated Mixing Valves for Hot Water Systems	Variable for system, not for final fixture control

ASSE – American Society of Sanitary Engineering
CSA – Canadian Standards Association

former supply to batteries and battery packs and now to water-powered generators, which recharge their own batteries. The battery-supplied units are easier to install and more economical at installation. Flushing devices have range and sizing recognition capabilities and can be controlled from remote locations, even based upon time. As you might expect, their memory-control modules provide information to the service technician concerning a multitude of factors. The bottom line is that the products vary a great deal from one supplier to the next.

Service with such a wide range of product differences begins with one basic concept—become familiar with the product as quickly as possible. That is best accomplished by familiarizing yourself with all of the product literature. The installation instructions may have been left on site; however, the best sources are the manufacturers and what they provide on their websites, such as troubleshooting procedures. It is not uncommon today to find service technicians with laptop computers that have wireless connections. Those individuals have instant resources for troubleshooting, finding part numbers, and ordering instantly. Several companies now specialize in "tough book" computers developed for construction use.

In the Field

It is often necessary to flush and purge the water supply to the electronically metered faucet at its filter and port locations. Exercise great caution here so you don't compromise your safety or damage the electrical components by water leakage.

Troubleshooting

Troubleshooting electronically operated metering faucets is accomplished by using a test multimeter. *Plumbing 301* provided background on electric circuit troubleshooting. The electric parts to analyze are the transformer or power source, the sensor, and the solenoid valve.

First, use the appropriate setting on the multimeter scale and confirm the building's power source (110 volts), and then move to the transformer outlet at the sensor and solenoid connections to measure the control voltage. The transformer label should indicate the operating control output voltage such as 12 or 24 volt. The batteries and their output wiring, which is commonly 6 volts, should also be checked. If the batteries or transformer are not operating, replace the transformer or battery pack.

Second, test the sensor output with the multimeter measuring at the top of the solenoid head connections. You of course have to activate the sensor recognition by presence and movement. If the sensor does not send a signal to the solenoid, the sensor must be serviced.

The last operation is to observe the solenoid valve. Actuate the sensor and listen for a click or buzzing to indicate the solenoid actuation and travel. If it is not present, the solenoid valve needs attention. The valves vary a great deal and at times may be disassembled for repair.

Sump Control Switches

The move from considerations of safe potable water faucet sensors to sewage pump switches would seem to be jumping from one extreme to another. Indeed it is, and this type of balancing act is common for the average service technician.

Sewage sumps, waste lifts, subsoil piping sumps, and ejectors have many different purposes with different terminology and one thing in common: they all need an evacuation pump that is controlled by a switching mechanism. There are two major switches to consider for these pumps—the pressure switch and the float switch. Each of these has various spin-offs, but the principles are the same.

Pressure switches are mounted low in the sump on the pump, generally on smaller sewage lifts, and are turned on as the height of the liquid increases, which increases the pressure on the switches. They are the least susceptible to fouling problems from solids or paper and require the pump to be pulled for switch servicing. Float switches have two types of activating methods. A smaller pump, which may have the motor above floor level, is connected to its submersed impeller by a long shaft. This set-up generally has a float switch on the motor housing with a rod running down to a float. This mechanical linkage turns the pump on when the rod reaches a predetermined point, tripping the mechanism. This type of assembly is very serviceable without pulling the pump. The float tipping up (floating) and closing the contacts within the float, thereby activating the pump, activates the other type of float switch. They have their own mounting and may be pulled easily but are subject to fouling. Troubleshooting the sump switches is a fairly simple procedure. First, check the power supply and then visually check the float mechanism to identify obstructions, which retard the float's movements.

REVIEW QUESTIONS

1. Explain when a vacuum relief valve is to be installed on a water heater.
2. Describe when tempered water might be supplied to public lavatory faucets.
3. Describe what the benefits of a sensor-controlled water closet flushometer valve would be.
4. Explain why a pressure-reducing valve without a built-in bypass valve may adversely affect the water heater relief valve.

CHAPTER 11

DWV Sizing Utilizing Blueprint Reading

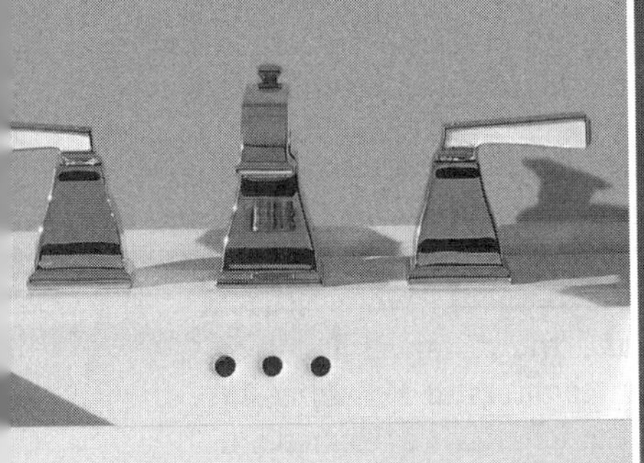

LEARNING OBJECTIVES

The student will:

- Describe how to size a sanitary drainage system.
- Interpret an isometric drawing to better understand a plan-view blueprint requirement.
- Explain how to size a drainage vent system.
- Discuss different piping layouts for a proposed venting isometric drawing.

PLUMBING DESIGN SIZING RELATED TO BLUEPRINTS

The next four chapters are arranged to provide sizing information for systems addressing sanitary drain, waste, and vent piping, storm drainage piping, potable water piping, and fuel gas piping with its venting. The system sizing will combine blueprint reading, which was addressed in both *Plumbing 201* and *Plumbing 301*, with code sizing requirements. Further, it will allow you to use your accumulated mathematics and practical training knowledge in the development of design/build systems using blueprints.

Your knowledge of how to install systems using the required itemized designs must advance to a level at which you can work in a fast-track design/build process in which you design the system in accordance with code requirements. In many cases, the design/build conceptualization takes place before the plans are in place, which is a fairly new practice and has the advantage of reaching a final product in a shorter time and at a lower cost than conventional methodologies. The key to this concept is implementing the desired result rather than going through the conventional process studied earlier to achieve that result. The philosophy is also characterized by the attitude of "Do not tell me what to do; I will provide the results you need." This type of delivery method is possible when the owner employs one company to serve as the designer, engineer, and builder with the builder's usual subcontractor team.

Your design and sizing will require you to use the architectural floor plans as the basis. The shop drawings you create for material take-off and installation procedures will commonly be in the form of isometric work sheets. This and subsequent chapters will use a very basic floor plan view and rough work sheet training method to aid in explanations.

Sizing Sanitary Drainage

Accurate sanitary drainage layouts are dependent upon consideration of three major factors: drainage fixture units, line sizing, and pipe slope. Venting will play an extremely important part in the layout because the type of layout used is dependent upon the type of venting method selected and the distances involved, such as the distance from the trap to the vent. The last chapters of *Plumbing 401* will detail code requirements for selected vent types, which have a tremendous effect on system layout. For training purposes in this chapter, venting methods will be utilized, named, and briefly described to aid the reader in understanding drainage table sizing.

In the Field

The codes may vary a great deal in their DFU tables. For example, many codes will have minimum trap size requirements built into the DFU tables, but some will not. The minimum size must be considered and understood in order to properly design a plumbing layout. Always defer to your local code requirements.

Drainage Fixture Unit Values

Drainage fixture unit (DFU) discharge values come into play when considering the load placed on a system based upon the type of plumbing fixture and its amount of discharge through the drainage system. That load amount must be considered to make adjustments for line sizing. Many years ago, 1 DFU was considered to be the amount a full lavatory discharged through a $1\frac{1}{4}''$ line in 1 minute. That old-fashioned lavatory contained 1 cubic foot, or approximately $7\frac{1}{2}$ gallons, of wastewater. Many changes have occurred since then to form a sophisticated methodology of determining discharge values. The greatest considerations are now based upon water conservation drainage reductions and use-group consideration. The following DFU table is composed of portions of the *National Standard Plumbing Code (NSPC)* that address buildings other than dwelling units. We will be using a single-story office building in our example, which will not require the entire NSPC table. See Table 11-1.

In the Field

The codes may also vary in how they provide DFU reductions when considering groups of fixtures, such as bathroom groups. Normally, reduced DFU values will be recognized. This makes sense due to the reduced number of uses for family groups. This consideration is referred to as probability of use. Again, always defer to local code requirements.

Building Drain and Sewer Sizing

The building drain and building sewer are the lowest portion of the sanitary drainage piping system. The minimum sizes of these pipes are based upon the number of DFUs

Table 11–1 Drainage Fixture Unit (DFU) Values

Type of Fixtures	Individual Dwelling Units	Serving Three or More Dwelling Units	Other Than Dwelling Units	Heavy-use Assembly
Bathroom Groups Having 1.6 GPF Gravity-tank Water Closets				
Half-bath or Powder Room	3.0	2.0		
1 Bathroom Group	5.0	3.0		
1½ Bathrooms	6.0			
2 Bathrooms	7.0			
2½ Bathrooms	8.0			
3 Bathrooms	9.0			
Each Additional ½ Bath	0.5			
Each Additional Bathroom Group	1.0			
Bathroom Groups Having 1.6 GPF Pressure-tank Water Closets				
Half-bath or Powder Room	3.5	2.5		
1 Bathroom Group	5.5	3.5		
1½ Bathrooms	6.5			
2 Bathrooms	7.5			
2½ Bathrooms	8.5			
3 Bathrooms	9.5			
Each Additional ½ Bath	0.5			
Each Additional Bathroom Group	1.0			
Bathroom Groups Having 3.5 GPF Pressure-tank Water Closets				
Half-bath or Powder Room	3.0	2.0		
1 Bathroom Group	6.0	4.0		
1½ Bathrooms	8.0			
2 Bathrooms	10.0			
2½ Bathrooms	11.0			
3 Bathrooms	12.0			
Each Additional ½ Bath	0.5			
Each Additional Bathroom Group	1.0			
BATH GROUP (1.6 GPF Flushometer Valve)	5.0	3.0		
BATH GROUP (3.5 GPF Flushometer Valve)	6.0	4.0		
Individual Fixtures				
Bathtub or Combination Bath/Shower, 1½″ Trap	2.0	2.0		
Bidet, 1¼″ Trap	1.0	1.0		
Clothes Washer, Domestic, 2″ Standpipe	3.0	3.0	3.0	
Dishwasher, Domestic, with Independent Drain	2.0	2.0	2.0	
Drinking Fountain or Watercooler			0.5	
Food-Waste-Grinder, Commercial, 2″ Min Trap			3.0	
Floor Drain, Auxillary			0.0	
Kitchen Sink, Domestic, with One 1½″ Trap	2.0	2.0	2.0	

Continued

Table 11–1 Drainage Fixture Unit (DFU) Values (Continued)

Type of Fixtures	Individual Dwelling Units	Serving Three or More Dwelling Units	Other Than Dwelling Units	Heavy-use Assembly
Kitchen Sink, Domestic, with Food-Waste-Grinder	2.0	2.0	2.0	
Kitchen Sink, Domestic, with Dishwasher	3.0	3.0	3.0	
Kitchen Sink, Domestic, with Grinder and Dishwasher	3.0	3.0	3.0	
Laundry Sink, One or Two Compartments, $1\frac{1}{2}''$ Waste	2.0	2.0	2.0	
Laundry Sink, with Discharge from Clothes Washer	2.0	2.0	2.0	
Lavatory, $1\frac{1}{4}''$ Waste	1.0	1.0	1.0	1.0
Mop Basin, 3″ Trap			3.0	
Service Sink, 3″ Trap			3.0	
Shower Stall, 2″ Trap	2.0	2.0	2.0	
Showers, Group, per Head (Continuous Use)			5.0	
Sink, $1\frac{1}{2}''$ Trap	2.0	2.0	2.0	
Sink, 2″ Trap	3.0	3.0	3.0	
Sink, 3″ Trap			5.0	
Trap Size, $1\frac{1}{4}''$ (Other)	1.0	1.0	1.0	
Trap Size, $1\frac{1}{2}''$ (Other)	2.0	2.0	2.0	
Trap Size, 2″ (Other)	3.0	3.0	3.0	
Trap Size, 3″ (Other)			5.0	
Trap Size, 4″ (Other)			6.0	
Urinal, 1.0 GPF			4.0	5.0
Urinal, Greater Than 1.0 GPF			5.0	6.0
Wash Fountain, $1\frac{1}{2}''$ Trap			2.0	
Wash Fountain, 2″ Trap			3.0	
Wash Sink, Each Set of Faucets			2.0	
Water Closet, 1.6 GPF Gravity or Pressure Tank	3.0	3.0	4.0	6.0
Water Closet, 1.6 GPF Flushometer Valve	3.0	3.0	4.0	6.0
Water Closet, 3.5 GPF Gravity Tank	4.0	4.0	6.0	8.0
Water Closet, 3.5 GPF Flushometer Valve	4.0	4.0	6.0	8.0
Whirlpool Bath or Combination Bath/Shower, $1\frac{1}{2}''$ Trap	2.0	2.0		

NOTES:

1. A bathroom group, for the purposes of this table, consists of not more than one water closet, up to two lavatories, and either one bathtub, one bath/shower combination, or one shower stall. Other fixtures within the bathing facility shall be counted separately to determine the total drainage fixture unit load.
2. A half-bath or powder room, for the purposes of this table, consists of one water closet and one lavatory
3. For unlisted fixtures, refer to a listed fixture having a similar flow and frequency of use.
4. When drainage fixture unit (DFU) values are added to determine the load on the drainage system or portions thereof, round the sum to the nearest whole number before referring to applicable tables for sizing the drainage and vent piping. Values of 0.5 or more should be rounded up to the next higher whole number (9.5 = 10 DFU). Values of 0.4 or less should be rounded down to the next lower whole number (9.4 = 9 DFU).
5. "Other Than Dwelling Units" applies to business, commercial, industrial, and assembly occupancies other than those defined under "Heavy-Use Assembly." Included are the public and common areas in hotels, motels, and multi-dwelling buildings.
6. "Heavy-Use Assembly" applies to toilet facilities in occupancies that place heavy, but intermittent, time-based loads on the drainage system, such as; schools, auditoriums, stadiums, race courses, transportation terminals, and similar occupancies where queuing is likely to occur during periods of peak use.
7. Where other than water-supplied fixtures discharge into the drainage system, allow 2 DFU for each gallon per minute (gpm) of flow.

that flow into the pipes and the slope or grade at which the pipes are installed. The line sizes are increased as additional fixtures are added in the direction of flow.

The drainage tables were developed based upon the lines being half full. It is important to understand, although controversial, that this half-flow concept enables the system to have some venting capabilities from the top half of the pipe, such as with low-flow, oversized lines in a combination drain and vent system. Another very important consideration, as shown in Table 11–2, is that it includes not only the building drain and building sewer but also the fixture branches connected to the building drain. A fixture branch is defined as a line serving two or more fixtures and discharging into another line. Do not be confused; these lower branches to the building drain can conduct a greater DFU value than the branches connected to stacks.

The table consideration of slope incorporates the aspect of having lines that build in flow velocities of 2 feet per second (fps) to ensure adequate flow. Several older plumbing codes increased the minimum line slope to ensure 4 fps for greasy waste such as food service sink discharges. The vast majority of codes will have a separate section to specify that drain lines will have at least a $\frac{1}{4}''$ per foot slope for 2-inch and smaller lines to ensure adequate velocity.

Building drains and building sewers are sized according to Table 11–2.

Sample Building System

The following example illustrates part of a fast-track process challenging you for a layout in an office building. Figure 11–1 shows a preliminary floor plan draft with several plumbing fixtures and appurtenances. The fixtures include the following:

5 – Flushometer valve water closets, flushometer valves, 1.6 gpf	Total 20 DFU
1 – Urinal, 1 gpf	Total 4 DFU
2 – Floor drains, 3-inch (auxiliary) = 0	Total 0 DFU
4 – Lavatories	Total 4 DFU
2 – Electric water coolers	1 DFU
1 – Service sink, 2-inch trap	3 DFU
1 – Residential dishwasher, drain to sink	—
1 – Residential kitchen sink (dishwasher connection)	3 DFU
	Total 35 DFU

Table 11–2 Building Drains and Sewers[1]

Maximum Number of Drainage Fixture Units (DFU) That May Be Connected to Any Portion of the Building Drain or the Buliding Sewer.				
Pipe Size Inches	$\frac{1}{16}$ Inch	$\frac{1}{8}$ Inch	Slope Per Foot $\frac{1}{4}$ Inch	$\frac{1}{2}$ Inch
2			21	26
3			42	50
4		180	216	250
5		390	480	575
6		700	840	1,000
8	1,400	1,600	1,920	2,300
10	2,500	2,900	3,500	4,200
12	3,900	4,600	5,600	6,700
15	7,000	8,300	10,000	12,000

[1] On-site sewers that serve more than one building may be sized according to the current standards and specifications of the Authority Having Jurisdiction for the public sewers.

In the Field

The codes may vary a great deal in their consideration of the number of water closets, which may discharge in the horizontal systems. For example, several codes may allow only two water closets on a 3-inch line, while others limit the number of water closets to 3-inch and larger lines based upon DFU allowable values. Readers are cautioned to follow local code requirements for their jurisdiction.

In the Field

Many years ago, very large lines for combination storm and waste municipal systems were egg-shaped. This accommodated large storm discharges but had a very small wetted perimeter area on the bottom of the pipe to improve flow-scouring action.

In the Field

The codes vary in their consideration of the minimum allowable line sizing for underground installations. For example, several codes require underground lines to be 2 inches or larger. This concept was based upon the lines' being less accessible and more difficult to service for cleaning. The larger line size would reduce service calls. Be sure to follow your local code requirements.

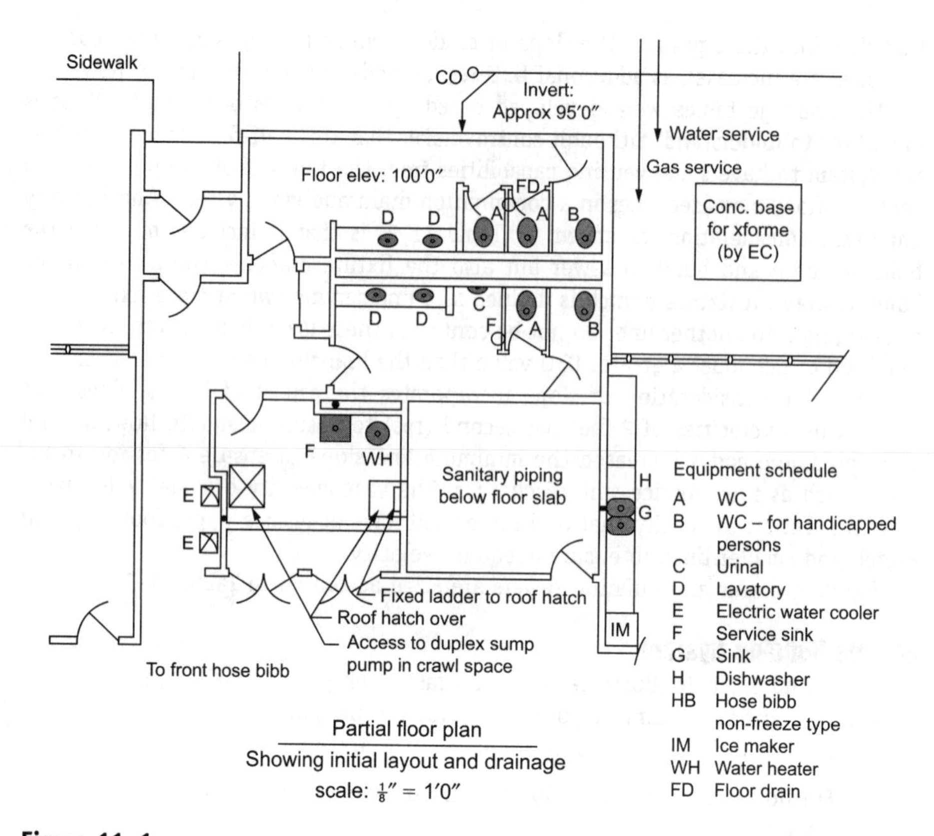

Figure 11–1
The partial plan view of an office building project illustrates the fixture locations and enables drawings to be created to assign drainage fixture units.

> **In the Field**
>
> Attention to details such as elevations in fast-track systems is a serious concern for two reasons. First, you must have sufficient depth to run lines as required by minimum slope and size requirements. Second, by considering only your needs, the strong possibility of interference with other mechanical systems may be overlooked, resulting in costly changes later.

> **In the Field**
>
> Some limited elevation concerns established by smaller lines requiring greater fall elevations ($\frac{1}{4}''$ rather than $\frac{1}{8}''$) can be solved by increasing the pipe diameter to allow less slope. For example, a 40-foot-long, 2-inch vented fixture branch requires 10 inches of fall, while a 40-foot-long, 3-inch vented fixture branch requires 5 inches of fall.

To provide a material take-off for ordering and installation purposes, a sanitary drainage piping drawing (isometric is best) is prepared by the plumbing contractor or its representative. This layout will indicate the fixture connections and fixture unit values. It is usually best to start at the uppermost fixtures and proceed downstream to the connections to the building drain and building sewer. An isometric sketch of this layout is shown in Figure 11–2. Because there is no continuous flow device, it is not necessary to compute an equivalent DFU value for such a load.

The line sizing in Figure 11–2 is self-explanatory using Table 11–2 and the notes, which indicate that the designer/technicians wish to use practical line oversizing to facilitate cleaning and possible changes in the structure at a later date. A 4-inch line is provided for the run serving the water closets in accordance with the *National Standard Plumbing Code* limit of four water closets to a 3-inch line; as the structure has five water closets connected to the branch, a 4-inch line is mandated. You should also note that the elevation information on the partial plan view allows sufficient depth for the design of double wyes in the vertical position for the back-to-back water closets. When technicians have difficulty digging in the branches, it is very common for them to bring the branches up to a higher elevation and use wyes on their side to pick up the fixtures.

Comments on "Final" Design

The codes represent minimum installation requirements. Often, because of cost considerations and competitive business pressures, licensed plumbers design and install systems per minimum code requirements. In addition, different concepts enter the plumbing industry along with new products. Caution and common sense should be exercised in your installation practices, such as maintaining past line-sizing principles. Maintaining a minimum pipe size underground of 2″ rather than matching some code allowances of $1\frac{1}{2}''$ may be a very good idea to service technicians who have repeatedly worked on the same underground line obstruction over a period of time from a kitchen sink or tub in a floor slab construction.

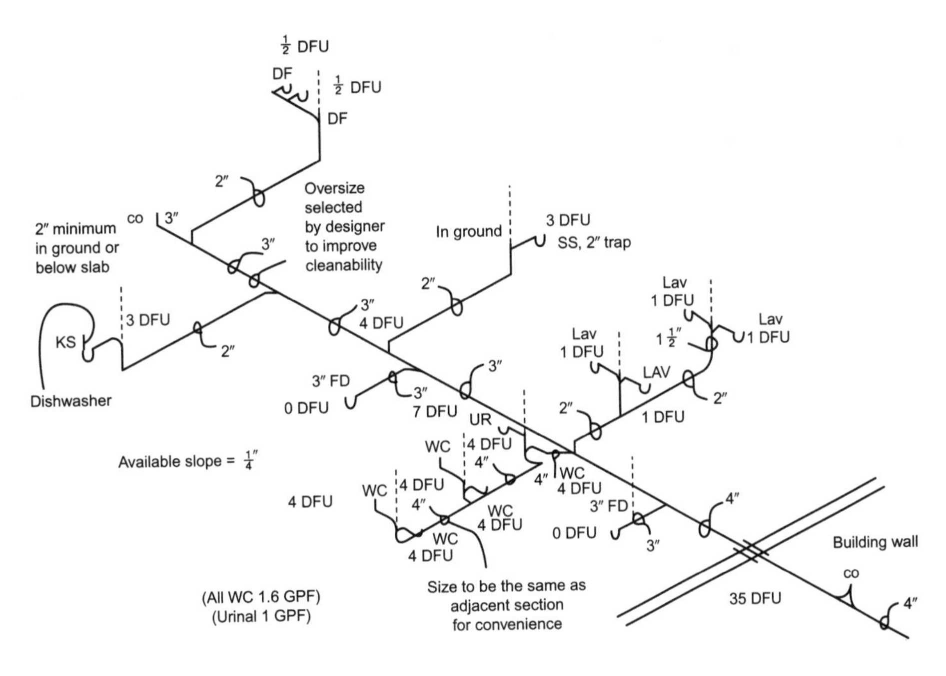

Figure 11–2
Creating an isometric view of an office building project provides the opportunity to size the line piping in accordance with the code-assigned, drainage-fixture unit values.

A few years ago, many designers believed that reduced-size automatic washer and shower lines were a wise engineering step, in response to a proposal that automatic washer standpipes be reduced from 2″ to $1\frac{1}{2}$″. When concerns were expressed about overflow spillage, the system advocates suggested increasing the standpipe height to compensate. Several years later, changes in some codes that incorporated air admittance valves seemed to warrant that automatic washer fixture branch discharges terminate into a 3-inch line. Later, another concern arose when it was discovered that automatic washer manufacturers had increased pump discharges, affecting line size concerns again. Not all codes allow air admittance valves, so be sure to check your local code before installing these devices.

The examples above are valuable, and yet new methods and products must be embraced. Many experienced plumbers remember the old days of trying to pour lead joints in wet sewer installations (a real problem) prior to gaskets and hubless bands. In addition, newer water piping materials, which replace cutting and threading galvanized piping along with fiberglass fixtures, were welcomed without any complaints.

Sizing Venting System The venting pipe sizes are selected after the plumbing fixture isometric layout is provided. The principal variables for selecting venting pipe sizes are the size of the stack servicing the fixtures, the drainage fixture units connected to the vent, and the total length of the vent pipe. The greatest consideration based on the type of venting was built into the drainage layout earlier. For example, when selecting the vent size, half the size of the branch served is mandated.

The basic vent consideration related to distance is the maximum distance a trap can be from its vent. Table 11–3 is provided to refresh your memory regarding trap arm distances. Another important consideration addressed in the table is the minimum slope the vent headers must have. Notice that the 3-inch vent line has a rating of $\frac{1}{4}$″ per foot, whereas the 3-inch waste line allowed a $\frac{1}{8}$″ per foot slope in drainage tables.

One of the most recognizable tables in the plumbing code industry is provided next to identify the vent sizes and maximum allowable travel. Vents with greater distances or lines with higher DFU values may be increased in size for proper operation.

Basic Venting Rules Table 11–4, as stated, is common to many codes across the country and is primarily used to identify the vent travel lengths based upon the

Table 11–3 Maximum Length of Trap Arm

Diameter of Trap Arm (Inches)	Length—Trap to Vent	Slope—Inches per Foot
$1\frac{1}{4}$	3′6″	$\frac{1}{4}$
$1\frac{1}{2}$	5′	$\frac{1}{4}$
2	8′	$\frac{1}{4}$
3	10′	$\frac{1}{4}$
4	12′	$\frac{1}{8}$

NOTE: This table has been expanded in the "length" requirements to reflect expanded application of the wet venting principles. Slope shall not exceed $\frac{1}{4}$″ per foot.

Table 11–4 Size and Length of Vents

Size of drainage stack or fixture drain inches	Drainage fixture units connected	Diameter of Vent Required (Inches) / Maximum Length of Vent (Feet): $1\frac{1}{4}$	$1\frac{1}{2}$	2	$2\frac{1}{2}$	3	4	5	6	8
$1\frac{1}{4}$	1	(1)								
$1\frac{1}{2}$	8	50	150							
2	12	30	75	200						
2	20	26	50	150						
3	10		30	100	100	600				
3	30			60	200	500				
3	60			50	80	400				
4	100			35	100	260	1000			
4	200			30	90	250	900			
4	500			20	70	180	700			
5	200				35	80	350	1000		
5	500				30	70	300	900		
5	1100				20	50	2200	700		
6	350				25	50	200	400	1300	
6	620				15	30	125	300	1100	
6	960					24	100	250	1000	
6	1900					20	70	200	700	
8	600						50	150	500	1300
8	1400						40	100	400	1200
8	2200						30	80	250	1100
8	3600						25	60	250	800
10	1000							75	125	1000
10	2500							50	100	500
10	3800							30	80	350
10	5600							25	60	250

(1) The length of the vent is unlimited

load, the DFUs connected to the drainage line. All codes have the same concept in that the minimum vent size is half that of the drain line size. This makes sense when considering that waste lines are sized for half full flow and stacks are designed for approximately $\frac{1}{3}$ full flow.

Distance considerations are addressed to ensure that proper amounts of air are provided to protect the fixture trap seals. It is often necessary to exceed the distances specified by the tables. Codes have different methods in dealing with this matter. Some of the major codes mandate an increase in pipe size after 40 feet, which allows for unlimited length. Other codes have a detail stating that the aggregate cross-sectional area of all vent terminals (through the roof) serving a sewer must not be less than the cross-sectional area of the minimum required size of the building drain that they serve at the point where it connects to the building sewer. This type of code requirement takes into consideration that fixtures scattered around a building generally have several different vent terminals totaling the required cross-sectional area.

The developed length of vents is increased as other vents are added with their DFUs. This process is called adding vent headers and is commonly used in venting layout planning. Individual vents through the roof would be wasteful and require several penetrations. Further, each individual penetration increases the likelihood of roof leaks.

Several different vent header methods are available for the isometric layout in Figure 11-2. The method shown in Figure 11-4 illustrates minimal sizes by utilizing $1\frac{1}{4}''$ headers. The common pipe used is of course $1\frac{1}{2}''$ based upon the selected pipe materials, such as hubless cast iron or PVC. The 2-inch vents off the water closet are the minimum half size of the 4-inch lines serving the water closets. The water closet double wyes could have been reduced to 3 inches in the layout; however, a 4-inch size was selected in the design.

The NSPC, like many codes, requires a vent terminal aggregate size through the roof (not including frost closure concerns) to be equal to the minimum required size of the building drain. See Table 11-5. The 35 DFU size in Figure 11-2 would allow a 3-inch building drain, meaning we are required to have a vent terminal equal to 7.1 square inches aggregate size as a vent terminal. Two 2-inch vent terminals would not be sufficient ($3.1 + 3.1 = 6.2$ rather than the required 7.1), which explains why a 3-inch vent terminal is used in Figure 11-4.

Later, the designer utilizes the plumber's isometric layout for plan review and trade coordination purposes and may create a normal blueprint plan view. This is an important consideration as the building code and plumbing permit process may require a plan review prior to a permit being issued.

In the Field

A new plumbing product has been introduced that reduces the number of roof penetrations by combining a roof drain strainer opening with a vent pipe riser. The product is illustrated in Figure 11-3. The product conforms to roof drain standards and may require acceptance by local authorities having jurisdiction.

Figure 11–3
A roof drain that has the ability to serve more than one function can reduce the number of roof penetrations, thereby reducing costs and the possibility of roof leaks. Courtesy of Froet Industries.

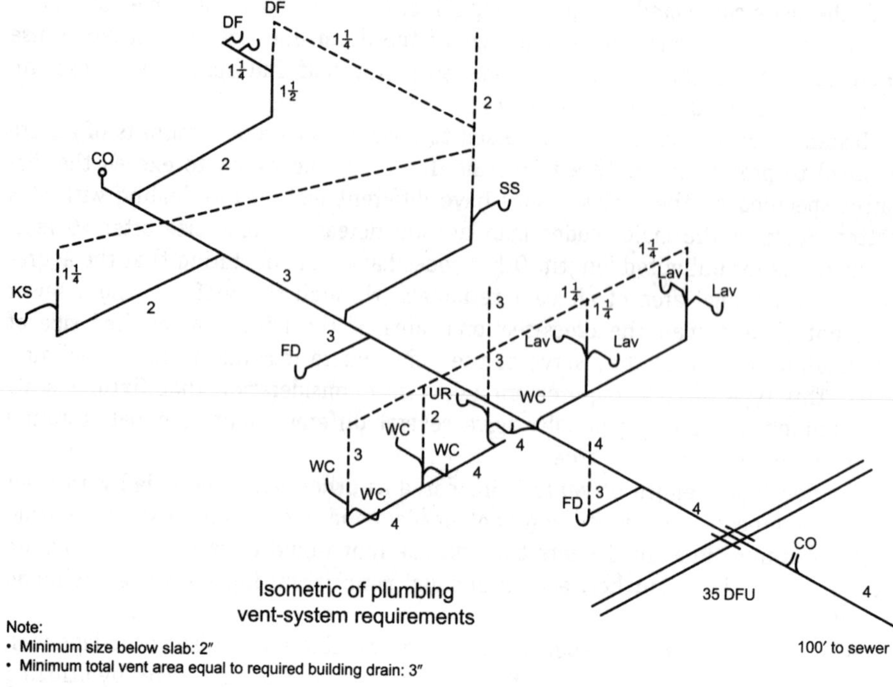

Figure 11–4
Creating an isometric view of the vents and exit locations on the roof provides the opportunity to size the line for later material takeoffs and enables technicians to consider locations of HVAC equipment.

Table 11–5 Nominal Pipe Cross Sectional Area (Sq. Inches)

Nominal Pipe Size (ID)	Cross Sectional Area (sq. in.)
$1\frac{1}{4}$	1.2
$1\frac{1}{2}$	1.8
2″	3.1
$2\frac{1}{2}$″	4.9
3″	7.1
4″	12.6
5″	19.6
6″	28.3
8″	50.3
10″	78.5
12″	113.1
15″	176.7

REVIEW QUESTIONS

1. Is the size of a vent based only on the fixture unit load on the drain?
2. Why is an isometric drawing of the plumbing drainage system developed?
3. Why are building drains and sewers capable of carrying heavier loads than branch drains?
4. It is possible to place a greater loading on a vent provided its distance to the terminal point is shorter?
5. What alternative design can be considered if sufficient fall is not available for a drain line 2 inches and smaller?

CHAPTER 12

Storm Drainage Sizing Utilizing Blueprint Reading

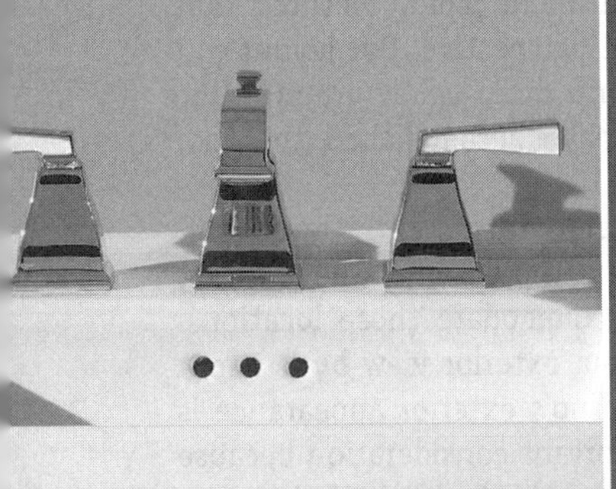

LEARNING OBJECTIVES

The student will:

- Calculate the sizes of primary and secondary storm drainage systems.
- Compare the various codes related to sizing methods for roof systems having parapet walls.
- Describe the added design considerations a secondary storm system requires of a structure.

STRUCTURE INFORMATION RELATED TO BLUEPRINTS

This chapter will provide instruction on the development and design of a storm drainage system for a flat-roof office building, which includes the previous chapter's restroom. A roof architectural plan view will aid in designing the system utilizing a design/build, fast-track process in accordance with code requirements. The shop drawings you utilize for storm drainage material take-off will commonly be in the form of plan view work sheets.

Storm drainage systems with their appropriate sizes are based upon the rainfall rates experienced at the location of the building and the roof area of the structure being considered. The building location's rainfall rate determines the volume of water (per unit area of roof) that must be carried away by the storm drainage system. Storm drainage components and mathematics information was provided in *Plumbing 301*. Designs in this chapter will begin with conventional storm drainage systems and close with a discussion of siphonic-flow systems, which require a great deal of engineering involvement and are often classified as alternative engineered design systems.

In the Field

Several plumbing codes provide information and a sizing table for exterior eaves troughs, referred to as roof gutters or semicircular gutters. The tables are helpful but are often deleted from the codes by the authorities having jurisdiction because local licensing requirements may not exclude exterior roofs eaves trough installers from installing roof drainage runoffs to eaves troughs.

Utilizing your skills as a technician incorporates an aspect of design, which is considered here. However, it is extremely important to understand that licensed design professionals have the skills and legal authority to require your common sense in following their direction. For example, in our consideration of sizing Figure 12–1, the designer has established the roof slope and location of the vertical conductors. Those considerations are based upon the interior of the structure, specifically, room locations and walls that could be sized to accommodate larger piping risers. This consideration is critical due to the many different piping methods that could be used. Simply put, the designer has the overall perspective. Our layout will use five different conductors from the roof to the storm drain below ground. One huge conductor or many smaller conductors could have accomplished the roof drainage.

Figure 12–1 requires your consideration of a building with its entry-level roof, which is 13,160 square feet in size with five conductors. The designers have established the downspout locations and roof slopes to accommodate those locations. Further, the building's HVAC equipment is shielded from exterior view by a 3-foot parapet wall around the complete structure. The building's exterior appearance is designed completely in glass. This is an extremely important consideration because the parapet walls could create a huge ponding effect on the building if the main storm drainage system became inoperable for some reason, such as a plugged line. This consideration now requires that the roof structure below be designed to carry this added weight, such as with added structural steel or a secondary drainage system. This secondary or emergency drainage system could utilize roof scuppers (openings in the parapet wall) sized to drain the roof in the event of the main storm system's failure. Furthermore, the owner and designer do not want the glassed exterior to have visible inconsistencies in appearance, so you must install a secondary system rather than scuppers or overflows. Our discussion of a secondary system will follow that of the entire conventional system due to its independence of the primary system.

Vertical Conductor Sizing

We will consider the building example in Figure 12–1 to be located in Paterson, New Jersey, in order to establish the area rainfall rate. Most plumbing codes have tables that provide the rates of rainfall for various cities based upon inches per hour for a 1-hour storm duration over a 100-year period. Paterson, New Jersey, has a 3-inch-per-hour rating. This information allows us to use Table 12–1, Size of Vertical Conductors and Leaders. The table identifies the area in square feet for various diameter piping sizes that will accommodate the 3-inch-per-hour discharge.

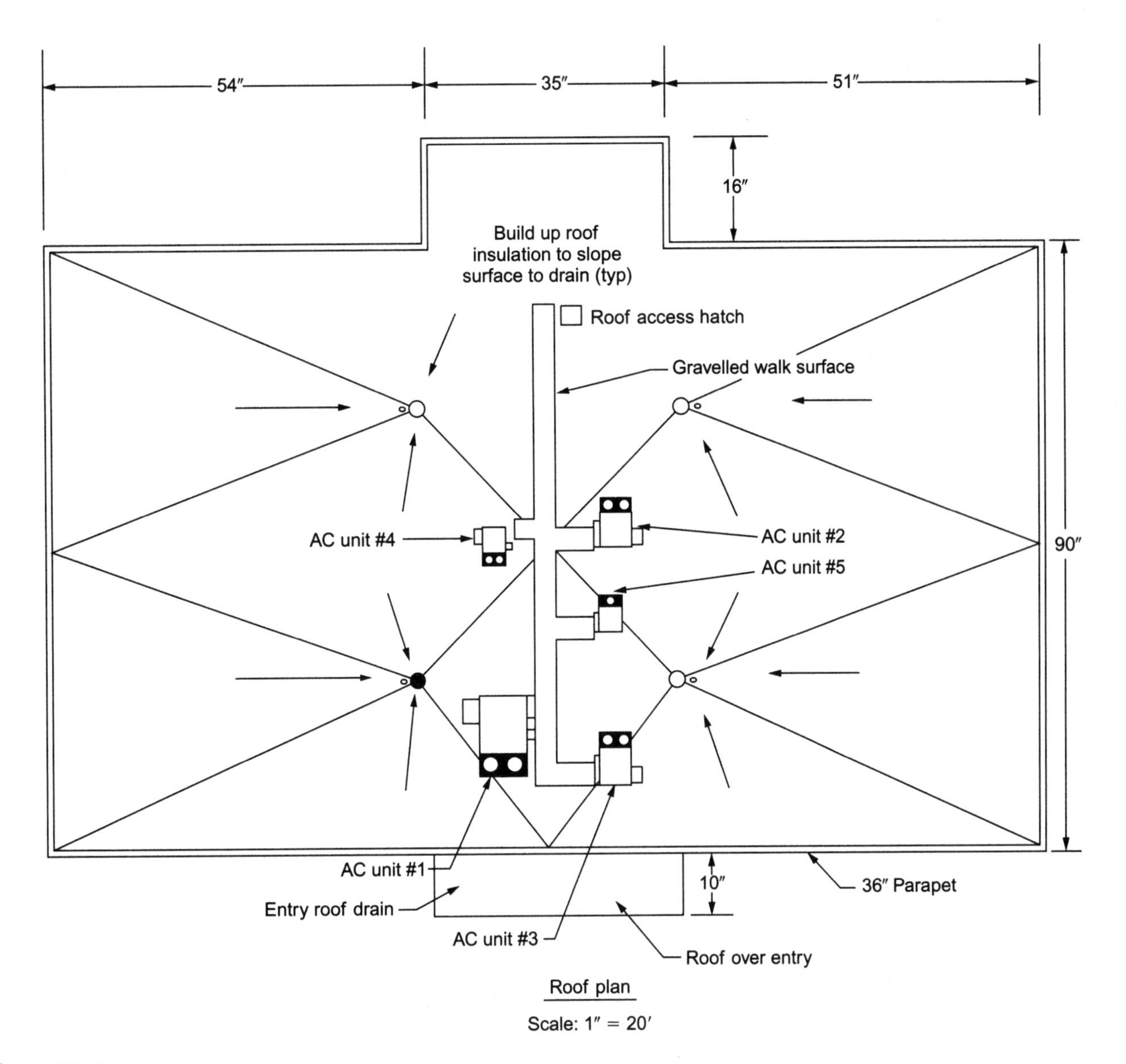

Figure 12–1
A roof plan illustrating roof drain locations is used to ensure that drains are placed correctly depending upon roof slopes to drain the roof properly.

The effective roof area to be considered when using Table 12–1 for conductor size and Table 12–2 for horizontal storm drains is the roof area plus $\frac{1}{2}$ the area of any vertical wall areas that can drain onto the roof. This concept takes into consideration that falling rain could be blown onto a vertical wall area and fall on the flat roof below, which of course would increase the load. Figure 12–1 shows a flat roof area with parapet walls. The roof is $140' \times 90'$ plus $35' \times 16'$. The parapet walls are $2(140' + 90' + 16') \times 3'$.

Thus, the equivalent roof area is calculated as follows:

$$\text{Main roof area}: (140 \times 90) + (35 \times 16) = 12{,}600 + 560 = 13{,}160 \text{ sq. ft.}$$

$$\text{Parapet wall}: 2(140 + 90 + 16)(3) = 2(246)(3) = 1{,}476 \text{ sq. ft.}$$

$$\tfrac{1}{2} \text{ the parapet wall area} = 738 \text{ sq. ft.}$$

Total effective main roof area: 13,160 sq. ft. + 738 sq. ft. = 13,898 sq. ft.

Table 12–1 Size of Vertical Conductors and Leaders

Nominal Diameter (inches)	Flow Capacity (GPM)	Allowable Projected Roof Area at Various Rates or Rainfall per Hour (Sq. Ft.)					
		1″	2″	3″	4″	5″	6″
2″	23	2,180	1,090	727	545	436	363
3″	67	6,426	3,213	2,142	1,607	1,285	1,071
4″	144	13,840	6,920	4,613	3,460	2,768	2,307
5″	261	25,094	12,547	8,365	6,273	5,019	4,182
6″	424	40,805	20,402	13,602	10,201	8,161	6,801
8″	913	87,878	43,939	29,293	21,970	17,576	14,646
10″	1655	159,334	79,667	53,111	39,834	31,867	26,556
12″	2692	259,095	129,548	86,365	64,774	51,819	43,183
15″	4880	469,771	234,886	156,590	117,443	93,954	78,295
		7″	8″	9″	10″	11″	12″
2″	23	311	272	242	218	198	182
3″	67	918	803	714	643	584	536
4″	144	1,977	1,730	1,538	1,384	1,258	1,153
5″	261	3,585	3,137	2,788	2,509	2,281	2,091
6″	424	5,829	5,101	4,534	4,080	3,710	3,400
8″	913	12,554	10,985	9,764	8,788	7,989	7,323
10″	1655	22,762	19,917	17,704	15,933	14,485	13,277
12″	2692	37,014	32,387	28,788	25,910	23,554	21,591
15″	4880	67,110	58,721	52,197	46,977	42,706	39,146

Source: *National Standard Plumbing Code*, 2006.

Because four roof drains with their associated conductors are to be used, each roof drain conductor will serve the following:

$$\text{Area to be served per roof drain} = \frac{13{,}898}{4} = 3{,}475 \text{ sq. ft. per conductor}$$

Table 12–1 indicates that a 3-inch conductor will serve 2,142 square feet and a 4-inch conductor will serve 4,613 square feet (for a 3-inches-per-hour rainfall rate). Thus, a 3-inch conductor is not large enough, and a 4-inch conductor is the correct choice. The calculations provide that four 4-inch conductors must be used.

The entry roof area utilizes the same process above identified in the following:

$$\text{The roof is } 10' \times 35' = 350 \text{ sq. ft.}$$
$$\text{The parapet wall is } 35' \times 3' = 105, \text{ half of which} = 53 \text{ sq. ft.}$$
$$\text{Area to be served by the roof drain is } 350 + 53 = 403 \text{ sq. ft.}$$

Table 12–1 indicates that a 2-inch conductor will serve 727 square feet for a 3-inch-per-hour rainfall rate. It is highly likely that a 3-inch conductor will be used when considering the availability of a 2-foot roof drain strainer assembly.

Underground System Sizing

Figure 12–2 requires your consideration of the building's underground storm drain rated as receiving a 3-inch-per-hour discharge rate of rain, as shown in Figure 12–1. As previously stated, the designer has established the downspout locations to accommodate the structure. The slope of the drainage system has also been established by the designer's consideration of the various elevations for the utility services. The structure's design will utilize a slope of $\frac{1}{8}''$ per foot.

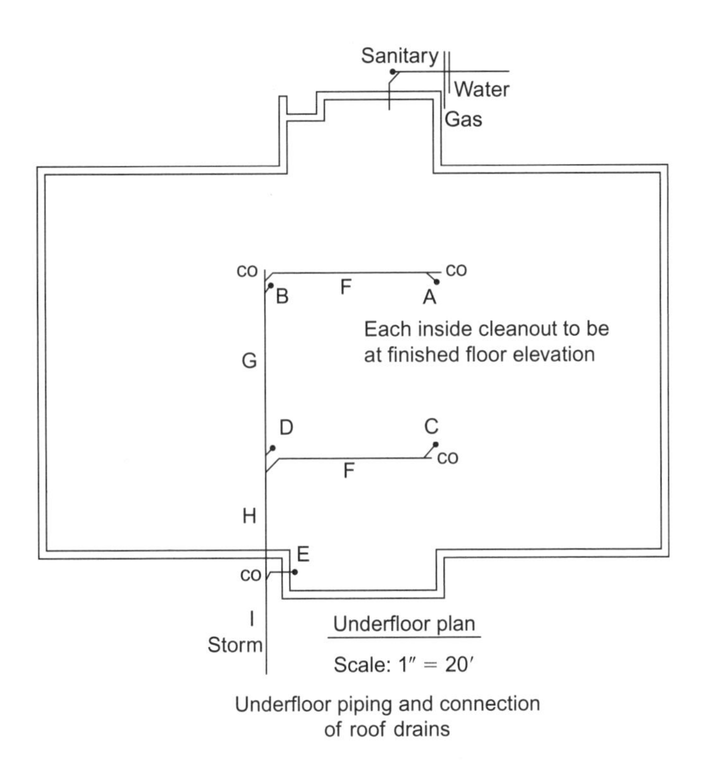

Figure 12–2
The underground storm drainage system plan is critical to ensure proper installation.

The horizontal piping in a storm drainage system is sized in accordance with a code table addressing the size of horizontal storm drains. The table will consider the various rainfall-rate-per-hour discharge and the pipe slope, which you will remember addressed the feet-per-second discharge rates covered in sanitary systems. The size of the horizontal storm drain is increased as the various downspouts are added into the system in the direction of flow.

We can now size the horizontal storm drain system shown in Figure 12–1 by using Table 12–2 and adjusting the pipe sizing based upon added roof area each time we include a new conductor. The pipe sizes are shown in Table 12–3 (all sizing is based on $\frac{1}{8}''$ per foot slope of piping). Each roof drain conductor, pipes A, B, C, and D, serves $\frac{13{,}898}{4} = 3{,}475$ sq. ft. of area.

Therefore, the building storm sewer must be 8 inches in size, laid at $\frac{1}{8}''$ per foot minimum slope. If some other slope is required, the appropriate values from Table 12–2 must be applied to the 3-inch rainfall rate and the pipe sizes selected accordingly.

Secondary (Emergency) Roof Drainage System The parapet walls require the structure to have a secondary/emergency roof drain system because the glass walls do not have scupper overflows or a roof constructed to hold the additional water weight should the primary system become inoperable. The majority of codes require this system to be completely separate from the primary. This structure will utilize two separate runs out the back to a wooded retention area rather than a single connection out the front like the main storm sewer with its connection to the public storm sewer at the property line.

The necessity of using two separate runs will be evident due to the drastic increase in the secondary storm line size in comparison to that of the primary system. While nearly all codes require separate systems, a few codes will double the line sizing capacity as in this illustration, which uses the NSPC requirements. The NSPC, Appendix A, provides secondary storm system sizing based upon 100-year, 15-minute storm duration. Our Paterson, New Jersey, location, with its 3-inch-per-hour primary, now considers 6.9 inches per hour secondary. We will round up the 6.9 to 7 inches per hour for calculation ease. Figure 12–3 illustrates the design/build plan for the secondary underground system.

Table 12–2 Size of Horizontal Storm Drains (for 1″/hr to 6″/hr Rain Fall Rates)

Size of Drain (inches)	Design Flow of Drain (GPM)	Allowable Projected Roof Area at Various Rates or Rainfall per Hour (Sq. Ft.)					
		1″/hr	2″/hr	3″/hr	4″/hr	5″/hr	6″/hr
Slope $\frac{1}{16}$ inch/foot							
2							
3							
4	53	5,101	2,551	1,700	1,275	1,020	850
5	97	9,336	4,668	3,112	2,334	1,867	1,556
6	157	15,111	7,556	5,037	3,778	3,022	2,519
8	339	32,629	16,314	10,876	8,157	6,526	5,438
10	615	59,194	29,597	19,731	14,798	11,839	9,866
12	999	96,154	48,077	32,051	24,039	19,231	16,026
15	1812	174,405	87,203	58,135	43,601	34,881	29,068
Slope $\frac{1}{8}$ inch/foot							
2							
3	35	3,369	1,684	1,123	842	674	561
4	75	7,219	3,609	2,406	1,805	1,444	1,203
5	137	13,186	6,593	4,395	3,297	2,637	2,198
6	223	21,464	10,732	7,155	5,366	4,293	3,577
8	479	46,104	23,052	15,368	11,526	9,221	7,684
10	869	83,641	41,821	27,880	20,910	16,728	13,940
12	1413	136,002	68,001	45,334	34,000	27,200	22,667
15	2563	246,689	123,345	82,230	61,672	49,338	41,115
Slope $\frac{1}{4}$ inch/foot							
2	17	1,636	818	545	409	327	273
3	50	4,813	2,406	1,604	1,203	963	802
4	107	10,299	5,149	3,433	2,575	2,060	1,716
5	194	18,673	9,336	6,224	4,668	3,735	3,112
6	315	30,319	15,159	10,106	7,580	6,064	5,053
8	678	65,258	32,629	21,753	16,314	13,052	10,876
10	1229	118,292	59,146	39,431	29,573	23,658	19,715
12	1999	192,404	96,202	64,135	48,101	38,481	32,067
15	3625	348,907	174,454	116,302	87,227	69,781	58,151
Slope $\frac{1}{2}$ inch/foot							
2	24	2,310	1,155	770	578	462	385
3	70	6,738	3,369	2,246	1,684	1,348	1,123
4	151	14,534	7,267	4,845	3,633	2,907	2,422
5	274	26,373	13,186	8,791	6,593	5,275	4,395
6	445	42,831	21,416	14,277	10,708	8,566	7,139
8	959	92,304	46,152	30,768	23,076	18,461	15,384
10	1738	167,283	83,641	55,761	41,821	33,457	27,880
12	2827	272,099	136,050	90,700	68,025	54,420	45,350
15	5126	493,379	246,689	164,460	123,345	98,676	82,230

Source: *National Standard Plumbing Code*, 2006.

Table 12–3 Storm Drain Sizes for Table 12–2

Rainfall rate 3 inches per hour; all horizontal lines at $\frac{1}{8}$" per foot slope			
Pipe Section	Roof Area Served, sq. ft.	Pipe Size Nominal (Inches)	Remarks
A	3,475	5	
B	3,475	5	Greater than 4-inch size of
C	3,475	5	2,406, shall be increased
D	3,475	5	
E	403	3	No provided consideration for
F	3,475	5	2 inches
G	6,950	6	
H	13,900	8	Greater than 6-inch size of
I	14,303	8	7,155, shall be increased

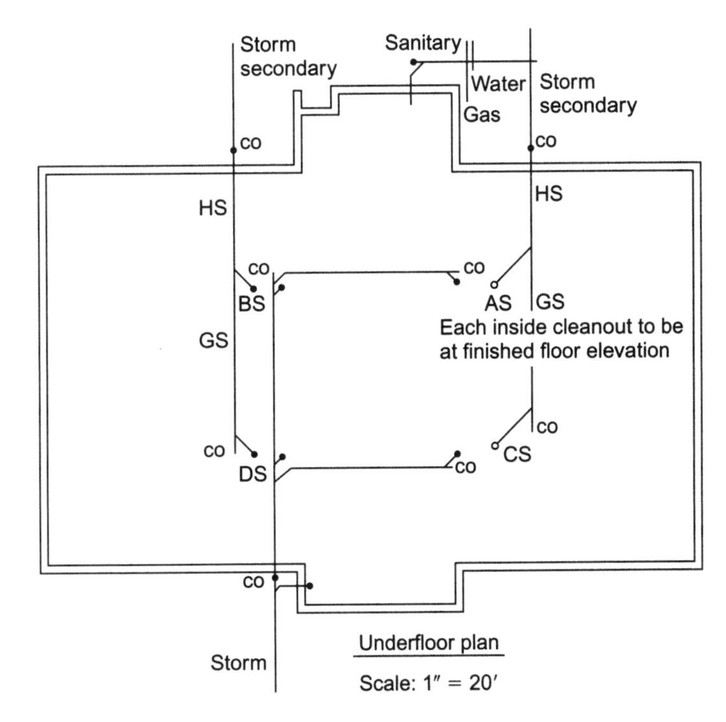

Figure 12–3
An underfloor plan of the secondary storm drainage system provides location and sizing information guidelines.

Table 12–4 provides line sizing for the secondary system from the previous NSPC vertical and horizontal tables using the 7-inch rate versus the previous primary system rate of 3 inches.

For comparison's sake, consider that if all four secondary drains were discharged to one common line like the primary, the secondary line would be 15 inches. This line size would be more expensive to purchase, and its size could create installation problems, including its larger size being in the way of other underground systems such as electrical cable troughs.

Siphonic Roof Drainage System Siphonic roof drainage systems are unique and have many cost advantages; however, they require a great deal of engineering design work and presently are not recognized by design standards in plumbing codes. The systems have been utilized as an alternative engineering design and conform to code criteria. Green Build and Leadership in Energy and Environmental Design (LEED), which are

In the Field

Water, snow, and ice are a major load factor to be considered in roof designs. One cubic foot of water weighs 62.48 pounds. Two inches of water over a flat surface equal to this building's 13,160 square feet have a total weight of 68.43 tons.

Table 12–4 Secondary Storm Sewer Sizes for Figure 12–3

Rainfall rate 7 inches per hour; all horizontal lines at $\frac{1}{8}''$ per foot slope			
Pipe Section	Roof Area, sq. ft.	Pipe Size for Vertical	Pipe Size for Horizontal
AS	3,475	6	–
BS	3,475	6	–
CS	3,475	6	–
DS	3,475	6	–
GS	3,475	–	8
HS	6,950	–	10

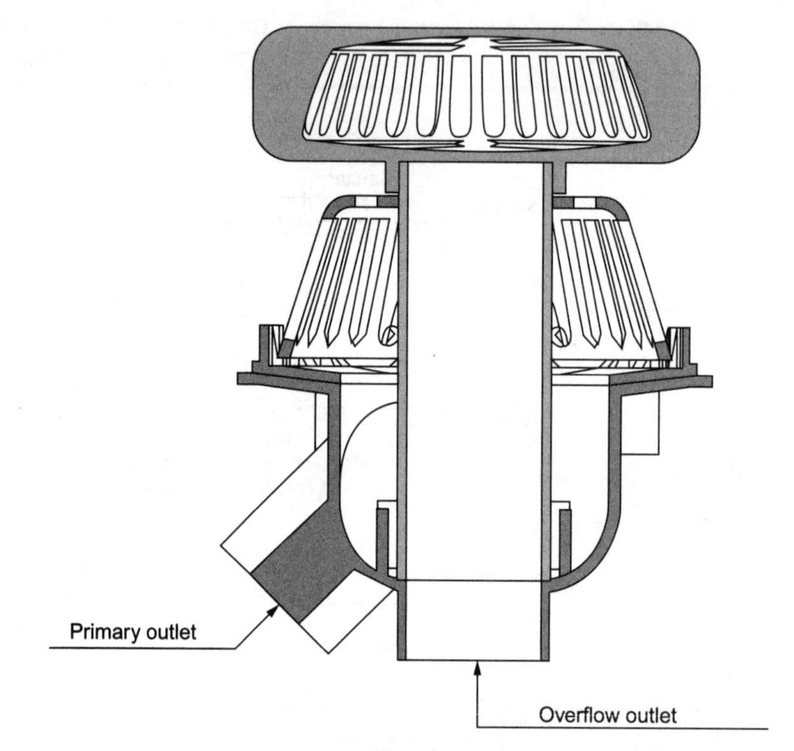

Figure 12–4
This is a new, innovative product with separate primary and secondary inlets and outlets. This product can reduce the number of roof penetrations and costs. Courtesy of Froet Industries.

building advocates, are especially supportive of the system. They have further incorporated storm water retention at the discharge in the interest of conservation.

The system starts with roof drains that have top baffles, which are designed to restrict air from coming into the system, thus promoting full-flow storm water conditions. Theoretically, this allows piping to be completely full of water, allowing higher velocities, capacities, and as some describe it, pulling (siphoning) higher rainfall discharges. See Figure 12–4. Proponents of the system refer to the benefits of reduced pipe sizes both horizontal and vertical and the lack of slope requirements, allowing greater-distance runs overhead. Opponents express concerns that the installation requirements specifically state that plumbing codes are not applicable, including required cleanouts and pipe reductions in the direction of flow.

Information on these systems is readily available from Jay R. Smith Co. at its website, http://www.jrsmith.com, and the American Society of Plumbing Engineers' site, http://www.aspe.com.

REVIEW QUESTIONS

1. Commonly, are roof gutters for sloped roofs to be installed by licensed plumbers only?
2. What is the reason primary and secondary storm systems are required to be completely separate?
3. Can a horizontal storm drain accommodate a greater load if it has $\frac{1}{2}''$ per foot slope rather than a $\frac{1}{8}''$ per foot slope?
4. What is the amount of load, in square feet, placed on a roof for a parapet wall that is 2 feet high on all four sides of an 80-foot-by-80-foot building?
5. Does the plumbing code require a precise number of roof drains for specific roof sizes in a conventional storm drainage system?

CHAPTER 13

Potable Water Line Sizing Utilizing Blueprint Reading

LEARNING OBJECTIVES

The student will:

- Explain why systems must employ an accurate sizing method in order to function safely.
- Describe several different methods of sizing potable water systems.
- Identify the minimum code requirements to ensure adequate performance and provide valuable insights into what is expected of potable water distribution systems.
- Summarize the operational consideration that demands supplies of adequate amounts of water through a system to its outlets.
- Demonstrate how to draw a diagram of a distribution system, utilizing fixture elevations, branches, lengths, and fixtures.

WATER DISTRIBUTION DESIGN

Having the correct size of water service and water distribution piping is critical in a building to address health and operational concerns. Undersized lines with low-pressure conditions have the potential to cause severe health hazards and improper operations. Oversized lines will result in high initial costs and can lead to adverse health conditions. Low pressures increase the potential of contamination through backflow and back-siphonage, and they will hinder the fixtures from flushing, filling, or operating properly in the required amount of time. Further, low pressures may result in higher amounts of scale build-up, which will accelerate the process to lower the pressure rating. Excessive pressures result in piping erosion and unacceptable piping noise levels.

This chapter will provide instruction on the development and design of a water distribution system utilizing the plumbing fixtures in the same office building of previous chapters. The sizing of water supply systems was also discussed in *Plumbing 201*. An architectural plan view, which includes the line locations, will aid in designing the system utilizing a design/build, fast-track process in accordance with code requirements. As stated in the previous chapters, the shop drawings you create will commonly be in the form of isometric work sheets to aid in material take-off and installation procedures

Utilizing your skills as a technician incorporates an aspect of design, which is considered here. However, as previously stated, it is extremely important to understand that licensed design professionals have those skills and the authority to legally design structures such as water distribution systems. This consideration is critical because of the many different pipe-sizing methods and variables that could be used. The methods can be classified generally into four different sizing styles. The major methods in alphabetical order are as follows:

A. *Parallel systems with manifolds* consider individual supply lines to each fixture from a central supply header.
B. *Segmented loss* uses calculations of friction loss for the pipe and fittings. This method is detailed and is one of the most accurate.
C. *Uniform pressure loss* considers that friction will result in a uniform pressure loss throughout the piping system. This method is conservative in its sizing approach.
D. *Velocity* considers the piping material used and the designed velocity of the water. It is one of the simplest methods and results in conservative sizing.

> **In the Field**
>
> The introduction and use of newer products such as PEX piping and newer delivery methods are not always clarified in documents used by technicians. The methods would include gridded and parallel water distribution system manifolds. Care should be taken to obtain up-to-date information when installing these systems. Manufacturers' installation instructions are readily available and are the best place to begin.

> **In the Field**
>
> Plumbers use what is commonly referred to as "normal design practices" for sizing water distribution systems. That is, they use proven methods such as 1", $\frac{3}{4}$", and $\frac{1}{2}$" lines in residences and smaller commercial structures and following the prints for larger industrial, assembly, and institutional structures. Other practices may include using $\frac{3}{4}$" line to the next to the last fixture (using a $\frac{3}{4}$" × $\frac{1}{2}$" × $\frac{1}{2}$" fitting), which may result in oversizing. Exercise caution here, because health/safety and performance are not always ensured by these procedures.

Design Expectations

Plumbing codes identify the minimum code requirements to ensure adequate performance and provide valuable insights into what is expected of potable water distribution systems. The basic requirement stated in several codes is that buildings with occupants must have adequate amounts of safe potable water at specified pressures. The codes then generally go on to state that the amounts are considered at peak demand and must conform to approved engineering practice. This allows the use of many designs and considers that precise predictions of demand use and equipment demand provide extremely difficult considerations.

Peak demand is defined as the maximum amount (rate of flow) of water required at any given time during 24 hours of operation. This peak demand is heavily dependent upon the type of structure (occupancy use group). In determining peak demand, it must be understood that not all fixtures are required to be in use but the probability of their use must be considered. Peak flow at system fixtures will consider the duration of use, frequency of each use, and the rate of water required. These calculations result in numerical assigned demand-load values called water supply fixture units (WSFU) and then are factored into sizing tables, which may use continuous flows from outlets such as hose bibbs for gallons-per-minute (gpm) calculations.

> **In the Field**
>
> Plumbing codes do not usually contain water distribution sizing method details within the body of the code because of the numerous methods available. However, the codes state the design expectations, such as minimum supply size to individual fixtures and required capacity at fixture outlets. Many codes provide a method for the technician in the book appendix that may be adopted by the authority having jurisdiction.

Design Considerations

Accomplishing the supply of adequate amounts of water through a system to its outlets demands consideration of several factors, which we will now discuss. While there are several methods for sizing, the factors listed are generally incorporated into the methods.

Characteristic Makeup of the Water Supply

The water composition will affect the system design in the choice of materials and sizing. Knowing that water is corrosive water, based upon its chemical composition, pH value, scale-forming tendency, and other factors, will aid a designer in choosing a proper piping material when considering the life expectancy of distribution systems. This information is generally available from the public water purveyor or from local health departments upon well-supply testing. Additionally, the water quality may warrant a water-conditioning device such as a softener or iron filter, which will cause additional pressure loss through the system supply.

Code Required Tables

It is extremely important for all aspects of plumbing that technicians be familiar with the applicable plumbing code in each area they are working. The major codes have different water-sizing requirement tables that must be considered, even though some do not address specifics. For clarification, we are not referring to the code appendices that might not be adopted by the authority having jurisdiction, but to the main body of the code dealing with water supply and distribution. The various table mandates are briefly addressed here for your consideration.

Some codes list various fixtures and provide the gpm and psi ratings for each that must be met at the outlet during peak flow, but others do not. Table 13-1 provides sample rates for a lavatory.

Some codes list the maximum flow rates and consumption at the outlet for the sake of conservation issues. Table 13-2 provides these numbers for a lavatory.

Some codes list the minimum sizes of fixture water supply pipes, while others will not. Table 13-3 addresses the size of pipes for a lavatory.

Some codes list the WSFU load values, as shown in Table 13-4 for a lavatory.

In the Field

Plumbing codes require that potable water be supplied to fixtures generally in the beginning of their water distribution chapters and later present tables that identify the objectives such as minimum fixture branch size, required pressure, and flow rate at each outlet. The *National Standard Plumbing Code* lists 22 basic principles that clearly identify its code objectives for sanitary and safety to protect health and safety matters. The principles define the code's intent and assist code officials and technicians in understanding the basic goals necessary for proper plumbing. The first four principles address water supply concerns.

Table 13-1 Required Capacity at Fixture Supply Outlets

Fixture Supply Outlet Serving	Flow Rate (gpm)	Flow Pressure (psi)
Lavatory	2	8

Table 13-2 Maximum Flow Rates and Consumption for Plumbing Fixtures

Plumbing Fixture of Fixture Fitting	Maximum Flow Rate or Quantity
Lavatory, private	2.2 gpm at 60 psi

Table 13-3 Minimum Sizes of Fixture Water Supply Pipes

Fixture	Minimum Pipe Size (Inches)
Lavatory	$\frac{3}{8}$

Table 13-4 Load Values Assigned to Fixtures

Fixture	Occupancy	Cold (WSFU)	Hot (WSFU)	Total (WSFU)
Lavatory	Private	1.0	1.0	0.75

Cost Should Not Unfairly Affect the Sizing Choice

Cost considerations will affect the performance of the water supply system. For example, a decision to select a backflow preventer with a higher cost that may have superior operation features and less friction loss could affect the overall budget. The choice will enhance the system's operations and affect the system's maintenance program by reducing costs. Structure designers, contractors, and technicians must make choices considering performance outcomes rather than installation costs.

Demand Load

Demand load and peak demand were discussed earlier, and although these are some of the most difficult aspects of the design, they are the most important. The building's use group, which affects the time of operations and number of projected occupants, is critical. The type of fixture and its application for public or private use is provided for in the code tables. For example, the code states that 1.6-gallon-per-flush water closets with a flush valve in a heavy-use assembly occupancy have an 8.0 WSFU rating, while a 1.6-gallon-per-flush water closet in a home with a conventional flush tank has a 2.5 WSFU rating. Projections of WSFU will be affected in the future with the introduction of other water conservation fixtures.

Developed Length of System

Consideration of the piping system length, which includes the water service and the distribution system from the water service to its farthest outlet, is critical in the system design. Pressure loss due to friction through pipe, fittings, and plumbing devices are part of this distance consideration. The uniform pressure loss sizing method tables are extremely dependent on the supply pressure provided and the developed length of the system.

Elevations

Consideration of the piping system elevation differences, including the water service and the distribution system from the public or well supply, is also critical in the system design. Pressure loss for end outlets above the supply or even pressure increases for outlets below the supply must be considered. The uniform pressure loss sizing method considers the pressure loss calculation in its design outcome. A reminder summary statement is "Pressure is a function of the weight and height of water."

Health and Safety

A commitment to ensuring that health and safety considerations are addressed when sizing systems is mandated by the codes. Undersized lines will not provide adequate amounts of water for fixture operation. Further, during periods of peak demand, inadequately sized lines will increase the likelihood of negative pressures, which contribute to backflow from sources of contamination. A potential cause for this would be line restrictions that cause negative pressures with excessive velocities and booster pump applications.

Materials

Consideration of the piping materials in the system is critical in the system design and the system's tolerance from velocity to erosion. Different piping products will vary in surface roughness, which affects pressure loss due to friction. In addition, the interior size, which affects capacity, will vary from one material to another.

Required Pressure at Fixtures

The proper operation of fixtures is dependent upon their water supply volume and water supply pressure. Most fixtures require a minimum of 8 psi, while water closets with flushometer valves require 25 psi. These numbers may sound small individually, but the reader is cautioned to consider the many additional factors, such as peak demand, that radically increase the pressure for all the fixtures involved in the system.

The codes limit the maximum system pressure to 80 psi and require a pressure-reducing valve for supplies in excess of 80 psi. Minimum pressures required for proper operation and identified by code tables are ensured by proper sizing and increased when necessary by booster pumps with various safety controls. Further, it is extremely important to consider the minimum pressures supplied from the source when performing sizing calculations. Supply pressures from a municipality may vary at peak demand times, and a well supply may provide lower pressures between pump operations.

The amount of pressure loss caused by friction is dependent upon the pipe roughness, diameter, length, fittings, and velocity. Friction is decreased as the size of the pipe is increased. This is attributed to the fact that an increase in pipe size allows a much greater volume of water to pass through a line in proportion to the increase in pipe wall area.

Pressure losses in the system can also be caused by equipment such as backflow preventers, meters, and valves. Pressure losses for meters and backflow preventers vary a great deal with the operating style of the meter or the degree of safety of the backflow preventer. The manufacturer of the product in question is the best source for obtaining pressure loss information.

Velocity

Consideration of the piping system design velocity is of extreme importance and is a contributing factor to each sizing method. Velocity, the speed of water moving through the pipe, is determined by the size of pipe and the volume of water passing through it in a given time frame, usually feet per second. The first consideration in the velocity discussion is that major plumbing codes limit the maximum velocity to 8 feet per second. Velocities higher than this would be detrimental to a system, causing wear, noise, and pressure surges that damage the system components such as valves.

The amount of friction is increased by higher velocity as the pipe diameter is decreased. The benefit of higher velocities is that the scale-forming characteristics of the water are less likely to cause build-up in the pipe and reduce pipe diameter.

The velocity method used for pipe sizing recognizes peak demand WSFUs and various pipe materials. The different piping products with various thicknesses are designed for velocities of 4 fps, 6 fps, and 8 fps.

Sizing an Example System

A sample plan view is provided in Figure 13–1 of the same structure considered in earlier chapters to aid in creating water piping sizing diagrams. This partial floor plan from the mechanical portion of the blueprints includes water distribution line locations and facts. Abbreviated code sizing tables will be provided and isometrics developed for installation purposes.

The following WSFU table is provided for your consideration in developing sizing figures. The table is from the NSPC and abbreviated for space.

Section 10.14.2 of the NSPC adds fixture branch-sizing criteria to the table above and addresses the minimum acceptable pressures for each fixture. It then addresses the various water closet and urinal pressures. Lastly, the section references water flow rates for individual fixture supplies. Hose bibbs and wall hydrants have a minimum rating of 5.0 gpm as potential continuous flows.

Table 13–5 is an abbreviated conversion table created from lengthy WSFU-gpm conversion information referenced in the NSPC to assist you in providing the gpm values in Figure 13–2 and Figure 13–3.

You could create Table 13–6 to clarify the individual fixtures' WSFU values and the minimum fixture branch sizes in preparation for filling in the isometric sizing sketches.

To size the water service and water distribution layout, proceed with the following methodology.

STEP 1. First, obtain the lowest probable pressure delivered at the site from the water utility as discussed earlier. In this example, a municipal supply pressure of 80 psi has been provided, and the specifications require you to use Type L copper.

In the Field

Technicians are reminded that code officials must ensure that the sizing has been properly designed. The code official may request the sizing methods and calculations.

In the Field

Hydraulic designs involve many variable circumstances that make specific rules for sizing difficult. The following method is often called the segmented loss method. Several simplified methods exist; however, this example is designed to consider most of the various sizing factors.

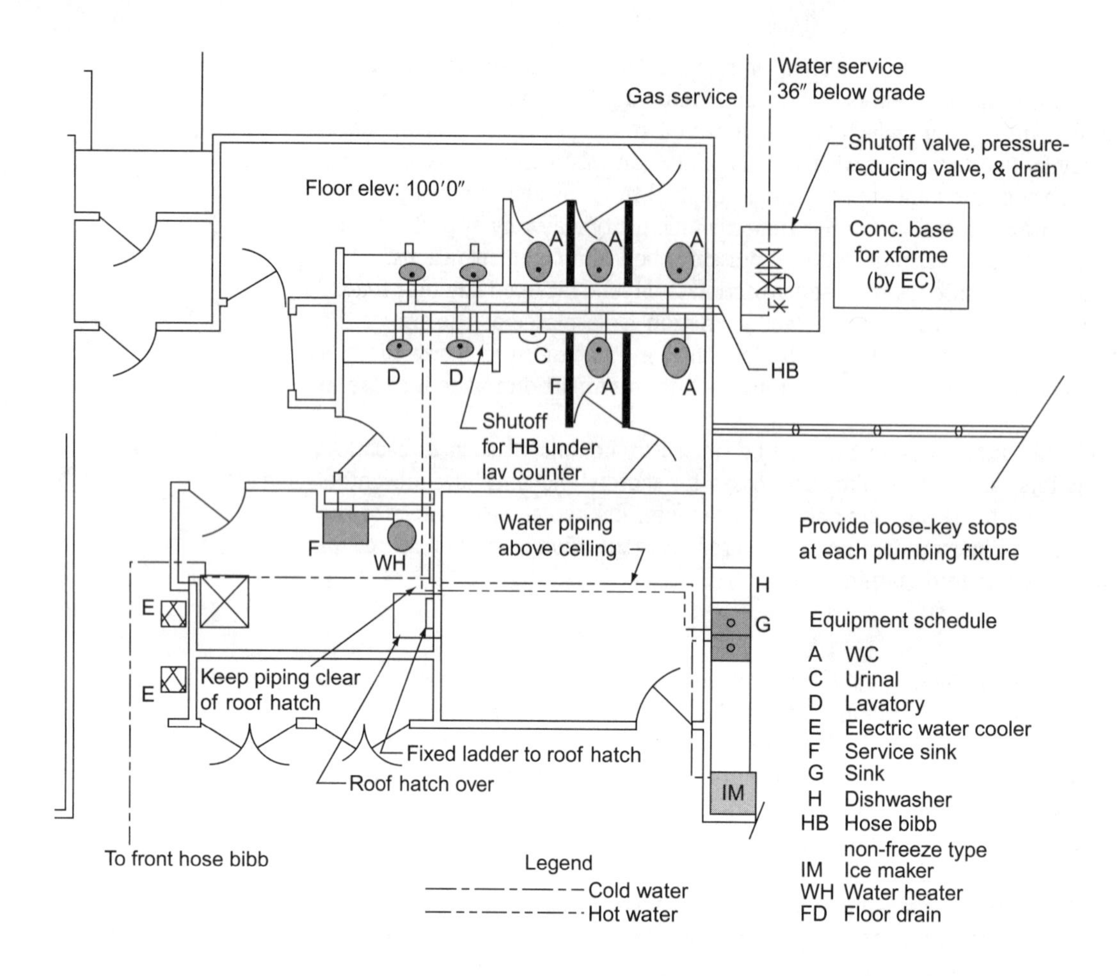

Figure 13–1
The partial plan view of plumbing fixtures with the locations of water distribution lines provides the contractor with necessary information for cost bidding and location.

STEP 2. Next, prepare a diagram of the distribution system as shown in Figure 13–2 and Figure 13–3, showing fixture elevations, branches, lengths, and fixtures. Determine the design load for the system by indicating the WSFU load at each fixture and each branch from Table 13–7 and the corresponding gpm values from Table 13–5. Insert this information into Figure 13–2. All individual fixture supply pipes to water outlets are sized on the isometrics from the minimum sizes information in Table 13–7.

Note that water closets and the urinal utilize flushometer valves, information that was provided in the design specifications. Your development of Table 13–6 indicates that upon entrance to the restrooms' pipe chase, you have 80 WSFU equated to 62 gpm from Table 13–5, plus 10 gpm (total) from the two hose bibbs.

STEP 3. Determine the available pressure drop for pipe friction loss by subtracting all the known pressure drops, the minimum pressure required at the highest fixture, meter pressure drop, pressure due to elevation, and the minimum probable inlet pressure supplied by the utility company.

The available pressure for pipe friction is calculated as follows:

Initial pressure = 80 psig
15 psig = Minimum residual pressure required at the farthest fixture
10 psig = Meter and PRV pressure drop
3.5 psig = Elevation pressure drop
28.5 psig

In the Field

Sometimes meters are one size smaller than the water service pipe based on flow rates determined by the water company or meter manufacturer.

Table 13–5 Table for Estimating Demand

Supply Systems Predominantly for Flushometer Valves	
Load (WSFUs)	Demand (gpm)
10	27
15	31
20	35
25	38
30	41
40	47
50	51
60	55
80	62
100	68
(The code continues on to) 10,000	790

Table 13–6 Plumbing Fixture Size Schedule

Plan Designation	Description	Fixture Unit Value	GPM Value	Cold	Hot
A	Water closet, flushometer	10	27	1″	—
C	Urinal, flushometer	5	—	$\frac{3}{4}$″	—
D	Lavatory	1	—	$\frac{3}{8}$″ or $\frac{1}{2}$″	$\frac{3}{8}$″ or $\frac{1}{2}$″
E	Electric water cooler	1	—	$\frac{1}{2}$″	—
F	Sink, service	3	—	$\frac{1}{2}$″	$\frac{1}{2}$″
G	Sink, kitchen, comb. fix	3	—	$\frac{3}{4}$″	$\frac{3}{4}$″
H	Dishwasher	1	—	—	$\frac{1}{2}$″
I	Wall hydrants	—	5	$\frac{3}{4}$″	—

NOTE: The water cooler and dishwasher are given the WSFU value of 1 because they are $\frac{3}{8}$″ size.

Initial pressure = 80 psig
−28.5 psig

51.5 psi = Available-pressure-for-pipe friction

STEP 4. This calculated available-pressure-for-pipe friction is divided by the equivalent length of piping from the water main to the most remote fixture, commonly called the basic design circuit. This value is the maximum allowable pressure drop per foot of pipe.

The equivalent length of pipe equals $1\frac{1}{2}$ times the actual length of pipe from the main or source to the farthest fixture. The calculations for the example problem, using the dimensions shown in Figure 13-3, are as follows:

1.5(270 + 24 + 4 + 5 + 14 + 7 + 16 + 14 + 8) = Equivalent length

1.5(362)

543′ = Equivalent length

Note: The remaining pipe length serves a lawn hydrant. Such pipe distances are not included in equivalent length.

The pressure drop available for pipe friction is calculated by the following method:

$\frac{51.5}{543} \approx 0.095$ psi/ft

Table 13–7 Water Supply Fixture Units (WFSU)[1] and Minimum Fixture Branch Pipe Size for Individual Fixtures

Individual Fixtures	Minimum Branch Pipe Size[2]		In Individual Dwelling Units			In 3 or More Dwelling Units			In Other than Dwelling Units			In Heavy-Use Assembly		
	Cold	Hot	Total	Cold	Hot	Total	Cold	Hot	Total	Cold	Hot	Total	Cold	Hot
Bathtub or Combination Bath/Shower	1/2″	1/2″	4.0	3.0	3.0	3.5	2.6	2.6						
Bidet	1/2″	1/2″	1.0	0.8	0.8	0.5	0.4	0.4						
Clothes Washer, Domestic	1/2″	1/2″	4.0	3.0	3.0	2.5	1.9	1.9	4.0	3.0	3.0			
Dishwasher, Domestic		1/2″	1.5		1.5	1.0		1.0	1.5		1.5			
Drinking Fountain or Water Cooler	3/8″								0.5	0.5		0.8	0.8	
Hose Bibb	1/2″		2.5	2.5		2.5	2.5		2.5	2.5				
Hose Bibb, Each Additional	1/2″		1.0	1.0		1.0	1.0		1.0	1.0				
Kitchen Sink, Domestic	1/2″	1/2″	1.5	1.1	1.1	1.0	0.8	0.8	1.5	1.1	1.1			
Laundry Sink	1/2″	1/2″	2.0	1.5	1.5	1.0	0.8	0.8	2.0	1.5	1.5			
Lavatory	3/8″	3/8″	1.0	0.8	0.8	0.5	0.4	0.4	1.0	0.8	0.8	1.0	0.8	0.8
Service Sink Or Mop Sink	1/2″	1/2″							3.0	2.3	2.3			
Shower	1/2″	1/2″	2.0	1.5	1.5	2.0	1.5	1.5	2.0	1.5	1.5			
Shower, Continuous Use	1/2″	1/2″							5.0	3.8	3.8			
Urinal, 1.0 GPF	3/4″								4.0	4.0		5.0	5.0	
Urinal, Greater Than 1.0 GPF	3/4″								5.0	5.0		6.0	6.0	
Water Closet, 1.6 GPF Gravity Tank	1/2″		2.5	2.5		2.5	2.5		2.5	2.5		4.0	4.0	
Water Closet, 1.6 GPF Flushometer Tank	1/2″		2.5	2.5		2.5	2.5		2.5	2.5		3.5	3.5	
Water Closet, 1.6 GPF Flushometer Valve	1″		5.0	5.0		5.0	5.0		5.0	5.0		8.0	8.0	
Water Closet, 3.5 GPF[4] Gravity Tank	1/2″		3.0	3.0		3.0	3.0		5.5	5.5		7.0	7.0	
Water Closet, 3.5 GPF[3] Flushometer Valve	1″		7.0	7.0		7.0	7.0		8.0	8.0		10.0	10.0	
Whirlpool Bath or Combination Bath/Shower	1/2″	1/2″	4.0	3.0	3.0	4.0	3.0	3.0						

NOTES:

1. The "total" WSFU values for fixtures represent their load on the water service. The separate cold water and hot water supply fixture units for fixtures having both hot and cold connections are each taken as $\frac{3}{4}$ of the listed total value for the individual fixture.
2. The fixture branch pipe sizes in Table B.5.2 are the minimum allowable. Larger sizes may be necessary if the water supply pressure at the fixture will be too low due to the available building supply pressure or the length of the fixture branch and other pressure losses in the distribution system.
3. The WSFU values for 3.5 GPF water closets also apply to water closets having flushing volumes greater than 3.5 gallons.
4. Gravity tank water closets include the pump assisted and vacuum assisted types.

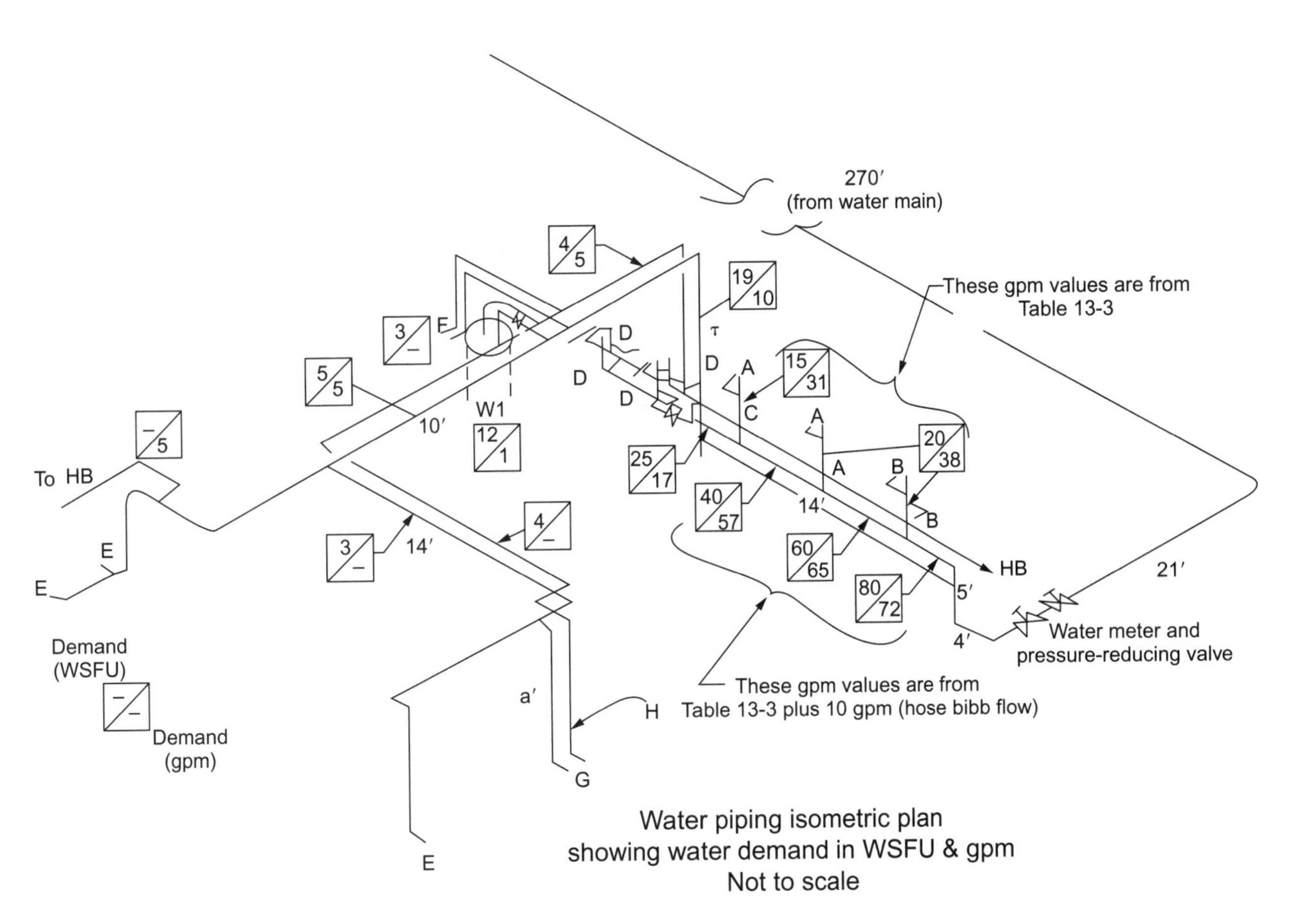

Figure 13–2
A water distribution isometric drawing enables the installing contractor to ensure code conformance by using accepted design factors, including water supply fixture units.

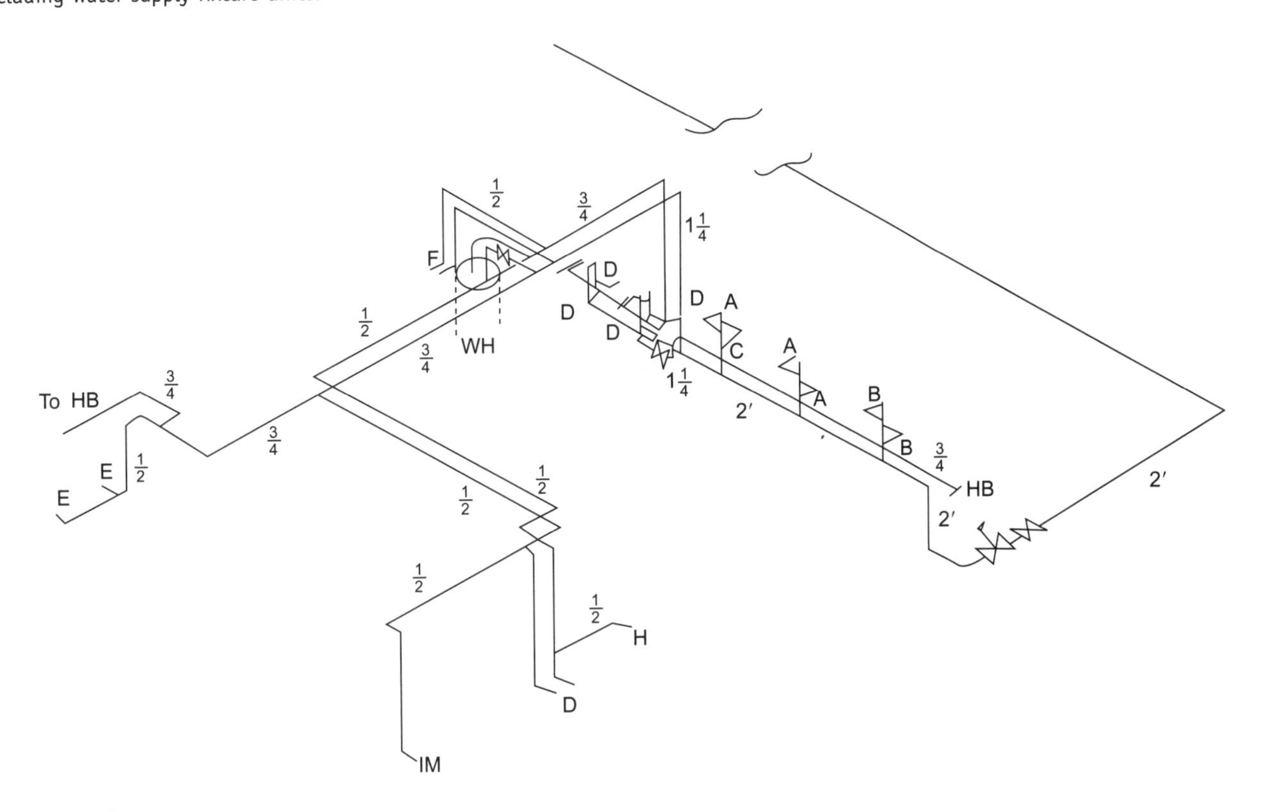

Figure 13–3
The water distribution isometric providing final line sizing enables the contractor to accurately order installation materials.

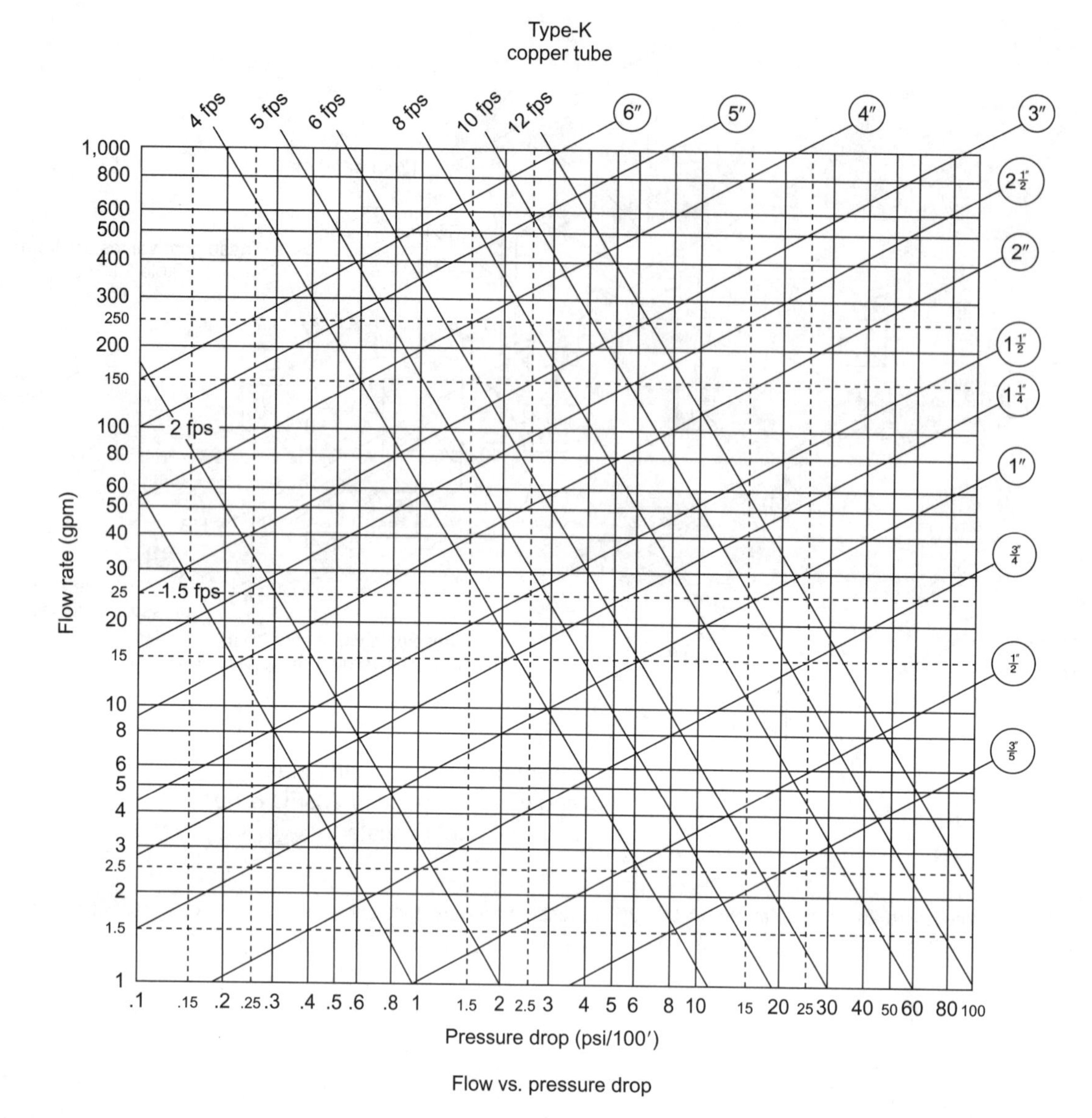

Figure 13–4
Copper Development Association chart for Type-K copper addressing the flow versus pressure drop.

> **In the Field**
>
> The hot water sizing is conducted in the same way as the cold. While the example is provided for basic understanding, real-life sizing would consider a pressure drop across the water heater and additional pressure loss from the pressure-reducing valve.

Because most pressure drop/flow charts present the pressure drop in psi per 100 feet, you should multiply the calculated drop per foot by 100 to convert it to the same units as the charts, which in this case equals $9\frac{1}{2}$ psi/100′.

STEP 5. To complete the exercise, select water pipe sizes for each portion of the layout in Figure 13–2 and Figure 13–3, and apply the table information and calculations using information from Figure 13–4 and the Copper Development Association chart for Type L copper addressing the flow versus pressure drop.

Select each size so as not to exceed *either* the maximum velocity limit or maximum allowable pressure drop per 100 feet of pipe (9.5) permitted by the calculations described above. An increase in size is applicable for the limitation of maximum velocity.

REVIEW QUESTIONS

1. What is the definition of peak demand?
2. Explain why water composition will affect the system design related to choice of materials and sizing.
3. Will undersized lines decrease the likelihood of negative pressures, which contribute to backflow?
4. Do the major codes leave all the water pipe sizing clarification to be addressed in the appendices only?
5. What term is applied to fitting pressure loss when calculating the pressure loss amount in comparison to pipe?

CHAPTER 14

Fuel Gas Pipe Sizing Utilizing Blueprint Reading and Category I Venting

LEARNING OBJECTIVES

The student will:

- Describe how to use sizing isometrics to create material lists for gas piping systems.
- Identify the maximum flow of gas that can occur in each section of pipe by calculating the gas flow for each appliance.
- Identify past appliance venting factors that contributed to efficient venting of gas appliances.
- Explain the use of a sizing table for Type I appliance venting.

In the Field

Fuel gas piping installations have been historically within the scope of licensed plumbers. Many licensing jurisdictions throughout the country have moved toward establishing mechanical licenses separate from plumbing licenses. The licenses are based upon established law and define the scope of work that may be performed by the individual. It is the responsibility of the technician, not just the contracting owner, to understand the licensing requirements in the area of construction.

DESIGN OF FUEL GAS PIPING VENT SIZING

The plumbing professional involved in the installation and design of fuel gas piping is well aware of the safety aspects necessary for the proper installation. Some technicians may be guilty of using abbreviated "rule of thumb methods" in sizing systems, which is unacceptable for both the safety of operating equipment and the satisfactory operation of gas-supplied equipment.

The same elements that are required to size other pipe systems are needed to determine the size of gas piping, including the amount of load, the length of the system, and the available initial and final pressure at the point of use. Other considerations are a reference source that shows the capacity of typical pipe sizes as a function of pressure drop and equivalent length. Commonsense consideration will dictate that sizing should be done to accommodate the worst-case scenario. That is, we will consider that all the appliances may be in use at the same time.

The sizing procedures, with respect to the drawings provided with the chapter, use tables from the *National Fuel Gas Code (NFGC)*, 2006 edition.

Sizing an Example Gas Piping System

Two fuel gas piping plan views of the same structure considered in earlier chapters to aid in creating a sizing diagram are provided in Figure 14–1 and Figure 14–2. The roof plan has five heating/cooling units and the supply gas piping that serves them. Information from the gas utility confirms that the heating value of the supplied gas is 1,025 BTU/ft.[3] Further, the material you have chosen for the installation is schedule 40 steel pipe conforming to ASTM A53.

Your efforts to size the system would begin by creating a table similar to Table 14–1, providing a schedule of gas appliances in the plan. The equipment's

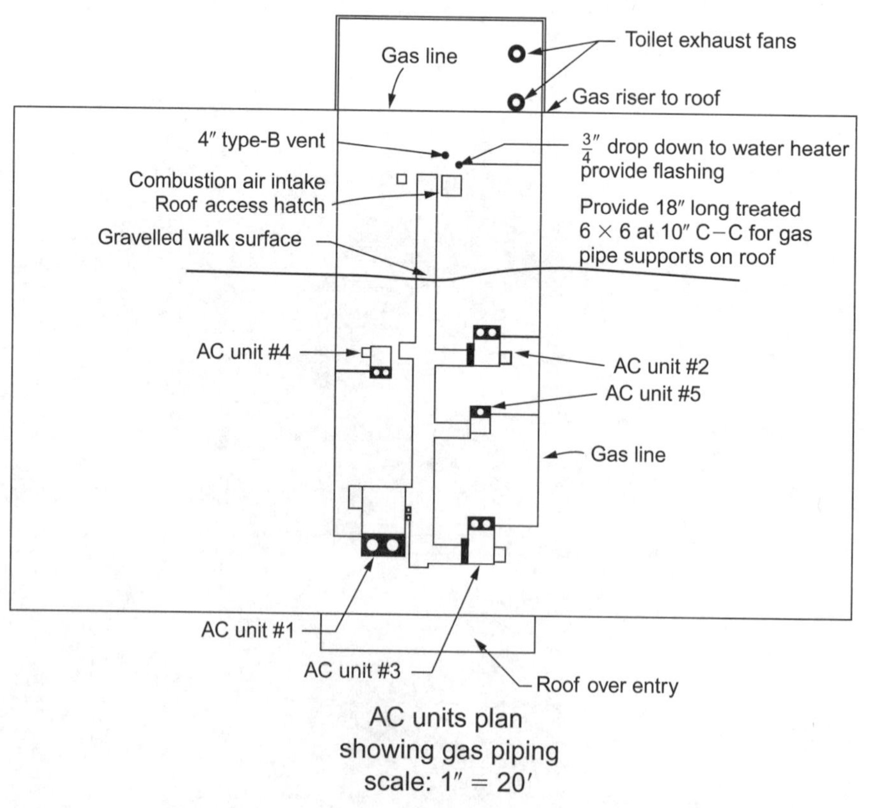

Figure 14–1
The plan view of an office building project illustrates the HVAC equipment locations.

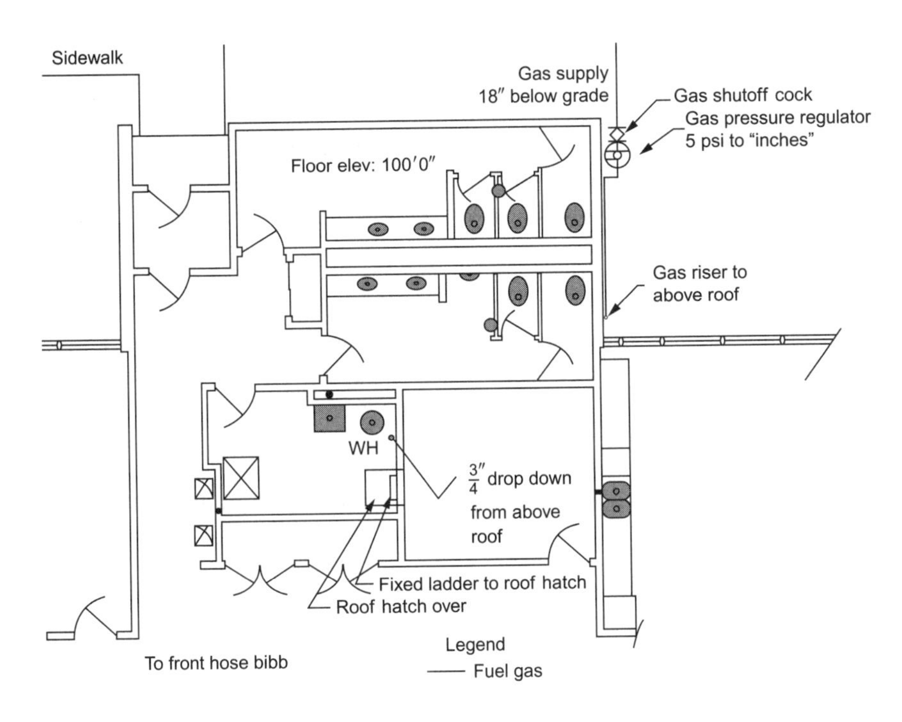

Figure 14–2
A partial plan view of the main floor mechanical room and other potential gas equipment provides location information that will be considered with the main rooftop units.

Table 14–1 Gas Appliance Schedule

Description	BTU Input Heating	Maximum Cubic Ft./Hr. Requirements*
AC heating unit #1	360,000	351.2
AC heating unit #2	225,000	219.5
AC heating unit #3	225,000	219.5
AC heating unit #4	1–70,000	165.9
AC heating unit #5	80,000	78.0
Water heater – 30 gal.	46,000	44.9

*Based upon heating value of 1,025 BTU/ft.^{3}

BTU value has been provided from construction documents, which must be obtained in order to properly size the system. You would then illustrate an isometric view to aid in designing the system and later in ordering material utilizing a design/build, fast-track process in accordance with code requirements.

The maximum cubic foot per hour values are obtained when you divide the BTU values by the gas heating values and, when used with provided distance allowance tables, indicate the required sizes. An example of the math for the first table value follows:

$$\frac{\text{Firing rate of the appliance in BTU}}{\text{BTU heating value of gas used}} = \text{Amount of gas burned in an hour}$$

$$\frac{360{,}000}{1{,}025} = 351.2$$

The partial floor plan of Figure 14–2 from the mechanical portion of the blueprints includes the water heater in the mechanical room below the rooftop units of Figure 14–1.

Figure 14–3 is an isometric sketch of all the gas loads and the gas piping. The lengths of the various sections of this piping layout with alphabetical location references are shown. Using this illustration, we can select the proper sizes for the piping from the tables.

Create a table to indicate the maximum flow of gas that can occur in each section of pipe by calculating the gas flow for each appliance (from Table 14–1) and for each section of pipe (heating value of gas is about 1,025 BTU/ft.^{3}). For reference, Table 14–2 includes the correct answers to this exercise. This table will later be expanded to indicate the line sizing using the alphabetical points of the isometric view in Figure 14–3.

The utility supplier often provides the gas service. We will provide a background discussion here without sizing tables to improve our understanding of the sizing. In a real-life installation, a site plan would have been provided and would indicate the location and distances of the various site utilities such as sanitary, storm, gas, and electric services. Our example would indicate that the customer's piping begins at the gas meter located 359 feet from the regulator, which is beside the building. The gas utility has installed a $1\frac{1}{2}''$ service line to the building's regulator with a gas delivery pressure from the meter to the regulator of 5 psig.

Calculating the effective length of the high-pressure section of the system from the site plan onto our isometric is provided in the following example:

The length of this section is $150' + 87' + 98' + 18' + 3' + 3' = 359'$

Thus, the effective length is $359(1.5) = 540'$, approximately.

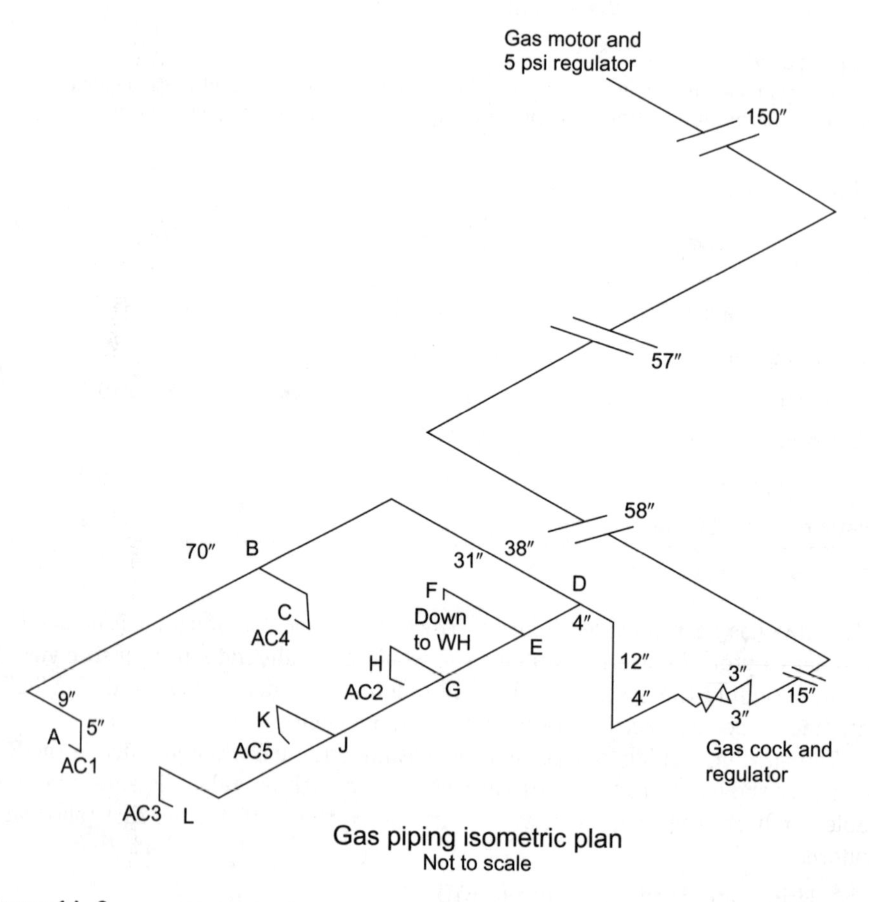

Figure 14–3
An isometric piping view enables the installing contractor to ensure code conformance by using accepted design factors.

Table 14–2 Maximum Flow Example Table

Section	Flow ft.^{3}/hr.
A–B	351
B–C	170
B–D	521
D–E	563
F–E	45
G–E	518
H–G	220
H–J–G	298
H–K–J	78
L–J	220
REGULATOR–D	1,080

Several design document tables and the *NFGC*, in previously printed editions, provide a table for high-pressure service lines. The heading fits into our application: Pipe Sizing Table for 5 Pounds Pressure Capacity of Pipes of Different Diameters and Lengths in Cubic Feet per Hour for an Initial Pressure of 5.0 Psig with a 10 Percent Pressure Drop and a Gas of 0.6 Specific Gravity. The table has a distance column for a 500-foot length, which of course is under the 540-foot distance we calculated. For that reason, we would use the 1,000-foot length column and obtain a size of $1\frac{1}{2}''$ for the schedule 40 steel pipe.

Calculate the effective length of the low-pressure section of the system, which is the line from the building's pressure-reducing valve to the farthest unit. The length of the system in the building is obtained by measuring the longer of the two parallel systems.

$2' + 4' + 12' + 35' + 76' + 9' + 5' = 143'$

The equivalent length is $143(1.5) = 215'$.

The equivalent length must be compensated for fitting considerations by an increase of 50%. This is accomplished by multiplying the actual pipe length by 1.5, as shown above.

The proper table to use for low-pressure gas (less then 1 psig), is from the *NFGC*, as it contains allowances for a normal number of fittings. The equivalent length above is greater than the 200-foot-length column, which warrants your use of the 250-foot-length column.

Use the 250 ft. row information above to complete the required sizes for Table 14–3.

Table 14–3, as completed, provides the necessary size information for the isometric view identified as Figure 14–3.

Sizing Category I Appliance Vents

The proper sizing of vented gas appliances is accomplished by using the *National Fuel Gas Code*, plumbing codes, and mechanical codes. The authority having jurisdiction will have selected the appropriate documents to address its installation and licensing concerns. Our discussion will be based upon National Fire Protection Association 54 and the *NFGC*, with a table from the 2009 edition. This code is commonly referred to as ANSI Z223/NFPA 54. The various documents involved are very similar in their application and makeup. A review of recent documents indicates the leadership content of the NFPA regulation, which influenced the other documents' contents. *Plumbing 301* provides insight into Category I appliance vents.

In the Field

There are several design document tables in the *NFGC*. You can select the proper one by determining the initial gas pressure and the allowable pressure drop and selecting the table that comes closest to matching the conditions you have selected.

In the Field

A Safety Reminder for Technicians: While this chapter's discussion has centered on the design aspect of fuel piping installations, a brief reminder of installation purging (bleeding the air from the system) is appropriately shared here. There have been occasions where technicians bringing fuel gas piping installations on line during purging processes have experienced dangerous conditions because they did not smell the odor, which is present in the fuel as an additive. The gas companies add an odorant called mercaptan as a safety feature and warning property. Under certain conditions, the odor may dissipate; for example, cutting oils may absorb the odor. Purging and allowing gas to accumulate presents the very dangerous possibility for unexpected combustion.

Table 14-3 Maximum Flow Example Table

Section	Flow ft.^{3}/hr.	Pipe Size Inch
A–B	351	$1\frac{1}{2}$
B–C	170	$1\frac{1}{4}$
B–D	521	2
D–E	563	2
F–E	45	$\frac{3}{4}$
G–E	518	2
H–G	220	$1\frac{1}{4}$
H–J–G	298	$1\frac{1}{2}$
H–K–J	78	1
L–J	220	$1\frac{1}{4}$
REGULATOR–D	1,080	$2\frac{1}{2}$

From the *NFPA Fuel Gas Code*, page 137 of their 2006 edition, Table 13.1 (b), Type B Double Wall. Courtesy of NFPA.

In the past, gas appliances that were intended to be vented adhered to the following assumptions:

1. The products of combustion leaving the appliance contained enough heat to warm the vent system quickly.
2. The amount of heat was adequate to warm the vent system sufficiently to produce the thermal lift necessary for the formation of draft.
3. The products of combustion were hot enough that condensation would not form after steady state conditions were obtained in the vent system.
4. The running time of the typical appliance and the total heat in the products of combustion were such that most vent systems delivered adequate draft most of the time.
5. Nearly all gas devices operated with negative pressure (i.e., draft) at the outlet, and the vent systems maintained negative pressure; thus, outflow into the building from the vent system could not occur.

During the past several years, with the implementation of high-efficiency appliances, these assumptions have not been correct. High-efficiency devices deliver relatively cool gases to the vent system, better construction of buildings means less heat is required in our furnaces or boilers (these space-heating products dominate the heat input in most equipment rooms), and less building heat loss means more short-time operating cycles during times of cooler outdoor weather. The fuel gas codes recognize the trends in equipment concepts and the effects of reduced design heat loss values. For these reasons, masonry chimneys have become a poor choice for venting the typical residential mechanical room.

Gas appliances are categorized in the following order for the purpose of matching the appliance to the required type of vent method. Previous editions of the *NFGC* recognized four categories of appliances, as follows:

I. Flue gases at negative pressure and at a temperature unlikely to form condensate.
II. Flue gases at negative pressure and at a temperature that is likely to form condensate.
III. Flue gases at positive pressure and at a temperature unlikely to form condensate.
IV. Flue gases at positive pressure and at a temperature likely to form condensate.

NOTE: Category II, III, and IV appliances must be vented according to manufacturer's instructions. Therefore, all the venting tables in the *NFGC* are for Category I appliances.

Sizing an Example Gas Appliance Vent

Gas appliance vent piping is sized by considering the BTU per hour firing rate, height of chimney, nature of chimney, and horizontal distance to the chimney.

A sizing example for a Category I appliance is taken from the plan view illustration in Figure 14–1 and Figure 14–2, the same structure considered in earlier chapters. The water heater, the only gas appliance located inside the building, will be considered in creating a sizing diagram. The roof plan includes five heating/cooling roof-mounted units that are gas-fired and probably Category III, which have integral power vent exhaust systems designed and provided by the manufacturer.

The water heater firing rate is 46,000 BTU per hour, as provided previously, and a double-wall Type B gas vent will be used in an adequately ventilated equipment room. Based on the vertical rise for the gas vent from the elevation plan (not included in this chapter), it is determined that the vent height is 15 feet above the top of the water heater. This 15 feet includes the vent portion above the roof starting at the flashing base. Table 14–4 provides the sizing capacity of Class B vents serving a single appliance as a function of vent diameter, vertical height, and offset length from the *NFGC*.

If the vent rises straight up from the heater without any lateral offset, the capacity of a 3-inch Class B vent (in the 15-foot height row) is 57,000 BTU per hour max, and a 4-inch vent has a capacity of 111,000 BTU per hour max. Consider that if a lateral offset of only 2 feet is included in the installation, the capacities are greatly reduced to 47,000 BTU per hour max for a 3-inch vent and 94,000 BTU per hour max for a 4-inch vent. Figure 14–4 from the *NFGC* further identifies the vent height (H) and offset (L) used in Table 14–3.

Consider two important limitations in this sizing process:

1. Do not connect an appliance flue outlet to a smaller vent, no matter what the venting capacity may be. For example, a 4-inch outlet should not be connected to a 3-inch vent. Appliance warranties may be voided if the flue pipe is connected to a smaller size vent pipe.
2. The vent should not offset horizontally more than 75% of the height of the vent. This is mandated in Section 12.7.3.2 of the *NFGC*. Vent tables indicate capacity values for greater offsets, but recent laboratory and computer calculations show that these longer offsets are unsatisfactory.

In the Field

Category II, III, and IV venting methods are dependent upon the manufacturer's recommended procedures.

In the Field

Note that the actual NFPA identification is included on the figures to aid in your later identification when you use *NFGC* Table 13.1(b).

In the Field

When working with a set of plans, most contractors or technicians do not check sizing unless an error is obvious. If your company policy is to check pipe sizing on all work you perform, you have additional incentive to become skilled at all the sizing problems described in this set of chapters.

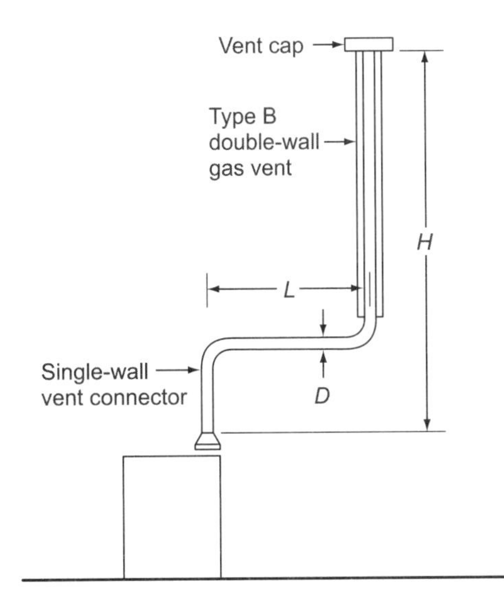

Figure 14–4
Vent height and offset are dictated by the *NFGC* and are defined as shown in this illustration. Courtesy of NFPA.

Table 14–4 Type B Double-Wall Vent

		Number of Appliances: Single																								
		Appliance Type: Category I																								
		Appliance Vent Connection: Single-Wall Metal Connector																								
		Vent Diameter — *D* (in.)																								
		3			4			5			6			7			8			9			10			12
		Appliance Input Rating in Thousands of BTU per Hour																								
Height *H*	Lateral *L*	FAN		NAT	FAN		NAT	FAN		NAT	FAN		NAT	FAN		NAT	FAN		NAT	FAN		NAT	FAN		NAT	FAN		NAT
(ft)	(ft)	Min	Max	Max	Min	Max	Max	Min	Max	Max	Min	Max	Max	Min	Max	Max	Min	Max	Max	Min	Max	Max	Min	Max	Max	Min	Max	Max
6	0	38	77	45	59	151	85	85	249	140	126	373	204	165	522	284	211	695	369	267	894	469	371	1118	569	537	1639	849
	2	39	51	36	60	96	66	85	156	104	123	231	156	159	320	213	201	423	284	251	541	368	347	673	453	498	979	648
	4	NA	NA	33	74	92	63	102	152	102	146	225	152	187	313	208	237	416	277	295	533	360	409	664	443	584	971	638
	6	NA	NA	31	83	89	60	114	147	99	163	220	148	207	307	203	263	409	271	327	526	352	449	656	433	638	962	627
8	0	37	83	50	58	164	93	83	273	154	123	412	234	161	580	319	206	777	414	258	1002	536	360	1257	658	521	1852	967
	2	39	56	39	59	108	75	83	176	119	121	261	179	155	363	246	197	482	321	246	617	417	339	768	513	486	1120	743
	5	NA	NA	37	77	102	69	107	168	114	151	252	171	193	352	235	245	470	311	305	604	404	418	754	500	598	1104	730
	8	NA	NA	33	90	95	64	122	161	107	175	243	163	223	342	225	280	458	300	344	591	392	470	740	486	665	1089	715
10	0	37	87	53	57	174	99	82	293	165	120	444	254	158	628	344	202	844	449	253	1093	584	351	1373	718	507	2031	1057
	2	39	61	41	59	117	80	82	193	128	119	287	194	153	400	272	193	531	354	242	681	456	332	849	559	475	1242	848
	5	52	56	39	76	111	76	105	185	122	148	277	186	190	388	261	241	518	344	299	667	443	409	834	544	584	1224	825
	10	NA	NA	34	97	100	68	132	171	112	188	261	171	237	369	241	296	497	325	363	643	423	492	808	520	688	1194	788
15	0	36	93	57	56	190	111	80	325	186	116	499	283	153	713	388	195	966	523	244	1259	681	336	1591	838	488	2374	1237
	2	38	69	47	57	136	93	80	225	149	115	337	224	148	473	314	187	631	413	232	812	543	319	1015	673	457	1491	983
	5	51	63	44	75	128	86	102	216	140	144	326	217	182	459	298	231	616	400	287	795	526	392	997	657	562	1469	963
	10	NA	NA	39	95	116	79	128	201	131	182	308	203	228	438	284	284	592	381	349	768	501	470	966	628	664	1433	928
	15	NA	NA	NA	NA	NA	72	158	186	124	220	290	192	272	418	269	334	568	367	404	742	484	540	937	601	750	1399	894
20	0	35	96	60	54	200	118	78	346	201	114	537	306	149	772	428	190	1053	573	238	1379	750	326	1751	927	473	2631	1346
	2	37	74	50	56	148	99	78	248	165	113	375	248	144	528	344	182	708	468	227	914	611	309	1146	754	443	1689	1098
	5	50	68	47	73	140	94	100	239	158	141	363	239	178	514	334	224	692	457	279	896	596	381	1126	734	547	1665	1074
	10	NA	NA	41	93	129	86	125	223	146	177	344	224	222	491	316	277	666	437	339	866	570	457	1092	702	646	1626	1037
	15	NA	NA	NA	NA	NA	80	155	208	136	216	325	210	264	469	301	325	640	419	393	838	549	526	1060	677	730	1587	1005
	20	NA	NA	NA	NA	NA	NA	186	192	126	254	306	196	309	448	285	374	616	400	448	810	526	592	1028	651	808	1550	973

Continued

		Number of Appliances: Single																								
		Appliance Type: Category I																								
		Appliance Vent Connection: Single-Wall Metal Connector																								
		Vent Diameter — *D* (in.)																								
		3			4			5			6			7			8			9			10			12		
		Appliance Input Rating in Thousands of BTU per Hour																								
Height *H*	Lateral *L*	FAN		NAT	FAN		NAT	FAN		NAT	FAN		NAT	FAN		NAT	FAN		NAT	FAN		NAT	FAN		NAT	FAN		NAT
(ft)	(ft)	Min	Max	Max	Min	Max	Max	Min	Max	Max	Min	Max	Max	Min	Max	Max	Min	Max	Max	Min	Max	Max	Min	Max	Max	Min	Max	Max
30	0	34	99	63	53	211	127	76	372	219	110	584	334	144	849	472	184	1168	647	229	1542	852	312	1971	1056	454	2996	1545
	2	37	80	56	55	164	111	76	281	183	109	429	279	139	610	392	175	823	533	219	1069	698	296	1346	863	424	1999	1308
	5	49	74	52	72	157	106	98	271	173	136	417	271	171	595	382	215	806	521	269	1049	684	366	1324	846	524	1971	1283
	10	NA	NA	NA	91	144	98	122	255	168	171	397	257	213	570	367	265	777	501	327	1017	662	440	1287	821	620	1927	1243
	15	NA	NA	NA	115	131	NA	151	239	157	208	377	242	255	547	349	312	750	481	379	985	638	507	1251	794	702	1884	1205
	20	NA	NA	NA	NA	NA	NA	181	223	NA	246	357	228	298	524	333	360	723	461	433	955	615	570	1216	768	780	1841	1166
	30	NA	NA	NA	NA	NA	NA	NA	NA	NA	NA	NA	NA	389	477	305	461	670	426	541	895	574	704	1147	720	937	1759	1101
50	0	33	99	66	51	213	133	73	394	230	105	629	361	138	928	515	176	1292	704	220	1724	948	295	2223	1189	428	3432	1818
	2	36	84	61	53	181	121	73	318	205	104	495	312	133	712	443	168	971	613	209	1273	811	280	1615	1007	401	2426	1509
	5	48	80	NA	70	174	117	94	308	198	131	482	305	164	696	435	204	953	602	257	1252	795	347	1591	991	496	2396	1490
	10	NA	NA	NA	89	160	NA	118	292	186	162	461	292	203	671	420	253	923	583	313	1217	765	418	1551	963	589	2347	1455
	15	NA	NA	NA	112	148	NA	145	275	174	199	441	280	244	646	405	299	894	562	363	1183	736	481	1512	934	668	2299	1421
	20	NA	NA	NA	NA	NA	NA	176	257	NA	236	420	267	285	622	389	345	866	543	415	1150	708	544	1473	906	741	2251	1387
	30	NA	NA	NA	NA	NA	NA	NA	NA	NA	315	376	NA	373	573	NA	442	809	502	521	1086	649	674	1399	848	892	2159	1318
100	0	NA	NA	NA	49	214	NA	69	403	NA	100	659	395	131	991	555	166	1404	765	207	1900	1033	273	2479	1300	395	3912	2042
	2	NA	NA	NA	51	192	NA	70	351	NA	98	563	373	125	828	508	158	1152	698	196	1532	933	259	1970	1168	371	3021	1817
	5	NA	NA	NA	67	186	NA	90	342	NA	125	551	366	156	813	501	194	1134	688	240	1511	921	322	1945	1153	460	2990	1796
	10	NA	NA	NA	85	175	NA	113	324	NA	153	532	354	191	789	486	238	1104	672	293	1477	902	389	1905	1133	547	2938	1763
	15	NA	NA	NA	132	162	NA	138	310	NA	188	511	343	230	764	473	281	1075	656	342	1443	884	447	1865	1110	618	2888	1730
	20	NA	NA	NA	NA	NA	NA	168	295	NA	224	487	NA	270	739	458	325	1046	639	391	1410	864	507	1825	1087	690	2838	1696
	30	NA	NA	NA	NA	NA	NA	231	264	NA	301	448	NA	355	685	NA	418	988	NA	491	1343	824	631	1747	1041	834	2739	1627
	50	NA	NA	NA	NA	NA	NA	NA	NA	NA	NA	NA	NA	540	584	NA	617	866	NA	711	1205	NA	895	1591	NA	1138	2547	1489

For SI units, 1 in. = 25.4 mm, 1 ft = 0.305 m, 1000 BTUH = 0.293 kW, 1 in.^{2} = 645 mm^2.
NA: Not applicable.

REVIEW QUESTIONS

1. What is the general sizing rule for Category II, III, and IV appliance vents?
2. What is the maximum percentage a horizontal offset may be of the vertical vent height?
3. What information is obtained when the firing rate of the appliance in BTUs is divided by the BTU heating value of the gas used?
4. Is schedule 40 steel pipe conforming to ASTM A53 the only material that can be installed to service HVAC rooftop units?
5. The rule-of-thumb method for obtaining piping equivalent length calculations is to multiply the total actual pipe length by what number?
6. Using Table 14–4, what is the recommended size of a Type B vent for a water heater with a 199,000 BTU per hour input rating and a vent 20 feet above the heater without any offsets?

CHAPTER 15

Indirect and Special Waste Installation Practices

LEARNING OBJECTIVES

The student will:

- Explain the importance of indirect waste connections and list the code requirements for their installation.
- Explain when indirect connections must be provided with an air gap or air break.
- Describe the various indirect receivers and explain code agencies' concerns, which may vary from jurisdiction to jurisdiction.
- Contrast special waste dilution and neutralization practices for special waste systems.

INDIRECT WASTE INSTALLATIONS

The sanitary drainage system is designed to remove the discharge from soil, waste, and drainage piping and convey this discharge to an approved point of disposal. Two methods of connecting fixtures or subsystems to the plumbing drainage system are called direct and indirect. Drainage systems with indirect waste systems are used to protect plumbing fixtures, appliances, and appurtenances from backflow contamination of the devices' contents, such as eating utensils. With direct connections, a sewage backflow can develop in a sanitary drain line when a blockage occurs. The indirect waste arrangement prevents this backflow from appearing in the protected drainage subsystem because the backflow condition simply causes sewage to spill onto the floor at the receptor that receives the indirect waste discharge. While this spillage may be unpleasant or worse, the protected refrigerator, storage tank, or food processing sink will not be affected.

Plumbing codes vary a great deal in requiring an extra trap close to the unit discharging into the indirect waste line. The trap at the equipment prevents odors from entering the fixture being drained. Most codes make this statement: "Traps shall be provided at fixtures and equipment connections where the developed length of indirect waste piping exceeds [a number of] feet." The codes range from 2 feet to 15 feet in this mandate. This is not a double-trapped fixture, which many codes specifically prohibit. The atmospheric break at the air break or air gap allows air to enter the indirect waste line for proper drainage operation.

> **In the Field**
>
> Several codes require healthcare institutions' central vacuum system discharges to be directly connected to the drainage system. This prevents aspiration of infectious diseases or contaminants into the open air to protect the habitable space.

Direct connections consist of any arrangement in which the plumbing fixture or appliance is continuously piped to the drainage system. These direct connections prevent the release of waste materials and odors into the habitable space from the drainage piping, building drain, building, and public or private sewer. The vast majority of plumbing fixtures in the drainage system have direct connections.

Indirect drainage piping connections are used to prevent backflow from the main drainage system into certain critical fixtures. An indirect connection consists of any piping arrangement in which the fixture, appliance, or subsystem discharges through open air into a trap, fixture, receptor, or interceptor, which is directly connected to the drainage system. The indirect connections prevent vermin invasion, provide additional potable water supply protection, prevent backflow of sewage into special fixtures, and prevent pressurization of the drainage system.

> **In the Field**
>
> Remember while considering the type of drainage connection that plumbing codes require a connection through a proper liquid-sealed trap that must be vented in accordance with the code.

A partial list of considerations regarding why equipment must be indirectly connected to the drainage system includes the following:

- Food located in or later placed in food preparation commercial sinks could not be contaminated from waste backing up in a malfunctioning drainage system.
- Ice located in a storage bin would not be contaminated from waste backing up in a malfunctioning drainage system.
- Insects or small animals could not enter the fixtures if they were indirectly connected.
- If a negative pressure develops on a water supply line to an appliance such as a water heater, the indirect waste maintains the integrity of the potable water system.

A partial list of equipment that must be indirectly connected to the drainage system includes the following:

- Clear water wastes
- Commercial food processing fixtures and appliances
- Condensate lines from air conditioners
- Drains from boilers and water heaters
- Drains from pressure tanks
- Overflow connections to potable water storage tanks
- Swimming pools
- Sterilizers
- Walk-in cooler drains and refrigeration equipment

Air Gaps

Indirect connections may be further broken down into the methods of their connection to an air break or air gap. These two methods vary in their degree of protection, with the air gap providing a higher level of protection. An air gap, according to the *National Standard Plumbing Code*, is the unobstructed vertical distance through the free atmosphere between the outlet of the waste pipe and the flood level rim of the receptor into which it is discharging. With this separation, it is impossible to have any backflow enter the waste pipe, regardless of what pressure or vacuum pattern exists. Figure 15–1 illustrates an air break discharge arrangement for indirect wastes.

The air gap is the ultimate protection against backflow under any circumstance. A thought-provoking consideration is the following:

1. If a vacuum can be developed in the device being drained, an air gap is required, such as in a relief valve discharge. In this way, the drainage system cannot be pressurized with a relief valve discharge, nor can a vacuum draw wastewater into the pressure vessel if there is an inlet pressure failure to the tank.
2. If a vacuum cannot occur, either an air break or an air gap is permitted, as in a food preparation sink compartment.

An air gap's vertical dimension must be at least twice the effective diameter of the discharge pipe. If the discharge opening is not circular, the effective diameter is the diameter of the circle of area equal to that of the actual opening. Thus, if an indirect waste discharge pipe is 1 inch, the air gap must be a minimum of 2 inches.

Figure 15–2 shows a reduced pressure principle backflow preventer with an air gap connection to the reduced pressure drainage chamber. The air gap is used to protect the potable water supply.

In the Field

A residential dishwasher must discharge through an air gap in some code jurisdictions, whereas other codes do not require this protection. The argument is that, without the air gap, simply restarting the dishwasher if a stoppage develops might cause a backflow condition. The possible spillage onto the countertop from an air gap is avoided. Further, most residential dishwashers have a float mechanism that acts as a drain from the inside washer compartment to the floor.

Air Breaks

According to the *NSPC*, an air break is a piping arrangement in which a drain from a fixture, appliance, or device discharges indirectly into a fixture, receptor, or interceptor at a point below the flood level rim of the receptor. The receptor may be a floor drain, standpipe, fixture, receptor, or interceptor. The advantage of the air break is

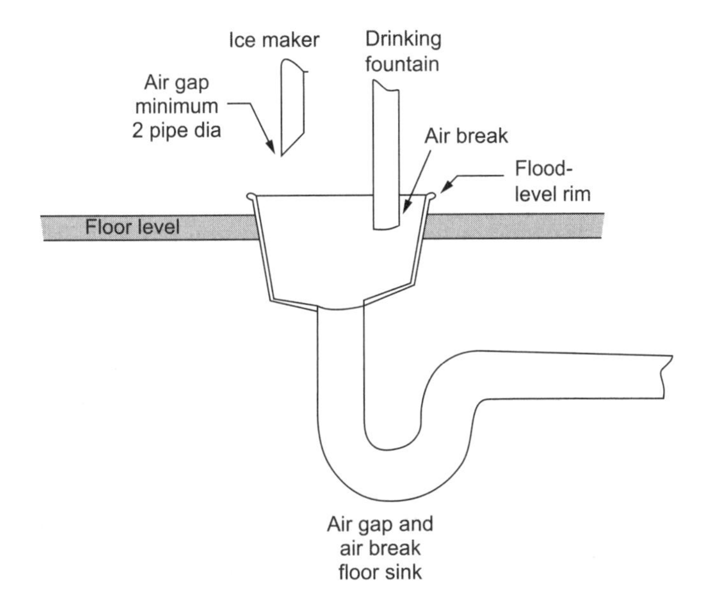

Figure 15–1
An air gap provides the best separation for food processing equipment, while an air break is acceptable for specific washing equipment.

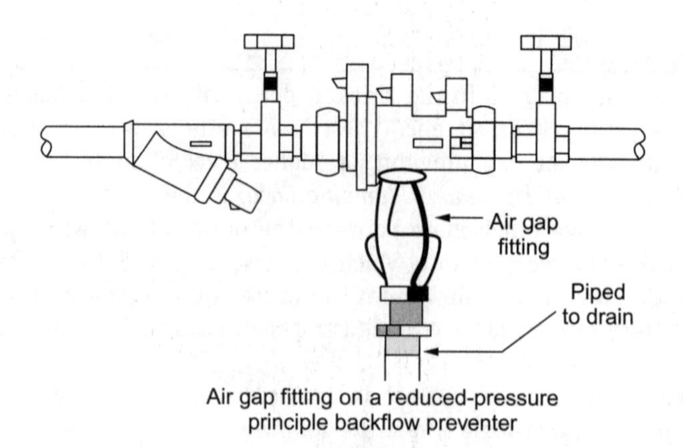

Figure 15–2
A code-required air gap that provides complete, fail-safe protection is necessary to ensure that a contaminant is not siphoned back into the potable water supply.

that there is less splashing with this arrangement than with an air gap, where splashing may be a nuisance. It should be noted that careful piping details could be employed to reduce the splashing problem of an air gap or air break, such as cutting the discharge pipe on an angle. Figure 15–1 illustrates air break discharge arrangements for indirect wastes as well as the former air gap.

Receptors

Waste receptors are provided to receive, accumulate, and conduct indirect waste discharges to the drainage system. They vary a great deal in design and must be adequately sized to prevent splashing and flooding from the indirect discharge. The receptor trap and connected piping must have a cross-sectional area not less than the sum of cross-sectional areas of all indirect waste pipes discharging into the receptor. The receptor must be an accessible floor sink, floor drain, service sink, standpipe, or other suitable trapped and vented fixture. Indirect waste receptors should not be installed in any toilet room or in any inaccessible or unventilated space such as a pipe chase.

The receptor should be equipped with a removable strainer to aid in cleaning it out. If the indirect waste carries clear water waste only, the strainer is not required. Clear water waste, according to the *NSPC*, is effluent in which impurity levels are less than concentrations considered harmful by the administrative authority, such as cooling water and condensate drainage from refrigeration and air conditioning equipment, cooled condensate from steam heating systems, and residual water from ice making processes. Plumbing fixtures used for culinary purposes should not be used as an indirect waste receptor. Practical consideration for fixtures receiving wastes are that residential sink traps or food waste grinders may receive a dishwasher hose connection, and a residential laundry tray may receive a clothes washer discharge.

> **In the Field**
>
> Several floor drain strainers are manufactured to have a funnel-type device protruding upward from the center of the strainer. This device is advertised as providing a combination floor surface area for drainage and a receiver area to prevent splashing for indirect connections. Exercise caution here in selection. Many health department officials believe the strainer area can become clogged and consider the device suitable only as an air break.

Condensate Waste

Condensate waste systems must have indirect connections and should be separately trapped close to the equipment. The piping should be easily reached for flushing and cleaning. Condensate from air conditioners is drained into trapped lines that are piped to convenient receptors. The trap in the condensate line prevents air movement through the line as a result of blower operation. Such air movement can cause condensate to hang up and not drain and can encourage algae growth and drain line stoppage. Slope the condensate pipe toward the drain receptor, and take as much vertical distance as needed at the outlet from the condensate-collecting pan.

Technicians are cautioned when dealing with different authorities having jurisdiction related to equipment pan drains. Code authorities have different opinions on where the drains may be connected. For example, large rooftop equipment may be regularly treated and purged and the pan used as a discharge accumulator drain. Numerous designers insist on draining condensate waste to the storm system, which may be more available to the location than the sanitary system.

Floor Sinks

Floor sinks (enlarged floor drain bodies) are very common receivers in commercial kitchens. They may have strainers at the discharge level rather than the top, as floor drains do. One of the major benefits of the fixture is that its expanded receiver area can easily accommodate various indirect lines. Several codes allow the omission of the strainer at the bottom if the elevation of the floor sink lip is higher than the floor. This higher lip would prevent kitchen crews from sweeping objects down the drain, such as French fries on the floor.

Standpipes

Several fixtures such as automatic clothes washers may discharge into a standpipe attached to a trapped and vented line. The standpipe could be connected to a floor drain strainer similar to the one mentioned in the previous sidebar. The vertical standpipe's height will vary from code to code but is commonly between 18 inches and 48 inches. The minimum 18-inch length is intended to prevent spillage, and the maximum 48-inch allegedly reduces the possibility of self-siphonage.

Relief Valves

Figure 15–3 shows how a relief valve must be discharged through an air gap, which is used to protect the potable water supply. If the receptor is in the same room, the relief valve drain must be air-gapped into the receptor. If the receptor is not in the same room, the relief valve discharge is air-gapped into a second indirect waste line, which should be selected according to gravity drain sizing methods.

Remember also that the pipe from a relief valve must be not less than the size of the relief valve outlet, and the discharge end of the pipe must not be threaded. Boilers and boiler blow-off tank piping must be piped to a receptor and discharged indirectly with assurances to discharge below the maximum allowable temperature of the code.

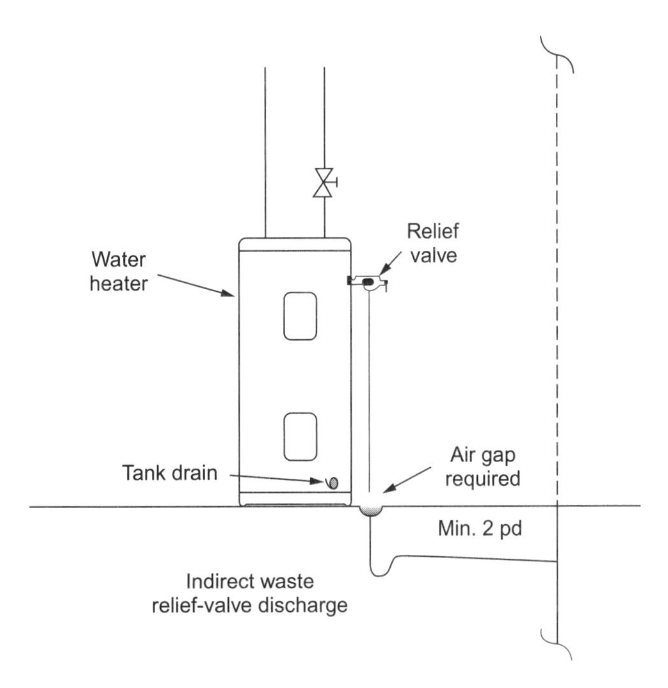

Figure 15–3
The code-required air gap protection ensures that a contaminant cannot be siphoned into the potable water supply when a relief valve is discharged.

In the Field

The subject of floor sinks and floor drains is a reminder to consider that code officials often require trap-seal primers on these devices indiscriminately. Most code officials do not require trap-seal primers on floor sinks as kitchens use these fixtures for discharges, which will replenish the trap seal. Discuss this matter with the authority having jurisdiction when you are involved in installations.

In the Field

Technicians must consider recent manufacturing changes that have increased the discharge volumes of residential automatic clothes washers. In some cases, standard sizing designs have proven to be incompatible with the greater discharge rates. Consider recent code changes that attempt to address these issues, such as one stating that the automatic washer branch must discharge into a 3-inch line, and venting types that may be affected by a pumping action discharge.

In the Field

Most codes require that if a relief valve discharge pipe leaves the room containing the water heater, it must discharge indirectly into a receptor before the pipe leaves the room. This also ensures that if it goes outside in colder climates, freezing the discharge will not render the draining pipe inoperable. In effect, you would be capping the discharge line.

Special Wastes

Special wastes are those wastes that require treatment prior to being discharged into the drainage system. Materials like steam, explosive liquids, gases, oil, grease, chemicals, and similar products are prohibited from entering the sanitary or storm drainage system. Such materials may damage, obstruct, or interfere with the normal sewage treatment processes or damage the piping that conveys the sewage. Special waste systems are designed to treat and render these problem materials harmless. Dilution and neutralizing systems including tanks may be made of materials like those used for fixtures. They are usually sealed and vented separately to the outdoors. If dilution water piping is connected to the tank, the water supply must be protected from contamination by suitable cross-connection protection methods.

Dilution and neutralization are the two methods employed to address problem materials. Dilution systems use water to dilute the problem waste materials to concentrations low enough to be considered harmless. Obviously, this method can work only on certain chemical types and may require the use of large amounts of water. Dilution may be used if the original waste is too hot. For example, a boiler blow-off tank, as in Figure 15–4, contains a quantity of water at room temperature. When boiler water is drained into the tank, the hot water is diluted by the water already in the tank. If more water is to be drained from the boiler than can be cooled by the original amount of water in the tank, cold water can be added to the flow. Most codes limit the maximum discharge temperature to 140°F.

Dilution tanks can be used to reduce the discharge or moderate the concentration of acids or bases, especially if the condition is intermittent. Some systems use a pH sensor that controls dilution water if the effluent pH varies by a set amount from pH 7.0, which is the neutral value for pure water. Figure 15–5 illustrates a dilution tank for a small chemical sink.

Neutralization systems use chemicals to change the harmful substances to products that are acceptable to the sanitary waste sewers and sewage treatment system. These systems must include methods or procedures for maintaining the treatment chemicals in sufficient quantity so that the process continues satisfactorily.

Neutralizing systems can handle more difficult problems and usually require less water, but they do require maintaining neutralizing chemical amounts dependent upon the system load. Very simple systems use a tile receiver with limestone, which neutralizes acid wastes before discharge into the building sewer. The limestone quantity must be checked regularly and increased as needed. Figure 15–6 shows a system with a pH sensor and equipment for adding chemicals. Such a system can handle wastes that vary in concentration of the objectionable material, but the added equipment means that more maintenance will be required.

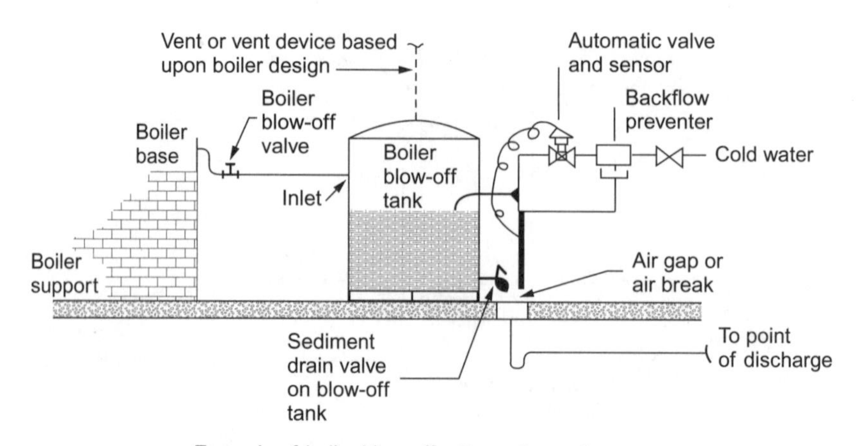

Figure 15–4
This illustration of a boiler blow-off with cooling water indicates one of several methods to lower temperatures of equipment discharge fluids before they enter the sewer system.

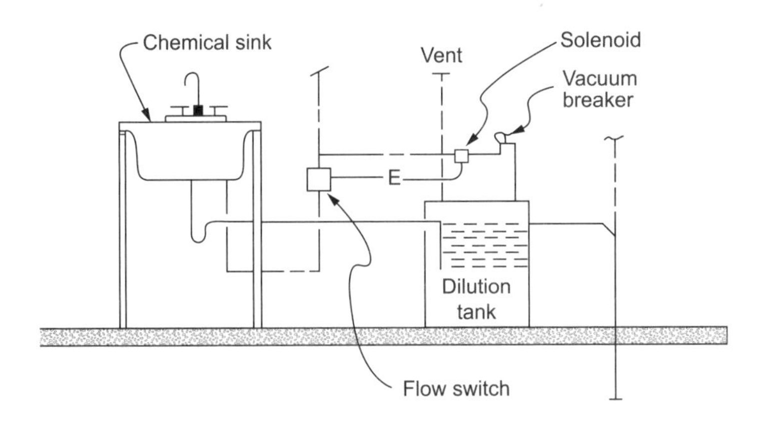

Figure 15–5
Small chemical dilution tanks are often used at locations where discharge from only one fixture requires treatment.

An important part of applying special waste systems is that they should be installed as close to the problem waste as possible. It is much easier to treat only the problem material than to treat all the building sewage to eliminate the problem component from the greater total stream.

All the components of special waste systems must be selected to be suitable for the waste stream that will be encountered. Careful thought must be given to probable temperatures and chemical concentrations so that a long-lived installation may be achieved. If possible, select pipe and equipment materials to include a safety factor. Failure or human error may produce more extreme conditions than would normally be present. Acids especially increase their chemical aggressiveness with temperature increases. Acid fumes are also usually very harmful, so the same factors are generally considered in selecting venting materials for acid waste systems.

It is especially important to obtain the advice of knowledgeable people when selecting these materials. Ask the manufacturers' representatives for their help. New materials and methods frequently appear in the marketplace to help address these problems. Study the problem and obtain expert advice in material system selection.

In the Field

Many years ago, technicians used a now-unacceptable phrase: "The secret to pollution is dilution." They used the concept in moderate acid waste systems such as school chemistry labs by discharging the lab waste immediately downstream of multiple-user restrooms, the idea being that large amounts of wastewater would dilute the lab discharges. Those systems had very short-term performance expectancies.

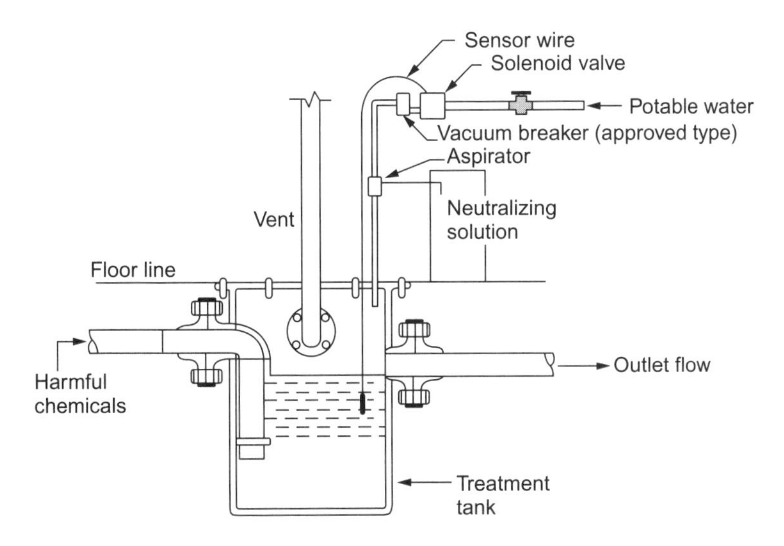

Figure 15–6
Large neutralization tanks with automatic sensors ensure proper maintenance and protect against harmful sewer discharges, which will damage sewer piping.

REVIEW QUESTIONS

1. What is the most reliable form of backflow protection?
2. What is the minimum height of a drainage air gap for an indirect waste line?
3. Is an air break an acceptable form of protection when a device is subject to a vacuum?
4. When circumstances permit, will an air break or air gap reduce the amount of splashing at a floor sink?
5. Storage tank drains and relief valve discharge pipes should be indirectly wasted by what connection?
6. Is it true that neutralizing systems require more monitoring and servicing than is needed for dilution systems?

CHAPTER 16

Interceptor and Backwater Valve Installation Practices

LEARNING OBJECTIVES

The student will:

- Recognize the role plumbing appurtenances play in the proper operation of drainage and treatment systems.
- Describe what type of interceptor is appropriate for the different commercial drainage systems.
- Recognize how to size oil interceptors related to structural conditions and floor area.
- Recognize different installation concerns for various code officials.

Interceptors and backwater valves are two types of plumbing appurtenances required by code for proper sanitary operations. A plumbing appurtenance is a manufactured device, a prefabricated or on-the-job assembly of component parts, that is additional to the basic piping system and plumbing fixtures. An appurtenance requires no additional water supply and does not add an additional discharge load to a fixture or the drainage system. It performs a necessary function in the operation, maintenance, servicing, economy, or safety of the plumbing system. Other appurtenances installed in plumbing systems would be water softeners and filters, and these devices are considered in other chapters.

INTERCEPTORS

Interceptors are used to keep foreign substances out of the drainage system; they are installed to collect sand, grease, oil, gasoline, hair, precious metals, plaster, and other substances. As with special waste treatment tanks such as chemical dilution tanks, interceptors should be located as close as possible to the point of introduction of the problem substance. This reduces the amount of piping exposed to the undesirable substance discharged and its effect on the system.

We will consider grease, oil, sand, and special purpose interceptors. Grease and oil interceptors are the most commonly encountered in the plumbing profession because grease is present in commercial kitchens and oil is present in service stations and other vehicle-related buildings.

In the Field

Grease interceptor is the presently accepted term for these collection devices. The plumbing profession, including codes, has also used the following terms:

1. Grease trap, meaning a small device located near the fixture discharging a greasy waste, which is referred to as a passive device and rated at 50 gpm or less. Grease traps are now sized by a flow-through capacity of up to 100 gpm with 200 pounds of retention capacity.
2. Grease interceptor, presently meaning all collection devices located inside and outside of a structure. When the term *grease trap* was used in earlier times, a grease interceptor referred to a large, passive device on the outside of a structure rated at 50 gpm or more.
3. Automatic grease removal device, which is an appurtenance that operates on a time- or event-controlled basis and accumulates grease automatically without assistance from the user, except for maintenance. The grease is accumulated in an additional holding container.

Grease Interceptors

Grease interceptors are an appurtenance designed to intercept and collect free-floating fats and oils in effluent discharges in and around food processing areas. Grease interceptors are necessary and mandated by plumbing codes. If left uncollected, the grease contained in kitchen waste discharges will adhere to waste lines and, if passed through waste systems, will adversely affect the performance of drains, sewers, and waste treatment processes. The waste treatment process could include going through a municipal treatment system or private septic and drain field system.

Interceptors must be located and installed so that they can be serviced readily. A major consideration is cleaning to remove the accumulated grease. Interceptors separate the grease from wastewater, based on the lesser density of the grease. The lighter material floats on top of the water where, in the case of grease interceptors, the grease is held by a series of baffles. Grease interceptors are equipped with a flow restrictor to limit flow to a rate that will not produce turbulence in the chamber that could dislodge the grease and cause it to travel down the drain. In addition to lowering the turbulence, the flow restrictor provides more time to allow separation in the discharged liquid. The flow control is commonly located outside the interceptor in the fixture discharge line, but some manufacturers design their flow controls for placement inside the interceptor.

Exterior flow controls are vented, which promotes air to be incorporated in the discharge, aiding separation in the interceptor. The vent is commonly an iron pipe $\frac{1}{2}''$ or $\frac{3}{4}''$ line above the $1\frac{1}{2}''$ or $2''$ waste inlet and out lines. Codes, manufacturers' installation instructions, and technicians have differing opinions on where the flow control vent is to be connected. Past-accepted methods are to have the control vent riser terminate above the flood level rim and turn the line downward. When the flow control vent is connected to the sanitary venting system, a fixture trap must be installed to prevent the passage of sewer gas from the venting system up through the fixture outlet and into the habitable space.

In the Field

Technicians are cautioned to read manufacturers' installation instructions closely. The instructions will point out that locating an interceptor below the fixture floor level, such as in the lower floor ceiling or lower level floor, will drastically reduce interceptor performance. Common sense here indicates that the increased elevation will increase flow velocities to render the interceptor less effective.

A grease interceptor acts as a trap, which in effect stops the passage of sewer gas from the sanitary system, through the supplying fixture drainage line, and into the habitable space. See Figure 16–1. Several codes consider and address that when lines reach a specified distance, drainage odors will be present between the fixture and the interceptor. The code will then require a fixture trap to be installed in addition to the interceptor (which serves as a trap). The distances vary and are commonly addressed in developed length, where the fixture-to-separator distance does not exceed 48 inches (some codes are 60 inches). These codes also limit the vertical distance from the fixture outlet to the interceptor to 30 inches. Consider that a double trap is not created here because the flow control vent adds air between the trap and the grease interceptor, which functions as a trap. Figure 16–2 illustrates proper installation of a grease interceptor.

Food waste grinders should not discharge directly into a grease interceptor but through a solids interceptor before discharging into a grease interceptor. Solids in a

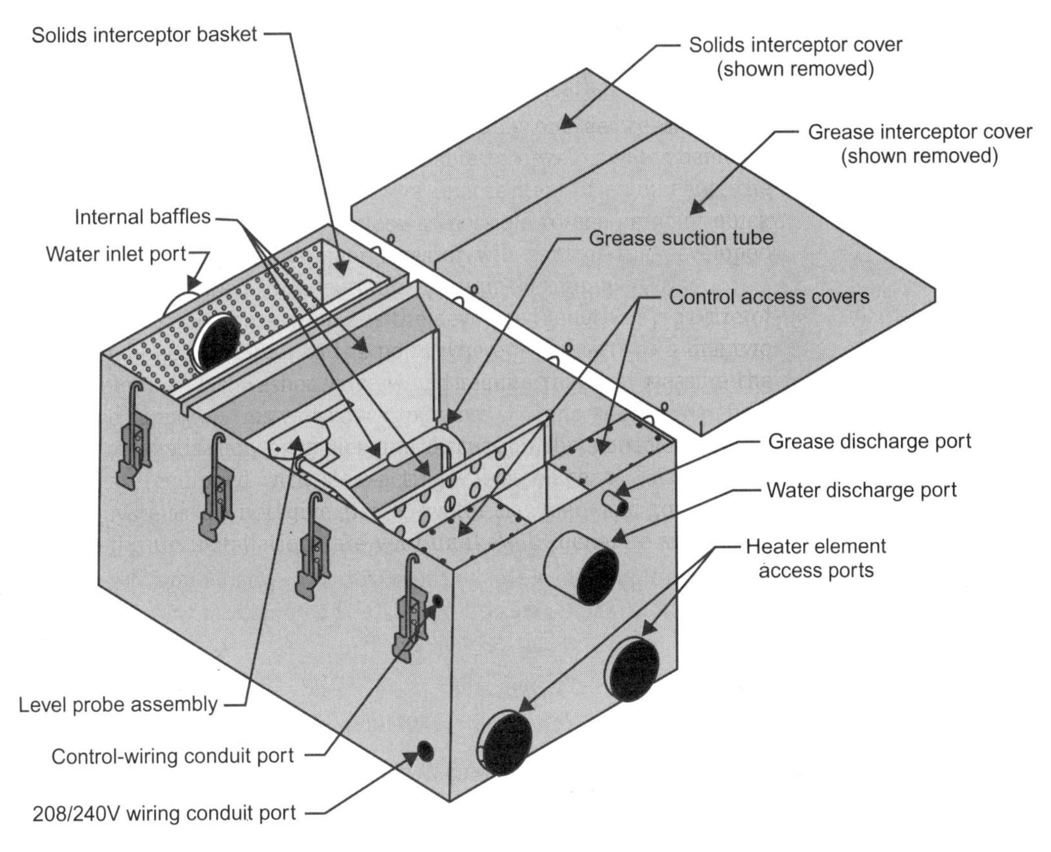

Figure 16–1
An automatic grease removal device is designed to meet the performance requirements of standard ASME A112.14.4. These have become sophisticated products with varying mechanisms.

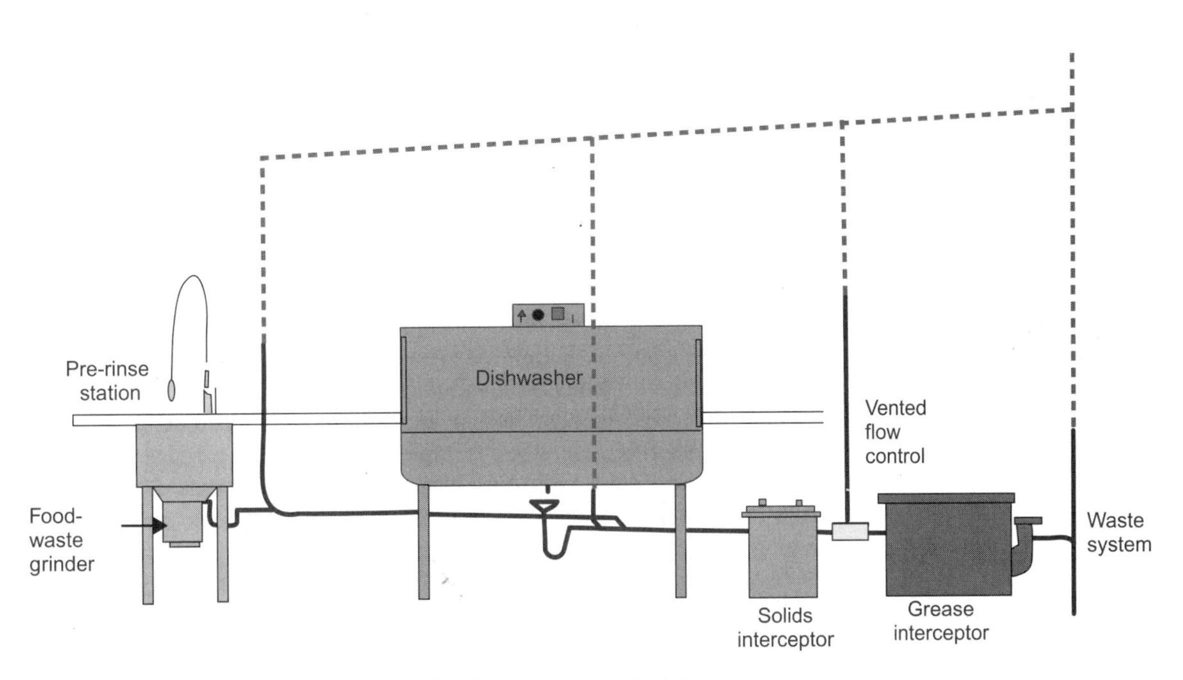

Figure 16–2
Grease interceptors are designed to meet the performance requirements of standards ASME A112.14.3 or PDI G1014. The vast majority of the interceptors have a vented flow control on the exterior of the device; however, in a few products the device is inside the unit.

In the Field

Consider guidance language here regarding code terminology. *Shall,* when used in the plumbing code, indicates that something is mandatory; *should* indicates that it is not mandatory but expresses a recommendation; and *may* implies neither requirement nor recommendation, only permission.

grease interceptor would take up space in the retention design area and would tend to dislodge collected grease. Additionally, dishwashers are seldom connected to grease interceptors, as the hot water discharge combined with the detergent would tend to negate the effective operation of the device.

Sizing

Newer plumbing codes require grease interceptors and grease removal devices to conform to certain standards, including Plumbing Drainage Institute, PDI G101, and American Society of Mechanical Engineers (ASME), A112.14.3 and A112.14.4. PDI G101 is the Testing and Rating Procedure for Grease Interceptors with Appendix of Sizing and Installation Data, which contains excellent information on sizing grease interceptors. Further PDI offers the entire standard free, and it is available on the Internet. ASME, A112.14.3, Grease Interceptors, and A112.14.4, Grease Removal Devices, also contain valuable information related to the performance requirements of interceptor products.

The PDI standard requires that the grease interceptor be able to retain $2\frac{1}{4}$ pounds of grease for every gallon per minute of flow capacity. These interceptors are rated in gpm flow rate, so the fixtures they serve must be checked for discharge rate. For an example, we will calculate the volume of water in a three-compartment sink, with each compartment being $15'' \times 24'' \times 12''$.

$$\text{Volume} = 15'' \times 24'' \times 12''$$
$$\text{Volume} = 4{,}320 \text{ cubic inches per compartment}$$

The volume for three compartments is as follows:

$$\text{Volume} = 3(4{,}320 \text{ cubic inches})$$
$$\text{Volume} = 12{,}960 \text{ cubic inches}$$

$$\text{One gallon} = 231 \text{ cubic inches}$$
$$\text{Sink volume in gallons} = \frac{12{,}960 \text{ cubic inches}}{231 \text{ cubic inches per gallon}}$$
$$\text{Volume} = 56 \text{ gallons}$$

The standard would advise us to consider that 25% of the sink volume is taken up by pans (items that are being washed) and will not be discharged to the drain. The practical adjusted discharge volume of the sink is calculated thusly:

$$\begin{aligned}\text{Discharge volume} &= (1 - 0.25)(56)\\ &= 0.75(56)\\ &= 42 \text{ gallons}\end{aligned}$$

Assuming that it will drain in 1 minute, the rate of discharge is 42 gpm. A 50 gpm device would be selected.

Note: Be sure to plan for the installation using adequate structural consideration. An interceptor may weigh as much as 2,000 pounds and must be set level and solid. Further, it must be readily accessible for servicing.

In the Field

The *Uniform Plumbing Code,* published by the International Association of Plumbing and Mechanical Officials, has excellent information related to large units. The information is contained in Appendix H, Recommended Procedures for Design, Construction, and Installation of Commercial Kitchen Grease Interceptors.

Grease interceptors vary in size from about a cubic foot in volume to field-fabricated units the size of septic tanks. Most plumbing codes in general do not address larger grease interceptors located outside structures. Commonly, the local health department having jurisdiction mandates requirements for large units outside structures. These sewage treatment municipalities and authorities having jurisdiction have established limits of fats, oils, and grease (FOG) through sewers to 100 ppm or 100 mg/L. This explains the monitoring stations in exterior manholes. Code requirements for methods of construction and performance standards are sketchy at best for these units. These large units are generally thought of as being septic tanks. They may be constructed of concrete, steel, fiberglass-reinforced polyester, or polyethylene. Concrete is affected by sewer gas degradation, and devices should have protections against leaks.

Oil Separators

Oil separators must be installed wherever a significant amount of oily waste is likely to be present. Locations such as repair garages, service stations, and industrial plants with an oily effluent require oil separators, but parking garages for vehicle storage do not. Figure 16–3 shows a typical installation. The oil separator must be located and installed so that it can be serviced readily. Each of these types is designed to separate the grease or oil product from water, based on the lesser density of the grease or oil. The lighter material floats on top of the water where, in the case of grease interceptors, the grease is held by a series of baffles. Oil can be drained off from an internal oil overflow located at the level of the top of the water in the separator to a retention container or pumped directly from the oil separator.

Sizing

Newer plumbing codes size oil separators based upon the floor area served rather than the amount of water containing oil, which may be discharge. The amount of water cannot be assumed to be provided from hose bibbs in the general area, such as washing facilities, when calculating separator sizing. The separators are required to gather planned accumulation such as oil in vehicle washing and unplanned accumulation such as a spill from vehicles or surrounding manufacturing.

Plumbing codes require oil separators to have a minimum container depth of not less than 2 feet below the invert of the discharge drain. They also require a minimum trap seal of not less than 18 inches. The 18-inch trap seal provides an extended protection time to guard against evaporation versus the typical 2-inch to 4-inch trap-seal depth requirement.

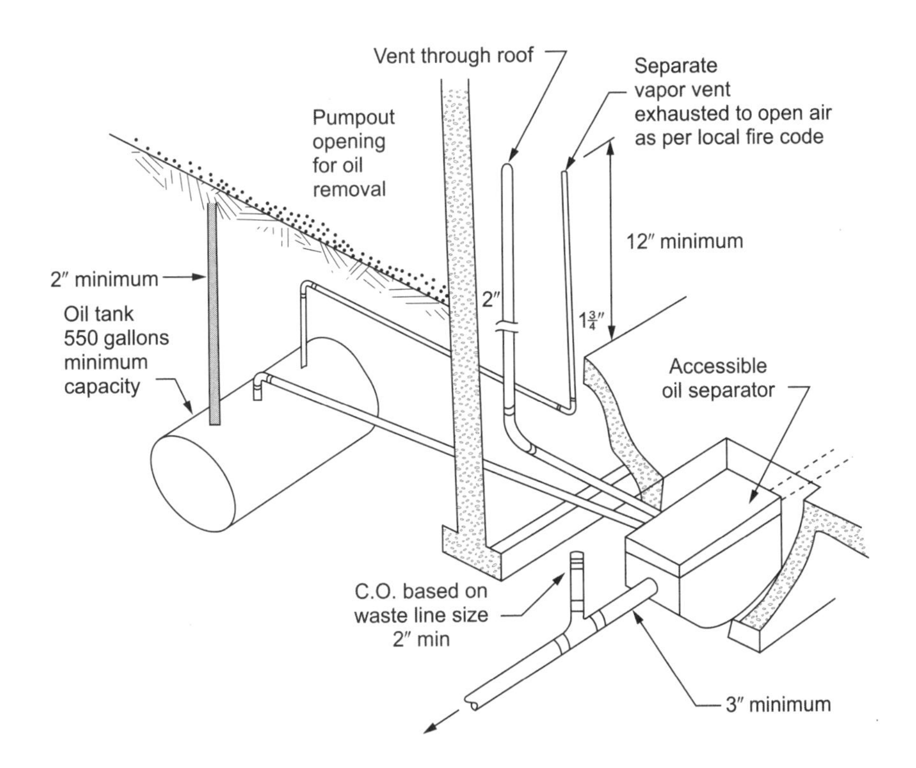

Figure 16–3
The typical oil interceptor shown here has many variations. Plumbing codes usually address the trap seal and sizing conditions but are not always specific in addressing the tank construction. Other valuable considerations for technicians come from the Environmental Protection Agency, local health departments, and the department of public works in specific jurisdictions.

In the Field

Some plumbing codes mandate that all car wash facilities have oil separators, while others mandate separators only for car washes that have the ability to clean areas of a vehicle having oil, such as engines.

In the Field

Several code officials require traps on all incoming floor drains or area drains serving an oil separator. These traps now require the waste lines to be properly vented with distance considerations according to code. The vents would have to be separate from the conventional drainage, waste, and venting system. The code officials base their requirements upon the fact that they do not want to provide an open conduit or conductor for explosive vapors from the separator to the habitable space.

The vast majority of oil separators accepted by code officials are factory-built devices provided in several sizes. Codes establish that a device must have 6 cubic feet of storage for the first 100 square feet of floor space to be drained. Another cubic foot of storage in the separator must be added for each 100 square feet of floor area. Common sense must be used in the floor design of the structures considered here. Typical floor sizes for these areas are very large, resulting in huge separators. For that reason, the designer will create smaller work floor areas by sloping particular work areas to the separator drains and not including parking areas. This commonsense planning will reduce floor space and translate to less required separator storage space. Designers could also have work areas not drain into any waste systems (no floor drain receivers), as when facility staff members have procedures for cleaning with a wet vacuum. There are several variables here because gasoline fumes near a conventional vacuum ensure explosive conditions.

In the Field

Oil separators will generally have two atmospheric vents to their storage areas. These vents are of course independent of the DWV system and when installed must have one vent terminal about 6 inches higher than the other. This common chimney effect promotes air movement to circulate combustible air out of the separator's storage container area.

Sand Interceptors

When dirt accumulation is associated with the process that produces the oil release for separator, a sand interceptor should be installed before the oil interceptor. Sand interceptors have been utilized for many years and are commonly referred to as sand traps, an appropriate term because most codes require at least a 6-inch deep trap seal and further require ready access for cleaning. The access, simply put, is provided by removing the grate covering the sand interceptor.

The essential design feature of a sand interceptor varies a great deal from one geographical area to another. Some areas use two compartments, the first compartment to collect solids (the sand) and the second to collect any carryover material (produced by turbulence). These interceptors are usually large (40–50 cubic feet minimum) and are located outdoors. The sediment storage space must be at least 24 inches below the invert of the outlet pipe. Other areas recognize a smaller circular-type basin in the garage area itself. These smaller basins are commonly clay tile with a poured concrete bottom. When the concrete is installed, shrinkage occurs, providing a leak path through the device to the surrounding ground area, which of course is not acceptable. However, they are generally accepted and forgotten about due to the sediment blocking the cracks and holding a water seal over a period of time.

In the Field

Environmental clean-up efforts have increased knowledge related to floor drains and separator problems. These efforts in some cases have prompted treatment facilities to establish policies that affect existing facility floor drains in a negative manner. For example, a code official reported a large environmental group "taking floor drains out of the system" by requiring them to be capped or plugged with concrete. Two serious problems occur here: first, the floor drains were installed for a specific purpose, and second, concrete is not accepted by the codes for connections due to shrinkage. Sewer gas from unprimed floor drains will escape into the habitable area.

Special-Purpose Interceptors

Special-purpose interceptors are available to catch precious-metal shavings or grindings from laboratories (such as dental labs), plaster scraps from medical treatment rooms where casts are made, hair from beauty shops, and other materials. Usually, the interceptors are placed close to the fixture served, but not always. For example, a large commercial darkroom could have a silver interceptor installed in the drain line. As with all interceptors, it is important that they be installed to make cleaning as convenient as possible.

Backwater Valves

Backwater valves are installed in sanitary systems to prevent soil and waste products from flowing back to and out of fixtures from blocked or restricted public sewers. The backflow is of course a contaminant whether we would be considering a habitable space or food processing fixtures. Most plumbing codes require the devices to conform to ASME A112.14.1, Backwater Valves.

The placement location of backwater valves is critical in sanitary faculties to ensure proper operation of all fixtures. Codes will clarify that the valves must protect all plumbing fixtures that have a flood level rim and are at elevations lower than the next upstream public sewer manhole. This requirement makes sense when you consider that an obstructed public system could build up and overflow,

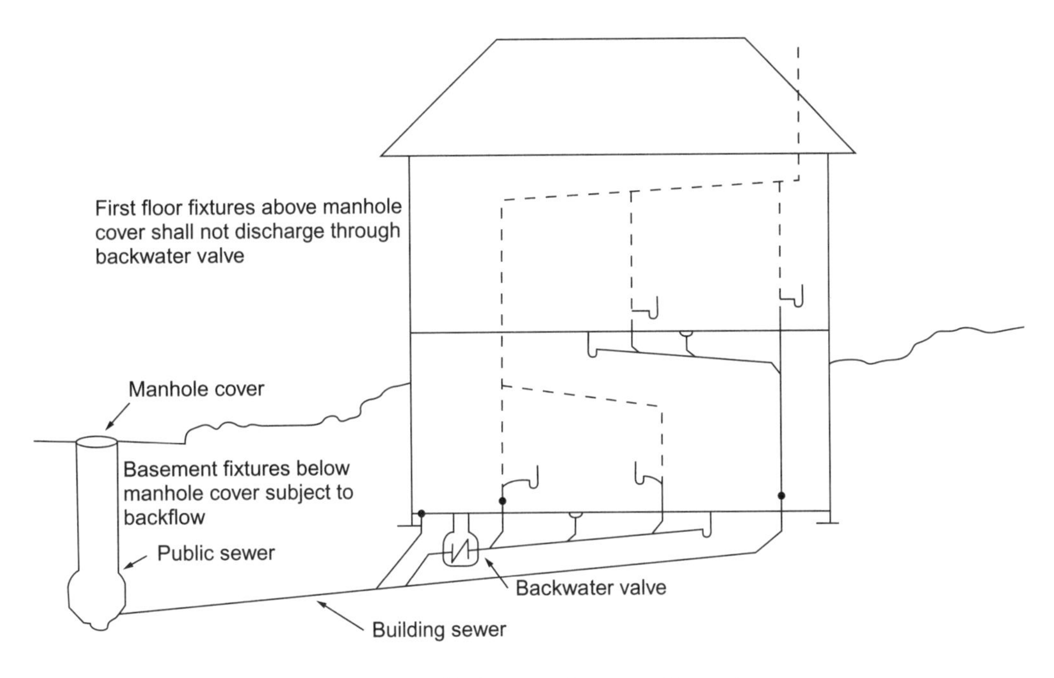

Figure 16–4
Backwater valve installations are required by code to ensure separation between systems above and below manhole overflow levels. This is a common-sense issue to allow only those fixtures needing protection from sewer backflow to flow through the devices. That will reduce the potential of fouling the backwater valve.

thus pouring out at the manhole. Manholes have holes in the cover to relieve sewer gas, and the holes would allow a visual indication of leaking to passersby. But sufficient pressure may not be present to lift the cover from its ring to relieve pressures. Lines that serve upper floor plumbing fixtures, which are not subject to backflow, must not be routed through backwater valves. This consideration is based upon the fact that the backwater valves may retard normal flow under some circumstances and solids in sewage flow may temporarily prevent the swing flapper from closing tightly. See Figure 16–4.

A typical horizontal backwater valve consists of a check mechanism. There is a slight drop downstream of the valve seat to aid in clearing any obstructing material from the seat. The valves must be accessible for repair and cleaning. Another version of the backwater valve is the combination horizontal backwater valve and manual gate valve, which includes a manually operated gate in addition to the check valve. This valve adds the security of the manual gate to make backflow prevention more nearly certain.

In the Field

The *National Standard Plumbing Code* requires a building that has backwater valves installed to have a posted warning including the backwater valves' locations. This will aid in troubleshooting drainage problems during service calls. Specifically, the notice will prevent service technicians from running drain cables through a backwater valve, which may ruin the valve or break the cable.

REVIEW QUESTIONS

1. Is it necessary to install all interceptors so that they are readily accessible?
2. Why are special-purpose interceptors installed close to the fixtures they serve?
3. What are the two main reasons interceptors are required to capture unwanted materials from entering the drainage system?
4. Are grease and oil interceptors designed to allow the materials to sink and accumulate in the separator, similar to septic tanks?
5. Should a backwater valve be installed on the building drain beginning at the lower level to prevent sewage backflow for all fixtures?

CHAPTER

17

Protection of the Water Supply

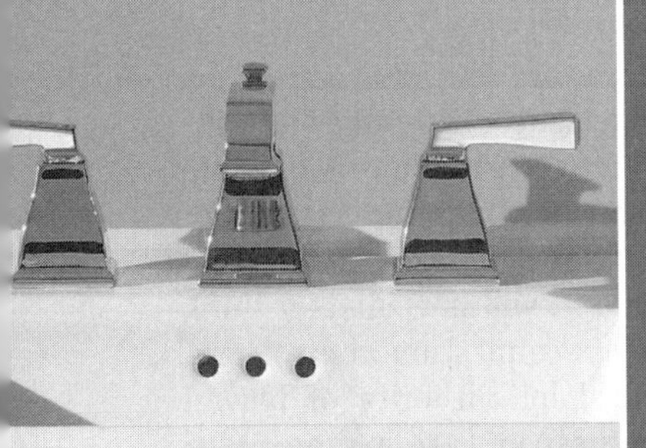

LEARNING OBJECTIVES

The student will:

- List situations that could compromise the safety of the potable water supply.
- Explain the different conditions that result in backflow conditions.
- Describe which devices to use to prevent backflow dependent upon the potable water system conditions.
- List the different prevention devices in order of performance and understand the devices' performance limitations.

In the Field

The *National Standard Plumbing Code*, in dealing with water supply and distribution, immediately summarizes the intent of an acceptable system in the following statement: "Only potable water shall be supplied to plumbing fixtures used for drinking, bathing, culinary use, or the processing of food, medical or pharmaceutical products."

PROTECTION OF WATER SUPPLY

The protection of the potable water system is of the utmost importance and the first responsibility of the plumbing professional. Isolating and keeping contamination out of each water outlet is the proper safety method rather than separating the building from the public water or well supply, which is called containment. Professionals understand that pollution affects the aesthetic quality of a water system, while contamination is a health hazard. Devices that protect against these compromises in the supply system are recognized for the degree of protection they can supply, such as pollution being a low hazard while contamination is a high hazard. For further discussion in this chapter, anything other than potable water will be referred to as contamination and unacceptable.

Numerous individual illnesses and major epidemics studied by industry researchers have been directly attributed to contamination of the potable water supply. The most famous of these situations occurred during the Chicago World's Fair in 1933, where it was discovered that submerged outlets on tub spouts led to contamination of the potable water system. Over 1,400 individuals became ill, resulting in 98 deaths. The common element in all these individuals' lives was that they had all stayed in the same hotel in Chicago while visiting the World's Fair. As a result of this case study and others, plumbing researchers developed installation standards that, if followed, prevent contamination of the potable water supply. Despite the efforts of code administrators, inspectors, and the plumbing industry to increase awareness of the possible dangers of cross-connections, contamination of the water supply continues to occur.

Cross-Connections

A cross-connection is any piping connection or arrangement where an objectionable gas, liquid, or material could enter the potable water system due to a pressure differential. Professional practices and codes establish guidelines for the use of devices to eliminate cross-connections. Plumbers and other professionals specializing in safe water, when hearing the term *cross-connections*, will immediately think of local water purveyor inspection programs called cross-connection control programs. These local rules or ordinances were established by authority of federal guidelines, the Safe Water Drinking Act passed by Congress in 1974 and amended in 1986 and 1996, to protect public health by regulating the nation's public drinking water supply.

The four methods through which cross-connections contaminate a system are direct connection, gravity, back-siphonage, and back-pressure.

Direct Connection

Direct connections are not permitted by code and would include hard piping between one system and another. An example of a direct connection is when a municipal water supply is connected to a structure with a well. The piping would have a control shut-off valve between the two systems. Any valve would be a potential leak path, and the whole concept is not acceptable.

Gravity Contamination

Contamination can enter a water distribution system or water main when it is opened during installation or for repairs. Simply put, the contaminant flows or settles into the open line. Flushing or chlorination will usually solve this type of problem, and most health agencies require assurance through testing prior to occupancy. While this gravity discussion is separate from the next backflow consideration, gravity and its effects are fundamental for all contaminations.

Back-Siphonage Backflow

Back-siphonage occurs when a negative pressure develops in a pipe because of a pressure reduction or failure at a lower level and a contaminant being available at a higher elevation. The contaminant is then drawn into the potable piping system because of reverse flow. The chance for that contaminant to get into the system comes of course through a cross-connection such as a submerged inlet.

The back-siphonage occurs when the water system falls below atmospheric pressure. Atmospheric pressure at sea level has the ability to push water up to 33.9 feet when a complete vacuum (absence of any pressure) exists on the other side. The siphonage takes place due to the difference of water levels. Other factors that could contribute to back-siphonage are created by the water or fluid in motion causing a venturi effect with an increased velocity, resulting in a lower pressure. If the contamination is available at the point of lower pressure, siphonage will occur. Pump intakes on the suction side also promote back-siphonage with undersized piping.

Back-Pressure Backflow

Back-pressure is created when a source other than the supply system obtains a pressure higher than the potable water supply pressure. This could be caused by an elevated source or a decrease in supply, resulting in a higher static head being present. A contaminant, which could be present at this other source (cross-connection), could then be forced into the potable system. Back-pressure is commonly created by a mechanical force such as a pump, an elevated tank, thermal expansion from a heat source, or even a steam connection, which could contain treatment chemicals.

Cross-Connection Possibilities

Cross-connections can occur anywhere in a potable water system, but certain appliances and configurations are more likely to cause problems than others are. Potential problem considerations and their remedies include the following:

Dishwashers

Dishwashers are usually provided with an air gap at the supply connection. If the air gap were not there, these appliances could deliver contaminated water to the potable system by back-siphonage or back-pressure.

Clothes Washers

These are also provided with an air-gap inlet like the dishwasher.

Water Heaters and Boilers

These devices must be air-gapped at the relief valve discharge, and boilers must be installed in such a way that boiler water cannot return to the potable system. Boilers with high operation pressures and those with toxic chemicals require a higher level of backflow protection device, such as a reduced pressure zone backflow preventer rather than a double check valve with an intermediate atmospheric vent port.

Plumbing Fixtures

Each water-using device must be installed to eliminate the risk of backflow. For that reason, faucet spouts are mounted so that an air gap is provided. Flushometer valves must be equipped with vacuum breakers, and flushometer tanks and flush tanks must also have vacuum breakers or air gap assemblies in the fill mechanism. Faucets with hose-thread outlets must be provided with a vacuum breaker at the hose-threaded connections because a hose attached to the outlet may be placed anywhere, in any solution, and thereby develop a hazardous cross-connection.

Miscellaneous Connections

Tanks and vats may be especially troublesome. Photographic processing tanks, etching tanks, plating tanks, and cleaning vessels are all examples of chemical tanks that require backflow-protected supplies from the water system. Heat exchangers with toxic transfer fluids (often referred to by the medical term *Gosselin Rating #3*), when connected to the potable water system, must be of the double-wall type with the area between the walls vented to the open atmosphere. This vent to the atmosphere ensures that a contaminant leak will not get into the second wall should a leak develop in the first wall. This will guard the potable water from a leak in the toxic system. Finally, drains, overflows, and relief valve discharge pipes must be indirectly connected to the waste by an air gap to ensure that sewage cannot enter any vessel.

Cross-Connection Prevention

Five common methods or devices are used to protect against cross-connections. Here, discussion and consideration will be in the order of the method that provides the best protection (an air gap) to the one that provides the least protection (a vacuum breaker).

> **In the Field**
>
> Backflow prevention devices are mechanical and must be designed, installed, and maintained to function correctly. A single error in any one of the three considerations could jeopardize the health and safety of an entire community.

Air Gap

The air gap is used on the water side to protect the potable supply and on the drain side to protect the fixture, appliance, or appurtenance. The air gap is the only absolutely fail-safe method of protection. The distance of the air gap is referred to as the "effective opening."

The essential component of an air gap is a physical, vertical separation between the outlet of the supply opening and the flood level rim of the fixture receiving the water. The air gap is the one fool-proof, maintenance-free cross-connection protection. It prevents any contaminants from crossing over from any vessel into the potable water system. Neither back-siphonage nor back-pressure backflow can produce a problem if an air gap is used for protection.

> **In the Field**
>
> All of the major plumbing codes in this country recognize the safety performance and function requirement standards developed by American Society of Sanitary Engineering (ASSE). These standards form the basis of cross-connection prevention by the codes.

Air gaps must be used with any spout or faucet to conform to the air gap standard ANSI A112.1.2. Sinks, lavatories, bathtubs, showers, and similar fixtures must be supplied by spouts through proper air gaps. Swimming pools, wading pools, spas, and hot tubs are also supplied by a fill pipe with an air gap. To be considered satisfactory for cross-connection protection, the air gap's vertical dimension must be twice the diameter of the effective opening of the faucet seat, supply pipe, or faucet spout. If the vertical dimension of the air gap is less than twice the effective diameter, it is possible for air entering the spout at high velocity to entrain water and carry it into the spout. It has been found that at twice the effective diameter dimension, it is not possible for any water to enter the spout. If the spout is not circular, the effective diameter is calculated by computing the cross-sectional area of the opening and then determining the diameter of the circle of the same area. Under some circumstances, the air gap must be greater than twice the effective diameter. If the spout opening is close to a wall (or two walls), the vertical dimension must be increased. See Table 17-1 for specific situations and information on where greater air gap dimensions are required to ensure protection.

> **In the Field**
>
> An air gap is defined in the *National Standard Plumbing Code* as the unobstructed vertical distance through the free atmosphere between the lowest opening from any pipe or faucet supplying water to a tank, plumbing fixture, or other device and the flood level rim of the receptor.

Another consideration in addition to the air gap is the flood level rim, which is the overflow level. The potential contaminant from a fixture or device will dissipate at the level where it can overflow on the floor or other area, thus not being available to be drawn or forced back to the water supply. Do not confuse the flood level rim with the overflow opening. The piping serving the overflow could become plugged, rendering it useless. Thus, the flood level rim measurement of the fixture is the level over which water would flow if the fixture were flooded. Exercise caution here because many containment devices will have overflows, such as a bathing

Table 17–1 Minimum Air Gaps for Plumbing Fixtures

Fixture Inches	Minimum Air Gap	
	When Not Affected by Near Wall[1] (Inches)	When Affected by Near Wall[2] (Inches)
Lavatories with effective opening not greater than $\frac{1}{2}''$ diameter	1	$1\frac{1}{2}$
Sink, laundry trays, gooseneck bath faucets, and other fixtures with effective opening not greater than $\frac{3}{4}''$ diameter	$1\frac{1}{2}$	$2\frac{1}{4}$
Over-rim bath fillers and other fixtures with effective openings not greater than 1-inch diameter	2	3
Drinking water fountains—single orifice not greater than $\frac{7}{16}''$ (0.437″) diameter or multiple orifices having total areas of 0.150 square inches (areas of circle $\frac{7}{16}''$ diameter)	1	$1\frac{1}{2}$
Effective openings greater than 1 inch	2 × Diameter of effective opening	3 × Diameter of effective opening

[1] Sidewalls or similar obstructions do not affect air gaps when spaced from the inside edge of the spout opening by a distance greater than three times the diameter of the effective opening for a single wall or by a distance greater than four times the diameter of the effective opening for two intersecting walls.

[2] Vertical walls or similar obstructions extending from the water surface to or above the horizontal plane of the spout opening require a greater air gap when spaced closer to the nearest inside edge of spout opening than specified in Note 1 above. The effect of three or more such vertical walls has not been determined. In such cases, the air gap shall be measured from the top of the wall.

tub trip waste overflow (not acceptable for safety's sake) or a potable water storage tank having a submerged inlet with an overflow.

In earlier years, bathtub spouts were located where overflows presently are installed. A tub could be filled to the spout, and any vacuum in the potable system would pull the tub water into the system. A vacuum can be developed readily by draining the potable system at an elevation below the submerged tub spout. When the spout is installed with an air gap, only air can be pulled into the spout from reduced pressure on lower floors. You should realize that an emergency demand on the city mains (e.g., fire fighting), an unexpected pressure drop (e.g., broken main), or maintenance operations in the building can produce these low pressures.

Reduced Pressure Zone Backflow Preventer

These backflow prevention devices are used to protect the potable water supply from back-pressure or back-siphonage and from the most serious contaminants. They are the next best thing to an air gap and must be tested regularly and maintained to ensure that there has been no component failure. The application of reduced pressure zone (RPZ) valves is usually considered to have the same effect as an air gap. An exception would be any application where regular maintenance of the RPZ valve cannot be ensured. In such cases, only an air gap is acceptable protection. Plumbing codes require the RPZ valves to conform to the performance requirements of ASSE 1003.

Pressure Drop

RPZ devices operate on a principle of reduced pressure zones. Traditionally, a significant problem with these backflow preventers was that they introduced a large pressure drop into the system, usually 8–15 psi. Designers and technicians should consult manufacturer's literature, which is generally available on the Internet, for data on the flow rate versus pressure drop.

In the Field

The RPZ valve is defined in the *National Standard Plumbing Code* as a backflow prevention device consisting of two independently acting check valves that are internally force-loaded to a normally closed position and separated by an intermediate chamber (or zone) and in which there is an automatic relief means of venting to the atmosphere that is internally loaded to a normally open position, including a means for testing the checks and openings for tightness.

In the Field

Several backflow prevention devices are identified with an additional name—detector assembly, or more specifically, reduced pressure principle back-pressure detector assembly. These devices have a water meter built into their assembly to detect low rates of flow, which aids the user in detecting leaks. It could also provide easier detection of unauthorized use such as illegal taps of water for equipment cooling off a fire-suppression system, which is commonly not metered by the water purveyor.

In the Field

Shut-off valves are required on RPZ assemblies and must be provided as part of the assembly. Ball valves are usually provided on pipe sizes of 2 inches or smaller. These specific valves are part of the valve performance during testing and an important part of the product listing.

Operation

An RPZ assembly consists of a pressure differential relief valve located in a zone between two positive-seating check valves. The back-siphon protection includes a provision to admit air into the reduced pressure zone by a channel separate from the water discharge channel, which is between the two check valves. The assemblies have two shut-off valves before and after the assembly. Test cocks are provided for testing, and manufacturers commonly require that a strainer assembly be installed prior to the first valve.

Figure 17–1 through Figure 17–4 provide a visual indication of how an RPZ acts under different circumstances. Figure 17–1, static pressure, illustrates the assembly with no flow and normal static pressure applied, say 60 psig. The reduced pressure zone would be 54 psig, and the outlet pressure would be 52 psig. The first and second checks are closed, as is the relief valve. Figure 17–2, full flow, shows the device with full flow. With 60 psig inlet, the reduced pressure zone is at 51 psig, and the output is at 46 psig. Figure 17–3, back-pressure, shows that partial or full relief valve opening occurs with any of the possible back-pressure levels. Finally, Figure 17–4, back-siphonage, shows that partial or full relief valve opening occurs with any of the possible abnormal conditions.

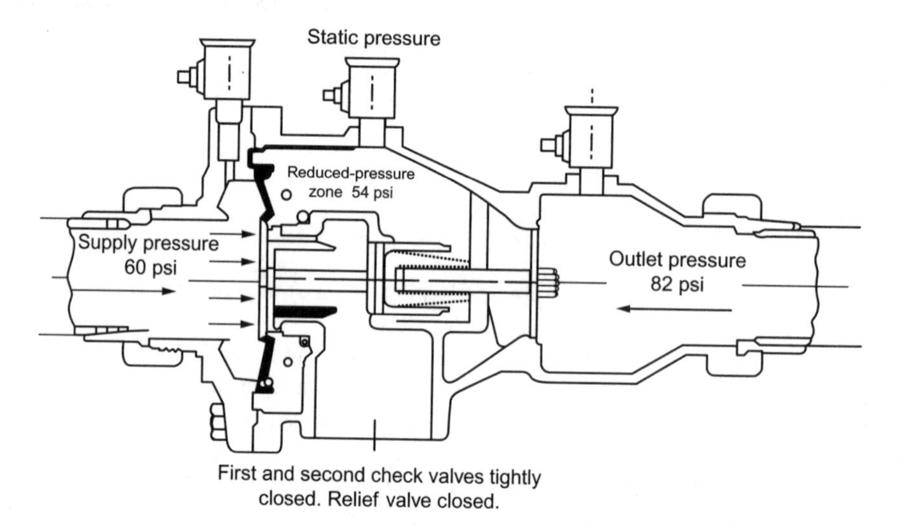

Figure 17–1
Static pressure cutaway view of a reduced pressure zone backflow preventer under various conditions.

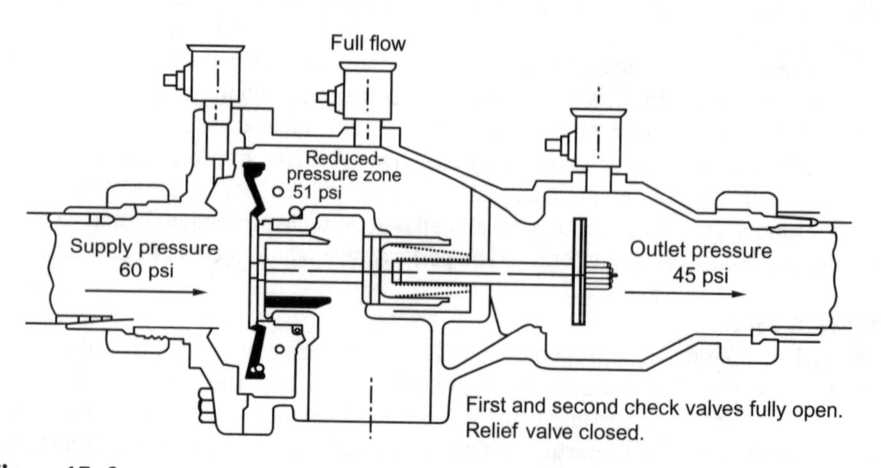

Figure 17–2
Full flow cutaway view of a reduced pressure zone backflow preventer under various conditions.

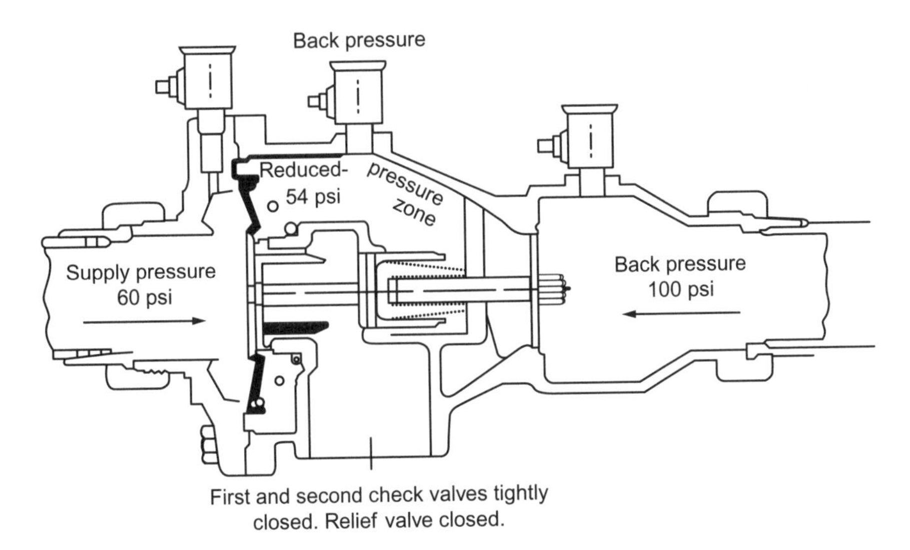

Figure 17–3
Back-pressure cutaway view of a reduced pressure zone backflow preventer under various conditions.

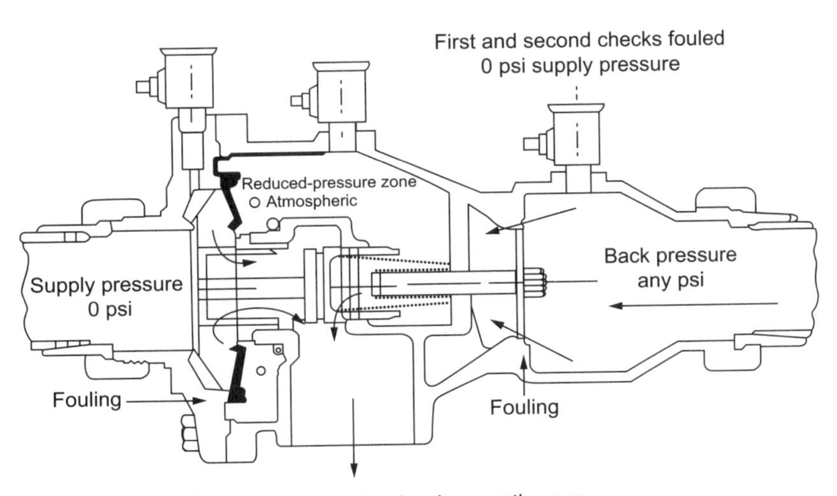

Figure 17–4
Back-siphonage cutaway view of a reduced pressure zone backflow preventer under various conditions.

Siphonage Installation

Installation considerations for RPZ valves include the following:

1. The units must be equipped and shipped with shut-off valves and test cocks.
2. The unit must be located to facilitate testing and repair and to protect against freezing and vandalism. Its location should allow for immediate notice of telltale discharge or other malfunction.
3. Observe the manufacturer's directions about orientation (horizontal, vertical, or otherwise).
4. If large flows are required, manifold RPZ valves should be installed in parallel. A manifold arrangement makes servicing easier because the water supply is maintained while the unit is isolated for testing or repairs.

5. The relief opening must be indirectly wasted to the building drainage system, and the RPZ valve should not be placed in a pit. Installing a backflow preventer in a pit or vault is not acceptable because flooding of the pit will cause a cross-connection at the relief port.
6. Prior to installation, thoroughly flush all pipes to remove any foreign matter.
7. The RPZ valve must be tested by a certified tester at the time of installation to ensure that the assembly is in proper working order and will protect the potable water supply as required by the code.

Testing

The testing frequency of the devices may be addressed in the applicable plumbing code or more likely by the water purveyor based upon its cross-connection control program. Often, the frequency is dependent upon the type of device, which considers the degree of hazard in the application. For that reason, you would expect to test devices in a mortuary before devices protecting beverage dispensers.

Commonly, a differential pressure gauge, as shown in Figure 17–5, is used to determine the condition of check valve 1, check valve 2, and the differential relief valve. Basic procedures are covered in the following general instructions in order to provide an initial level of understanding. Detailed test procedures may be found in ASSE Series 5000, Professional Qualification Standard for Backflow Prevention Assembly Testers, Repairers, and Surveyors.

The basic procedures are as follows:

1. Close the outlet ball or gate valve (valve 2).
2. Open test cocks 2 and 3.
3. Open vent valves to vent air from all lines.

Test 1 (Test Relief Valve)

4. To test the differential relief valve, open valve A and close valve C.
5. Open valve B very slowly until the gauge reading starts to drop.

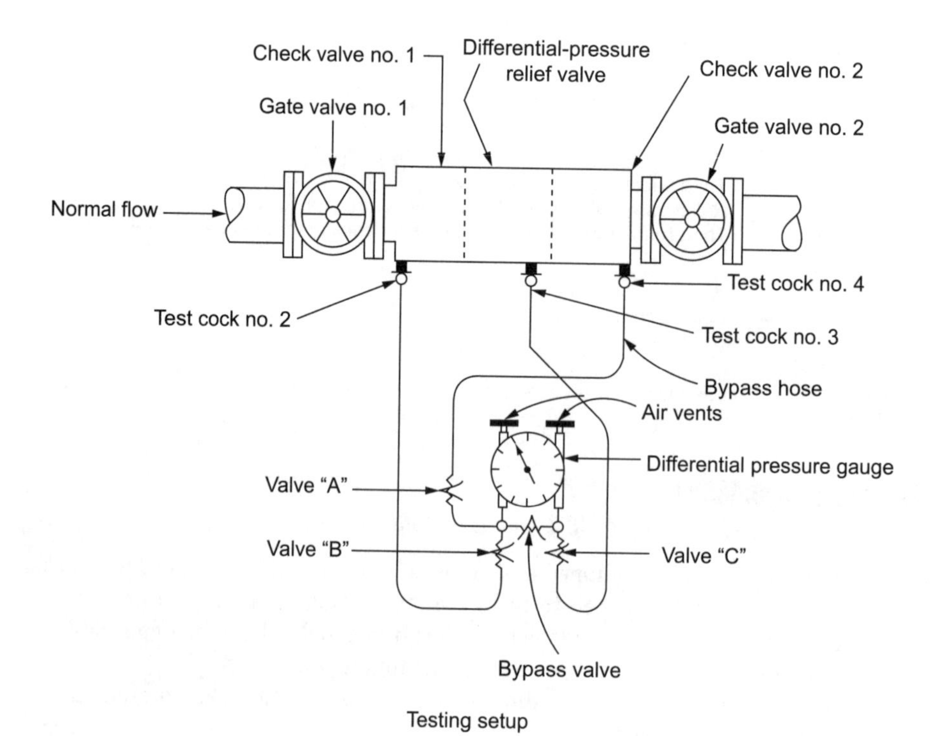

Figure 17–5
A differential pressure gauge test illustration for checking a reduced pressure zone backflow preventer.

6. Hold the valve at this position, and observe the gauge at the moment the first discharge is observed from the relief valve.
7. This value is the pressure differential of the valve and should be the designed value of the spill differential.

Test 2 (Test Check Valve 2)

8. To test the second check valve, open valves A, B, and C.
9. Open test cock #4 to bleed water from the zone by opening the low-pressure side bleed needle valve.
10. The gauge reading should stabilize and stay constant.
11. If it drops continuously to the relief valve opening point, the second check must be repaired.
12. If the indicated pressure differential decreases but remains above the relief valve opening point, the check valve is tight and does not need repair.

Test 3 (Test Shut-Off Valve 2)

13. Close test cock #2.
14. The differential pressure should drop slightly then hold steady.
15. If the differential pressure continues to drop, shut-off valve 2 is defective.

Test 4 (Test Check Valve 1)

16. Close valve A.
17. Open test cock A.
18. Vent air out of all connections.
19. The pressure difference should remain constant; otherwise, check valve 1 needs to be rebuilt.

In the Field

If the device must be repaired, be sure to use parts approved by the manufacturer. When the repairs are completed, retest the device to be sure that it operates as designed. Remember, if major repair is needed, it is best to replace all parts subject to wear. Be sure to follow special procedures and use any special tools as recommended by the manufacturer.

Vacuum Breaker

While air gaps are the ultimate backflow protection and RPZs follow when air gaps are impractical, cost and simplistic operating devices can be considered next. The next device to consider would be the atmospheric (Figure 17–6) or pressure-type (Figure 17–7) vacuum breaker, which protects against the siphoning effects caused by a vacuum or negative pressure on the potable water supply line. These devices will not protect against back-pressure backflow.

Figure 17–6
An atmospheric vacuum breaker. Courtesy of Watts Water Technologies, Inc.

Vacuum Breaker Design

The basic design of a vacuum breaker includes a check valve to close off the water line and a port that is simultaneously opened to atmosphere. Thus, if a vacuum develops in the supply piping, the check valve closes the water flow path and admits the atmosphere into the fixture connection space. This admittance of normal atmospheric pressure stops the back-siphonage. If the vacuum breaker is located at a higher elevation than the fixture protected, it is impossible to draw contaminated water into the water supply line. The height consideration is very important for all vacuum breakers because a head or outlet that is higher than the vacuum breaker would allow the higher elevated water pressure coming back (back-siphonage) to overcome the atmospheric pressure. For that reason, codes and standards discuss such things as device critical level and being 1 inch, 6 inches, or 12 inches above the outlet or flood level rim.

Two main design concepts of vacuum breakers are available:

1. Atmospheric (nonpressure) type – Installed on lines that operate at line pressure only when water is flowing through the vacuum breaker. These valves conform to ASSE 1001, Performance Requirements for Atmospheric Type Vacuum Breakers (pipe applications). They are recognized as protecting against a high or low degree of hazard. Several other hose-connected vacuum breakers with their appropriate standards exist.

Figure 17–7
A pressure-type vacuum breaker. Courtesy of Watts Water Technologies, Inc.

In the Field

While some hose-connected vacuum breakers are allowed to have nonflowing conditions for a few hours, most are not. No-flow conditions could allow them to become inoperable (set up). For that reason, hose-connected devices such as chemical dispensers and shut-off nozzles are viewed as being very hazardous.

2. Pressure-type – These vacuum breakers are designed to withstand line pressure when there is no water flow as well as when flow is occurring. This means a valve or valves may be installed downstream of the device, such as zone valves on a sprinkler system. These valves conform to ASSE 1020, Performance Requirements for Pressure Vacuum Breakers. They are recognized as protecting against a high- or low-degree hazard. Used to protect a supply line subjected to line pressure, these devices must be mounted higher than any opening that may become contaminated. Pressure vacuum breakers contain a check valve that is spring-loaded to close when the supply line pressure is negative (i.e., less than atmospheric). The pressure-type devices now also have an additional product feature that makes them spill-proof, which are governed by the ASSE 1056, Performance Requirements for Spill Resistant Vacuum Breaker. The spill-proof vacuum breaker is similar to the pressure-type device except for an additional feature that eliminates the water discharge through the air inlet vent each time the device is pressurized.

Atmospheric Types: Pipe Applied

Pipe-applied vacuum breakers can be visually checked. The protective cap or shroud must first be removed, and the control valve upstream of the valve can then be activated. When the pressure is removed, water should drain from the unit and air should enter the device. If the checking member does not move, pry it from its seat. If it continues to stick, replace the device.

A sticking check valve is usually a signal that the unit is pressurized constantly, a condition that is likely to occur when a valve is located downstream of the vacuum breaker, which is not acceptable.

Atmospheric Types: Hose Thread

Hose-connected vacuum breakers are designed to be throwaway items and are usually not testable in the field. Some more sophisticated wall hydrants allow you to field-test checking members using a spray nozzle and a length of hose.

Vacuum Breaker Installation

Unless listed or approved for a lesser clearance, vacuum breakers are usually installed at least 6 inches above the flood level rim of the fixture. (Some deck-mounted vacuum breakers may be installed as low as 1 inch above the flood level rim.) The point of measurement on the vacuum breaker is the critical level, which is the bottom of the device unless the mark "C-L" is cast on the side of the device. An atmospheric vacuum breaker must be installed on the discharge side of the last control valve. Pressure-type vacuum breakers must be installed 12 inches above the flood level rim of the device served. Exercise caution here as not all codes recognize the 12-inch height requirement.

Vacuum breakers should never be installed in locations where hazardous fumes could be present. If the vacuum breaker were to operate in such locations, these fumes would be drawn into the potable water piping. This consideration is usually centered on laboratories with fume hoods where the water outlet has a hose connection. For that reason, the vacuum breaker is mounted on the outside of the cabinet.

Testing and Servicing

Vacuum breakers require periodic inspection to ensure that they are functioning properly. The previous discussion concerning frequency of inspections and ASSE 5000 applies here also.

Pressure-Type Vacuum Breakers Pressure-type vacuum breakers can be checked with the sight tube method by using the following procedure as illustrated in Figure 17–8:

In the Field

Protection of the water supply considers both waterborne contaminants and airborne contaminants.

Air Inlet Test

1. Remove the top canopy.
2. Close valve B on the outlet.

3. Close valve A on the inlet.
4. Open test cock 2, and watch the air inlet poppet.
5. The poppet must be completely open before the water stops running from test cock 2.

Check Valve Test

1. Connect sight tube to test cock 1 and install it plumb.
2. Open valve A to fill the device.
3. Open test cock 1 and allow the water to fill the sight tube to a height of 30 inches above the centerline of the outlet pipe.
4. Close valve A (valve B is still closed).
5. Open test cock 2.
6. The water in the sight tube may drop about 3 inches. If it drops more, the check assembly needs repair. If the valve holds 1 psi (27-inch water column) in the forward direction, it will also hold 1 psi in the reverse direction.

In the Field

If any repairs are made on the pressure-type vacuum breaker, perform the complete test program on the valve before returning it to service.

Double Check Valve Assembly

Double check valve assemblies, as shown in Figure 17–9, are used for low hazard applications such as fire suppression installations. They also must be tested regularly to be certain of satisfactory operation. They are capable of operating under continuous pressure applications.

These devices may be used to guard against back-pressure or back-siphonage backflow but only where nontoxic substances are involved. Because there is no provision to divert reverse-flow liquid to a drain, it is possible with certain failure combinations for reverse flow to appear in the inlet of this device. For that reason, they are only acceptable under low-hazard applications.

It is important to note that these assemblies are more than just two check valves piped together. They are a single device containing two check mechanisms and a means for testing. Two check valves installed in series are not a substitute for double check valve assemblies. These devices should conform to ASSE Standard 1015, Double

In the Field

A double check valve assembly is defined in the *National Standard Plumbing Code* as a backflow prevention device consisting of two independently acting check valves that are internally force-loaded to a normally closed position between two tightly closing shut-off valves and that have a means of testing for tightness.

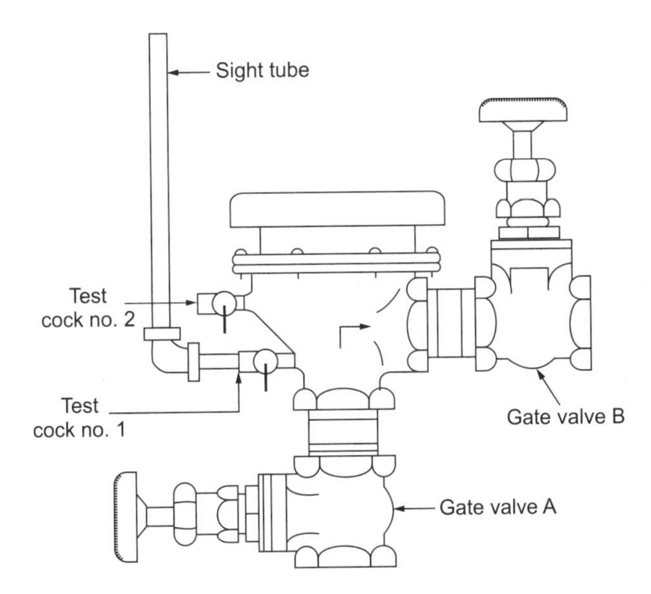

Figure 17–8
Testing a pressure-type vacuum breaker with the sight tube method. Methods using test gauges are recommended and the only acceptable methods to water purveyors. However, this is used to provide a simple explanation.

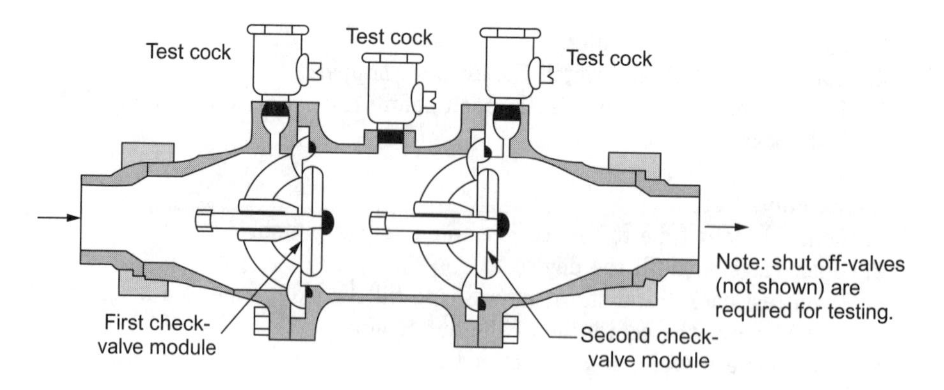

Figure 17–9
A cutaway example of a double check valve assembly. Courtesy of Plumbing–Heating–Cooling–Contractors—National Association.

Check Valve Backflow Preventers. The products are made of materials suitable for the applications, usually bronze or epoxy-coated iron bodies, bronze seats, and synthetic, specially formulated gaskets. Usual operating high-temperature ratings are about 160°F, with temporary excursions permitted to 210°F. If high temperatures are anticipated, select internal parts for the intended service.

The spring-loaded checks are closed under no-flow conditions. When water flows through the device, there is a pressure drop across the product, which increases as flow increases. Significant pressure drops (20–30 psi) may occur, especially if a strainer is combined with the unit. The three test cocks placed on the valve permit testing to verify operating condition.

Typical applications for double check valve assemblies include these:

1. Pumps, tanks, and lines that handle nontoxic substances
2. Supply lines to heating boilers that contain nontoxic fluids
3. Lawn sprinklers with elevated heads, where no toxic chemicals are used
4. Valve outlets or fixtures with hose thread outlets
5. Connections to fire protection systems
6. Steam generators, water boosting systems, solar systems, and beverage dispensing equipment

In the Field

Exercise caution when making changes to existing systems. For example, a fire suppression system with a double check valve assembly being changed to include a room with toxic antifreeze (a high-hazard application) would require an RPZ valve. Installing an RPZ valve on the system could result in a greater pressure drop across the device, thereby reducing the flow. The system would not operate as designed and could have an inadequate supply volume. Your actions would result in serious consequences. Design professionals are to be consulted in these situations.

Double Detector Checks

On large-diameter systems, which may supply fire protection systems or other approved applications, double detector checks are often specified. These devices consist of two double check valve assemblies on which a larger assembly is fitted with a smaller bypass line. The bypass line is equipped with equivalent backflow protection (a double check valve assembly) plus a small water meter. This combined assembly allows the building owner to assess whether water flow is occurring in the piping. Any small demand on the line (legal or otherwise) causes flow on the bypass line, which is registered on the meter. When a large demand that is significant enough to force the check members from their seats on the larger assembly is required, there will be more than adequate flow to meet the system demand if the valve is properly sized.

Double Check Valve Assembly with Intermediate Atmospheric Vent

These devices are also capable of operating under continuous pressure applications and under backflow conditions because the device spills to atmosphere. This device must conform to ASSE Standard 1012, Double Check Valve with Intermediate Atmospheric Vent, and is not testable. The relief port must discharge by an air gap and must be prevented from being submerged, which would cause a cross-connection.

Installation

Usual installation details for double check valve assemblies include the following:

1. Test cocks and shut-off valves must be accessible.
2. The valve body must be a minimum of 12 inches above the floor (codes may vary on this).
3. Sufficient clearance for service, including disassembly, must be provided.
4. Union or flanged connections should be used on each end to aid installation.
5. The permitted installed positions of the device must be verified (vertical, horizontal, diagonal).
6. Installation in a pit that could flood may be prohibited.
7. The installation must be protected from freezing and vandalism.

Testing

Regular testing must be performed to be sure that the device is operating properly. The tests are designed to determine that the check valve assemblies are holding tight. The previous discussions concerning frequency of inspections and ASSE 5000 apply here also.

Double check valve assemblies can be checked by the following method using Figure 17–10:

Test check valve 1 as follows:

1. Flush the test cocks.
2. Install a sight tube in test cock 2. Fill the tube with water to 28 inches, and then shut off the test cock.
3. The pressure at the base of the tube is 2.33 (0.434) = 1.8 psi, which is the requirement for the valve to hold.
4. Close line valves A and B.
5. Open test cocks 2 and 3. The water should remain in the sight tube. If it leaks out, the first check needs to be serviced. If water stays in the tube and a continual flow comes out of cock 3, then shut-off valve A is leaking.
6. Repeat the test for check valve 2 by installing the test tube in cock 3.

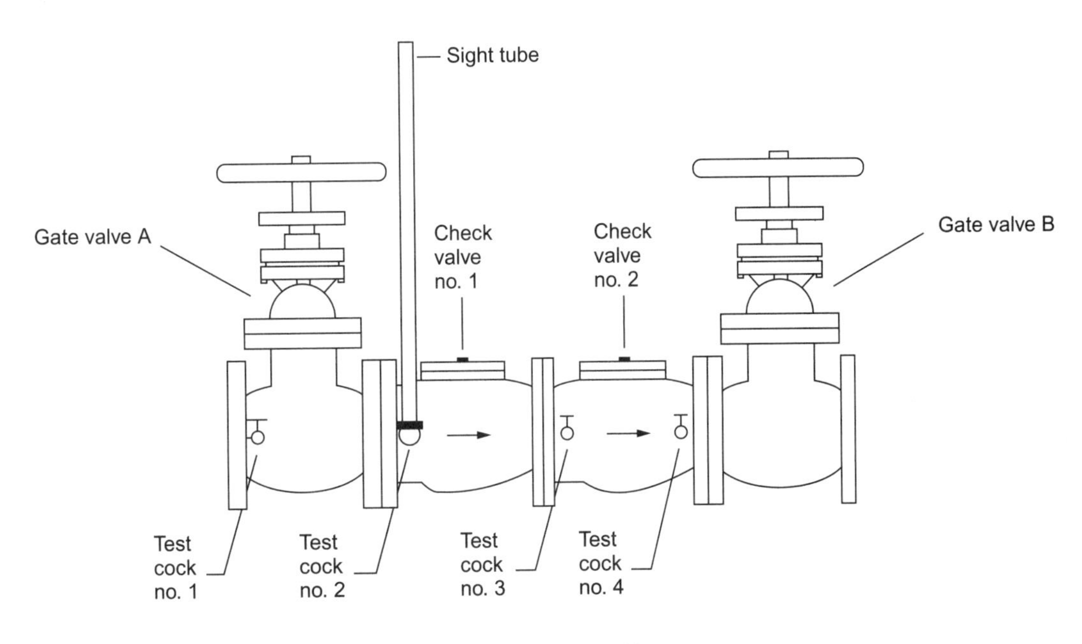

Figure 17–10
Testing a double check valve assembly with the sight-tube method. Methods using test gauges are recommended and the only acceptable methods to water purveyors. However, this is used to provide a simple explanation.

If either check assembly needs to be serviced, be sure to examine the other one to be certain it is in good condition. If replacement components are needed for one, it would be wise to replace both with similar new parts. Care should be taken to use the proper kit for the application involved. Different temperatures and pressures may require different materials.

Barometric Loop

The barometric loop is another method of backflow prevention identified in some plumbing codes that was recognized long before devices were manufactured. The barometric loop is generally impractical and covered by some codes as a matter of theory. The piping must extend vertically 35 feet above the highest installed system piping to the connection in question. Even then, it would serve only as an atmospheric or pressure-type vacuum breaker.

REVIEW QUESTIONS

1. What is the benefit of installing backflow preventers in parallel?
2. What is the main focus in installing proper plumbing according to codes?
3. What is the best protection against backflow that a potable water system can have?
4. Are all backflow preventers capable of protecting against back-pressure conditions?
5. List the five major methods used to protect against contamination of the potable water supply.

CHAPTER 18

Steam Systems, One- and Two-pipe

LEARNING OBJECTIVES

The student will:

- Identify the components of a heating system.
- Recognize and list the classifications of heat transfer media and the different types of fuel.
- Compare steam piping methods and their correct operations.
- Recognize the types of steam condensate return systems.

HEATING SYSTEMS

The next six chapters in *Plumbing 401* examine heating systems that run on steam, hot water, forced air, and solar systems. Understanding the fundamentals of these systems, through an analysis of their basic designs, application and equipment, is necessary for new installation and service/repair. Steam systems include a boiler, piping, devices, and controls. Hot water systems also include a boiler—designed to heat the transfer solution rather than turn the solution to steam—piping, devices, and controls. Forced-air systems, commonly called warm air furnaces, include humidification, and are also used to heat the air with their various system components. Finally, we will address solar systems, which can use the same piping distribution system or ducts as forced air systems.

Heating systems are required in most habitable building structures. They provide heat to a building both to ensure human comfort and to protect the integrity of the building and its internal systems. In order to survive, human beings in some climates require heated areas in the winter. The physical comfort range is a very small spread (three to five degrees Fahrenheit) of ambient temperature conditions. In addition, building structures and contents usually must be maintained in a narrow temperature range (50°F–80°F) to maintain building integrity.

The basic components of a heating system for structures in habitable climates are a power plant source, a distribution system to conduct the heat medium, and the terminal. Heating systems are classified according to the medium and the fuel used. The medium is the fluid that travels through the distribution system. The medium may consist of air, water, steam, or special fluids such as antifreeze or refrigerants. Additives are commonly placed in these mediums. The additives may include humidification in the air or corrosion-prevention products in the water. If the heat transfer medium is water or steam, the heat source is called a boiler. If the transfer medium is air, the heat source is called a furnace. There are two types of furnaces: forced air or gravity. Gravity furnaces are seldom used today because they are significantly larger and more expensive than forced-air furnaces of equivalent rating. Gravity furnaces were often converted to forced-air service by adding a blower, in an attempt to attain the features and advantages of a forced-air system. The components of a furnace are a combustion chamber and heat exchanger.

Classifications for the type of fuel include coal and coke, electricity, fuel gas, solar, oil, and wood. The selection of fuels depends on the fuel cost, fuel availability, and system design. All the elements of a heating system must be selected with care by a knowledgeable person to develop a coordinated, balanced system that will heat the space satisfactorily and efficiently.

In the Field

The basics for a heating system can be organized in outline form as follows:

A. System components
   1. Power provider
   2. Distribution system
   3. Terminal
B. Transfer medium used
C. Type of fuel used

STEAM SYSTEMS

Steam has been used for a very long time as a source of power, for example in steam locomotives and steamships. Its presence as a power source in early manufacturing areas provided the opportunity for steam to be used as a heating source. Steam, in conjunction with piping and devices used to transmit heat into various portions of a structure, has been widely accepted for many years.

Steam is an invisible gas generated by heating water to a temperature called the boiling point. At standard atmospheric pressure, heating water to 212°F (100°C) changes its physical state to steam in a process called vaporization. The liquid turns to gas and expands to 1,700 times its original volume. When the gas is cooled, steam condenses back into a liquid. The resulting liquid is called condensate. The temperature at which condensation takes place is known as the dew point. To raise the temperature of one pound of water one degree Fahrenheit takes 1 BTU up to 212°F. Forming steam at that temperature requires an additional 970 BTUs for the "latent heat of vaporization." Therefore, steam has (970 + (212 − condensate temperature)) BTUs per pound.

Example

When the dew point is 160°F, steam has (970 + (212 − 160)) = 1,022 BTUs per pound. This clearly shows why steam has more energy content than hot water.

Steam boilers, often described as power plants, are available in a multitude of sizes for a multitude of applications. The pressure and delivery amounts vary depending on the size and application needed. These boilers are pressure vessels and, as such, may suffer violent vessel failure explosions due to high internal pressures. The steam boiler industry is unique in that it was the only mechanical/construction industry to be regulated before the plumbing industry in many areas. Our understanding of necessary plumbing regulations for health and safety developed after devastating boiler explosions in the mid-1800s. Accidents in the late 1800s occurred at a rate of one every four days and peaked in 1905, with approximately 400 reported that year. A few examples follow:

- March 2, 1854. Fales and Gray Car Works, Hartford, Connecticut. 21 dead, `50 seriously injured.
- April 27, 1865. The steamboat Sultana, Memphis, Tennessee. 1500 of the 2200 passengers were killed.
- March 10, 1905. A shoe factory in Brockton, Massachusetts. 58 dead, 117 injured.

The American Society of Mechanical Engineers (ASME) has established boiler and pressure vessel standards for safety. Their first standard for the construction of boilers was developed in 1915. An organization known as the National Board of Boiler and Pressure Vessel Inspectors was established in 1919 to commission boiler inspectors. The National Board web site, http://www.nationalboard.org, contains a wealth of information on topics from accident investigation to repair and registration of boilers. Boiler construction, components, and installations are closely regulated now, with exceptionally safe results.

In the Field

Many boiler inspectors in the United States who have been commissioned and employed by governmental jurisdictions and insurance providers gained experience in the United States Navy. The Navy, with its vast utilization of steam boiler systems, allowed these individuals the opportunity to gain valuable on-the-job experience.

Power Plant Boiler

The heat source for a steam system is a boiler, referred to as the power plant. If the heat transfer medium is water or steam, the heat source is still called a boiler.

Boilers may be water tube or fire tube designs. They can be constructed of cast iron sections assembled together, of steel in either water tube or fire tube arrangements, or of copper tubing with extended-fin secondary surfaces. Figures 18–1 and 18–2 show different types of steam boilers.

Smaller boilers are generally rated in horsepower (500 hp and less). Larger units are generally rated in thousands of pounds of steam. One boiler horsepower is about 42,000 BTUs of input. One pound of steam is about 1,200 BTUs of input. Low-pressure steam boilers are generally operated at pressures less than or equal to atmospheric pressure (14.7 psi), while high-pressure units can greatly exceed atmospheric pressures, reaching 100 psi and higher.

Distribution Systems

A heating distribution system conveys the heat from the source to the terminal units. This distribution system may be piping, ductwork, or a combination of the two. In this chapter, the distribution system under discussion will be basic steam piping, which are one- and two-pipe systems. Older steam piping systems, which you may have to repair, will be classified for our consideration according to this one- or two-pipe system.

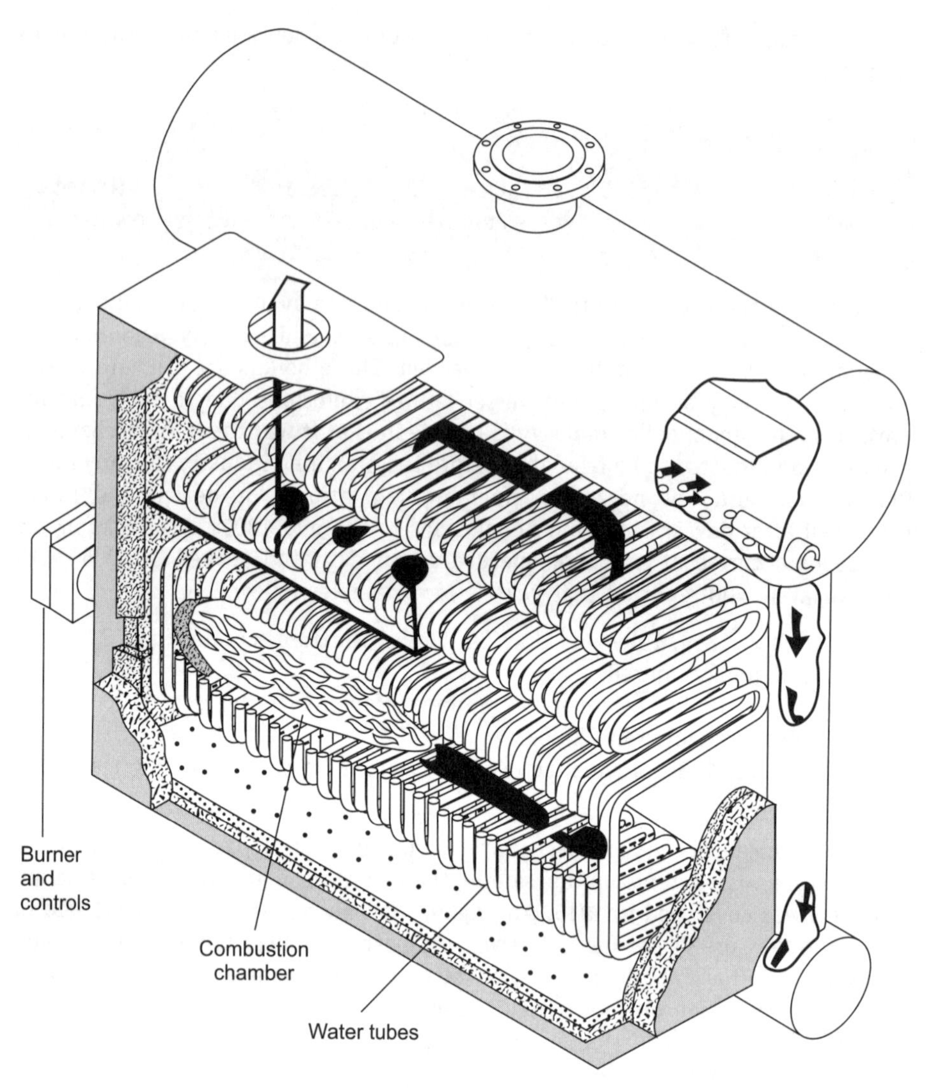

Figure 18–1
Water tube steam boiler. Courtesy of Bryan.

Distribution systems may be insulated to retain heat and to reduce the risk of burns from extremely hot surfaces. When insulation must be removed from an existing installation, first determine whether the insulation contains asbestos, as this material must be removed using very strict procedures, and only by authorized persons. Violations of asbestos abatement laws in many states will result in plumbing licensing actions and fines. These laws are necessary to safeguard the life and safety of all individuals who might be exposed to the danger of disturbing asbestos at construction sites. Asbestos in plumbing construction was primarily utilized in pipe installations until 1972.

Terminal Unit

The terminal unit is the device that delivers the heat from the distribution system to the space to be heated. In common language, it conveys heat from the source to the structure. The heat is delivered to the space at a terminal such as a radiator or convector for hot water or steam systems, and an air grille for air systems. Radiant panels can also be used with either air or water systems.

Terminals for general heating systems include the following examples:

- Air handler coils
- Baseboard terminals: electric, forced air, or hydronic fin tube or cast iron

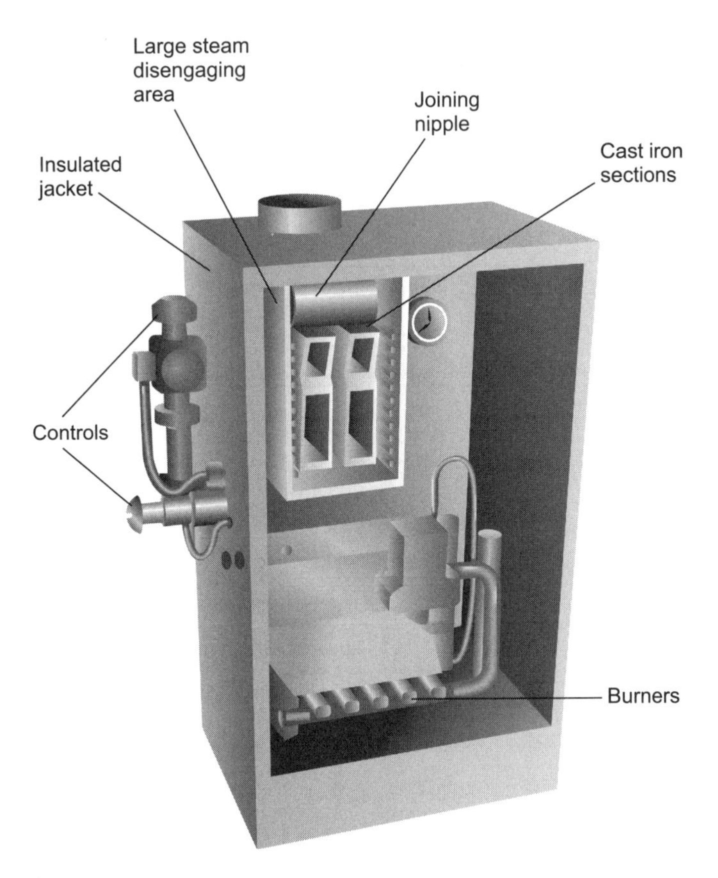

Figure 18–2
A Burnham cast iron steam boiler. Courtesy of Burnham.

- Convectors
- Radiant panels
- Radiators
- Registers
- Unit heaters

ONE-PIPE STEAM SYSTEMS

Steam is used in heating systems to transfer heat from the boiler to terminals in the building.

The principal advantage of steam as a heating medium is that it diffuses by itself throughout the system. For simple steam systems, the steam itself is the motivating agent that delivers the heating medium to the terminals. No pump or other device is required as long as a few conditions are observed. The most basic design of a steam heating system is referred to as the one-pipe system.

In order to understand the basic principles of steam heating and steam heating design, the following terms must be defined:

Essential Terms

Air Vent

An air vent is a device that permits air to pass out of the system (see Figure 18–3).

Boiler (Steam)

A steam boiler is a device in which energy is added to water to change it to steam (see Figure 18–4).

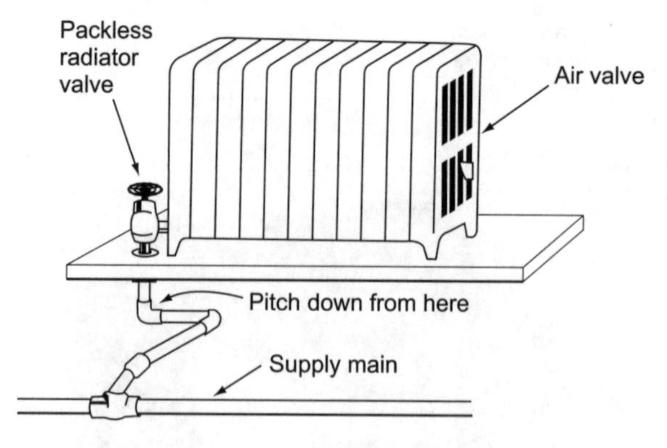

Figure 18–3
A steam radiator using an air vent.

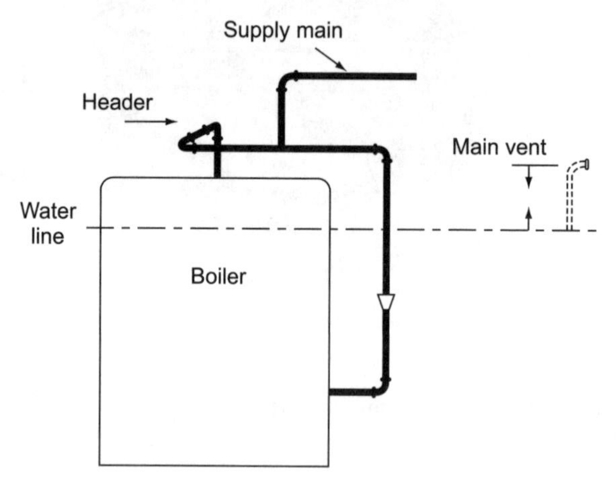

Figure 18–4
A steam boiler illustration indicating the critical water level.

Condensate

Condensate is the liquid resulting from steam giving up its latent heat and cooling back into a liquid.

Drip Connections

A drip connection is a piping arrangement between a steam pipe and return pipe to remove condensate from the steam pipe and convey it to the return pipe.

Dry Return

The dry return is the part of the return piping located above the water line of the boiler, or above the condensate receiver tank to which it connects (see Figure 18–5).

Hartford Loop

The Hartford loop is a piping arrangement at the boiler designed to equalize pressure between the boiler supply and return. The Hartford loop prevents boiler pressure from pushing boiler water into the return system, which significantly lowers the boiler water amounts. Firing the boiler with insufficient water amounts could permanently damage the boiler (see Figure 18–6).

Header

The header is the horizontal pipe connecting the outlet(s) of the steam boiler to the building main(s). The plural indicates that multiple boiler units could be tied together (see Figure 18–4).

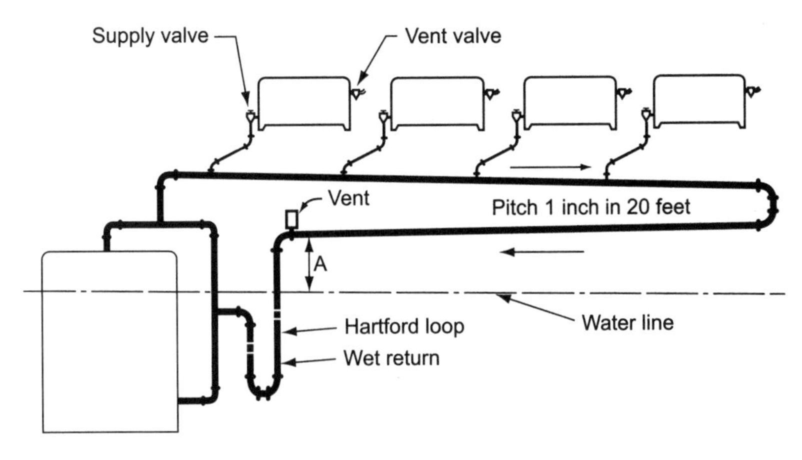

Figure 18–5
A one-pipe steam parallel unit flow system with a dry return.

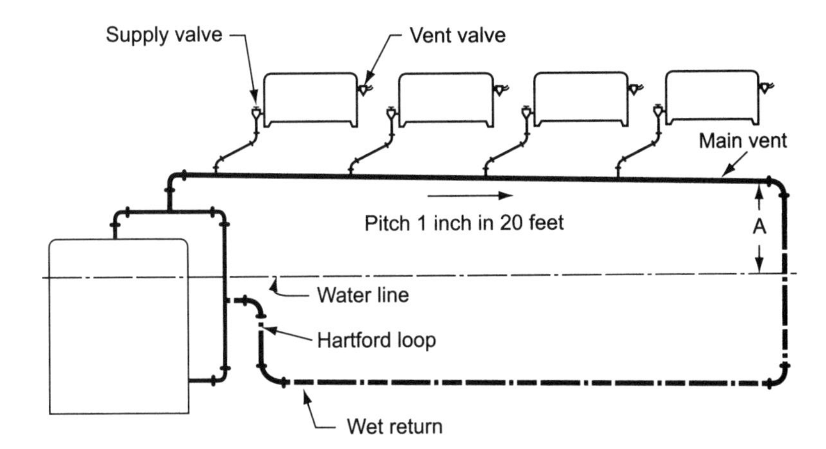

Figure 18–6
A one-pipe parallel steam unit flow system with a wet return and Hartford loop.

Heating Unit

A heating unit is any terminal unit (see Figure 18–3).

Radiator Valve

A radiator valve is a globe valve used to control steam flow to a heating unit (see Figure 18–3).

Riser Supply Main

A riser supply main is any vertical steam or return pipe.

Steam Supply Main

The steam supply main is the piping that carries steam from the boiler header to the branch piping serving the heating units.

Trap

A trap is a device, usually thermostatic or float and thermostatic, that prevents the discharge of steam from a heating unit and allows air and condensate to discharge to the condensate return lines.

Wet Return

The wet return is return piping located below the boiler water line or the condensate receiver to which it connects (see Figure 18–6).

Additional steam facts are listed below to aid in your understanding:

- The latent heat or heat of vaporization (a change from water to steam or steam to water) is generally 980 BTU at low pressure.
- The temperature of steam increases as the pressure increases. Thus, high-pressure boilers for commercial use have drastically increased energy potential over smaller, lower-pressure units.
- The volume of steam decreases as the pressure is increased. Conversely, as the pressure is decreased, the volume of steam increases.

One-pipe System Concepts

The one-pipe system is illustrated in Figure 18–3. The steam migrates to the radiator as the condensate flows back out the only pipe connected to the radiator. This migration is because the pressure at the radiator or convector is less than that of the boiler and as the pressure decreases the volume increases, forcing the condensate into the pipe. The condensate or water moves because of gravity or another form of assistance that will be discussed later. These facts will also apply to other piping systems, discussed later in this book.

One-pipe systems have less pipe length than other systems. However, the pipe size must be larger to accommodate both steam being supplied to the radiators and condensate running back to the boiler. The steam usually leaves the boiler at less than 5 psi, runs to radiators at the bottom of the room, which give off heat, and condenses back to a liquid. This condensed liquid drains back to the boiler. In poorly designed systems, water hammer often occurs because of the condensate flow against the steam head. Each radiator is provided with an air vent (see Figure 18–3) to let air escape from the radiator so that steam can enter. A main air vent is also connected at the end of the main. The condensate returns to the boiler via a dry or wet return. These air vents prevent the system from becoming what is commonly referred to as "air bound."

In the Field

The easiest way to understand a one-pipe system is to recall that the radiator (terminal unit) only has a feed and no piped return: one pipe.

In the Field

Troubleshooting on service calls for older boiler systems should always include pipe hanger observations. Condensate pockets caused by missing hangers will lower system performance and increase noise complaints.

It is best to install the steam main so it helps the collected condensate return to the boiler. This requirement means that the steam main must continuously slope downward until it drops vertically to return to the boiler. This is best illustrated by Figure 18–5, where a one-pipe system allows the condensate to come back to the boiler via a dry return. Pitch, commonly referred to as fall, is an important factor in the proper functioning of these systems. Often, the supply is pitched down and away from the boiler, which is called a wet return and allows the condensate to more or less run with the steam supply. This method reduces water hammer.

For small systems to operate satisfactorily, the supply pressure must be $\frac{1}{8}$ psi to $\frac{1}{4}$ psi above the return pressure. This pressure difference requires that the end of the supply pipe elevation must be at least

$$\frac{\frac{1}{8}\text{ psi}}{0.434\text{ psi/ft}} = 0.28\text{ ft}$$

above the beginning of the return. The return system is normally designed to have the same pressure drop, and a safety factor equal to the drops is usually added.

Thus,

Steam main drop ($\frac{1}{8}$ psi) $= 0.28$ ft
Return main drop ($\frac{1}{8}$ psi) $= 0.28$ ft
Safety factor $2(\frac{1}{8}) = 0.56$ ft
Total $= 1.12$ ft
Total $=$ approximately 14 in.

The usual standard is 18 in. elevation for small systems and 28 in. for larger one-pipe systems. This dimension is the elevation of the end of the steam/return main above the water line of the boiler. (Dimension "A" in Figures 18–5 and 18–6). If these minimum values cannot be obtained, it will be necessary to drain the

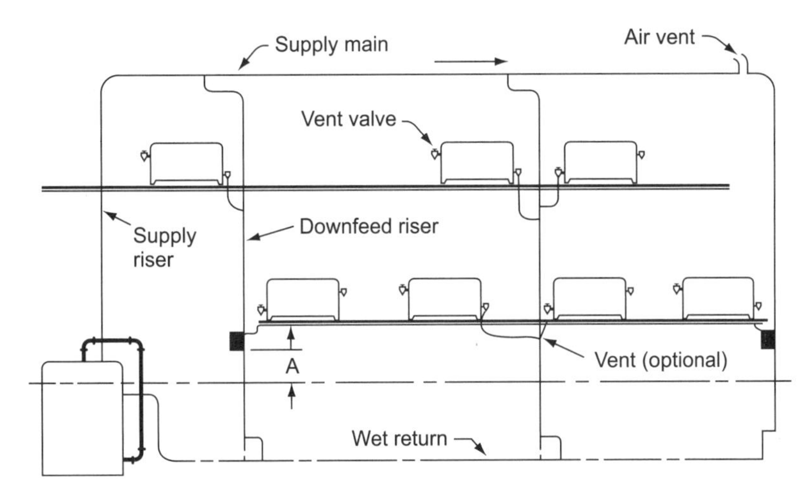

Figure 18–7
A multistory steam one-pipe parallel flow downfeed system.

return line into a condensate receiver and pump, which will return the water to the boiler.

Figure 18–7 shows the best arrangement for a multistory building: an attic steam main and downfeed risers to the floors below. Steam and condensate travel in the same direction. Each riser connects at the base to the wet return by a dirt pocket arrangement.

You may encounter one-pipe steam systems in service/repair work, but they have undesirable characteristics that make new installations impractical. Modern radiation equipment has a very small cross-section of internal passages, which means one-pipe applications will not work. One-pipe systems require large pipe sizes. The smaller sizes of two-pipe systems are more economical.

Finally, radiation cannot be throttled (partially closed off) in one-pipe systems. Attempting to do so causes the radiator to fill with water and/or to hammer when the boiler is firing.

TWO-PIPE STEAM SYSTEMS

The two-pipe system is a design that separates the steam supply and condensate return at the heat transfer terminal (commonly a radiator). Two-pipe systems are now more commonly used for building heating than the one-pipe system previously explained. Steam heating systems are almost always installed in two-pipe configurations, which allows smaller line size than the one-pipe system. Steam from a boiler with low pressure is conveyed to the terminals in a supply pipe system, and condensate is carried away by a separate return system. This arrangement operates well with any type of terminal unit and with any pressure pattern in the supply and return piping.

Steam enters the two pipe terminals at the top. Condensate exits at the bottom. A regulating valve at the top controls the amount of heat. The steam turns to condensate, which exits out a steam trap after giving up its heat. This thermostatic trap opens and closes depending on the difference in temperature between the steam and condensate at the trap. The trap closes to the hotter steam and opens to the cooler condensate, allowing condensate to drain and allowing new steam to come into the radiator.

The various two-pipe systems differ in the method of condensate return. Those methods include the gravity return system, the mechanical return system, and the vacuum return system. As the names indicate, the mechanical and vacuum systems require mechanical equipment assistance, which improves operation.

In the Field

The easiest way to understand a two-pipe system is to recall that the radiator (terminal unit) has a feed and a condensate return pipe (two pipes).

Two-pipe Gravity Return

A gravity return system has a steam main above the boiler water line. Steam flows in this main to the terminals. Condensate returns by gravity to the boiler in a return main, which may be a wet or dry return. Each terminal unit must be equipped with a thermostatic steam trap to keep steam in the terminal and condensate in the return piping. One example is illustrated in Figure 18–8: the float and thermostatic (F&T) trap. The return main is equipped with one or more air vents, which release the air in the system as steam fills the system. The condensate returns to the boiler as long as the return connections and drip connections are at least 18 inches above the boiler water line to overcome the boiler operating pressure.

Just as we saw for the one-pipe system, the two-pipe system must retain the elevation range of 18 inches for small systems (up to 200 MBH, approximately) and 28 inches for larger systems (dimension "A" in Figure 18–8). If the air vents contain check valve mechanisms, the systems can operate below atmospheric pressure under certain conditions.

Two-pipe Mechanical Return

A mechanical return system uses a steam main for distribution to the terminals. The return main conveys condensate to a vented receiver tank. The condensate is pumped from this tank into the boiler. Thus, a wider pressure range can be accommodated with this system. As with the gravity design, each terminal unit must be equipped with a trap to keep steam from entering the return piping system. One such trap is illustrated in Figure 18–9 as an F&T trap.

Two-pipe Vacuum Return

A vacuum return system is a mechanical return system in which the return pump is capable of developing a vacuum in the return piping of the system. The effect of producing a vacuum on the return side of the system is to increase the pressure difference in the system. This greatly increases the diffusion rate of the steam, so the system is more responsive to changes. Also, the mean steam temperature may be reduced (without increasing steam transfer time), which also improves the controllability of the system. The return pump must be capable of removing air from the system as well as moving water from the receiver to the boiler. Figure 18–10 shows a two-pipe vacuum system.

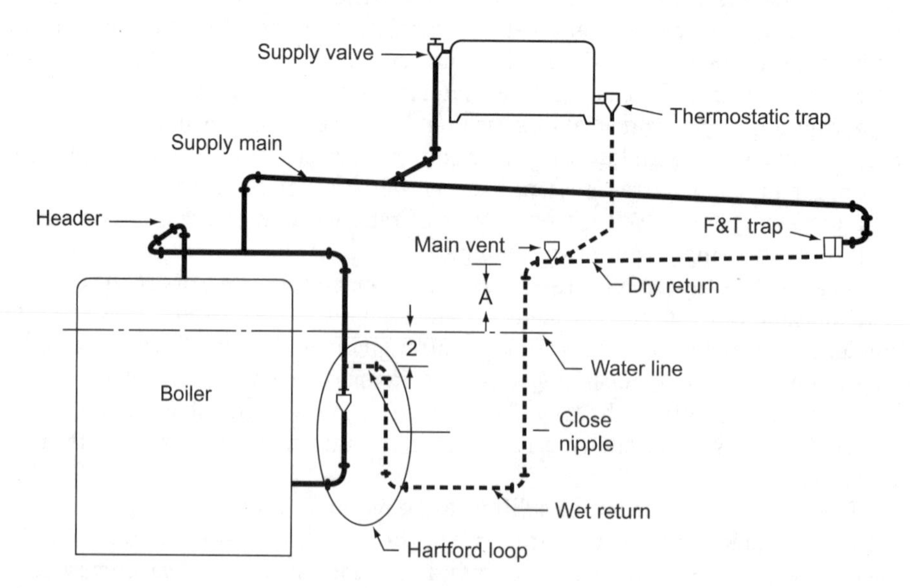

Figure 18–8
A two-pipe steam gravity return system.

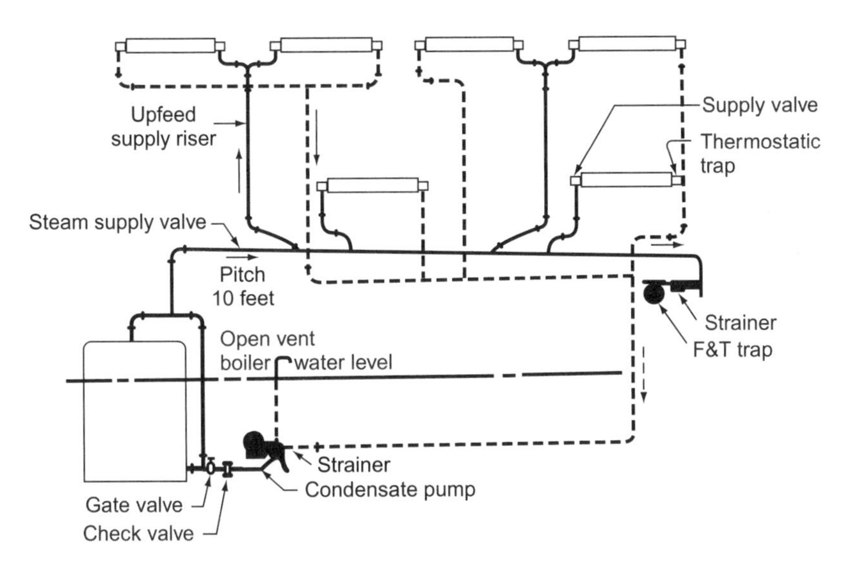

Figure 18–9
A two-pipe steam upfeed system, mechanical return system with a condensate pump.

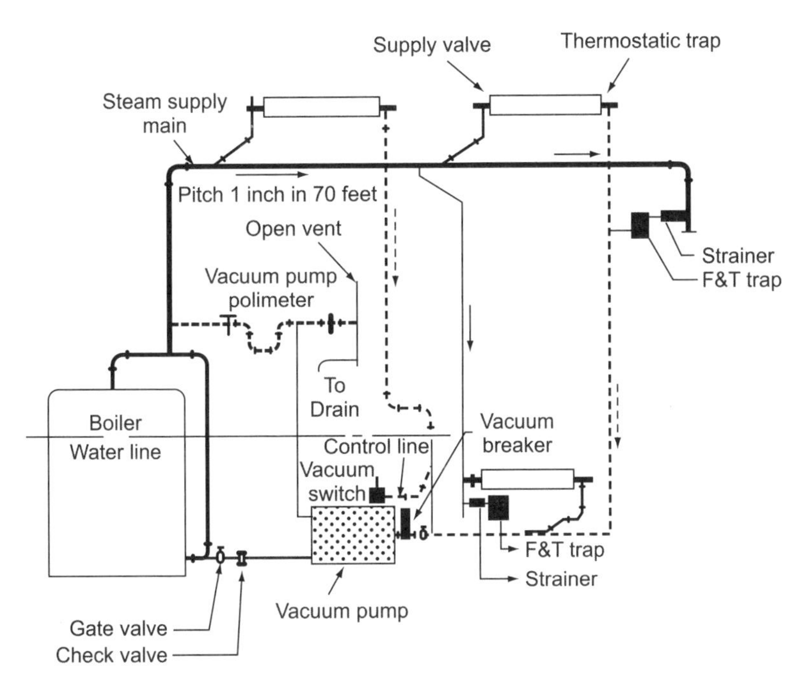

Figure 18–10
A typical steam two-pipe vacuum return system.

Vacuum return systems are especially suited to large buildings where the vacuum permits very rapid changes in the heating fluid in the system, making it possible for the system to follow changes in outdoor conditions more quickly. It must be noted that vacuum pumps for steam heating service are expensive and require considerable maintenance. Most vacuum-designed systems that you will encounter operate with the vacuum pump as a condensate pump only.

Figure 18–11 shows a piping arrangement that permits lifting of condensate and air from low points in the system. Condensate collects in the trapped piping until the vacuum lifts the accumulated condensate to the higher level. By making the height of these lifts less than the vertical distance corresponding to the vacuum, condensate can be raised from a lower elevation to the principal return line elevation in stages.

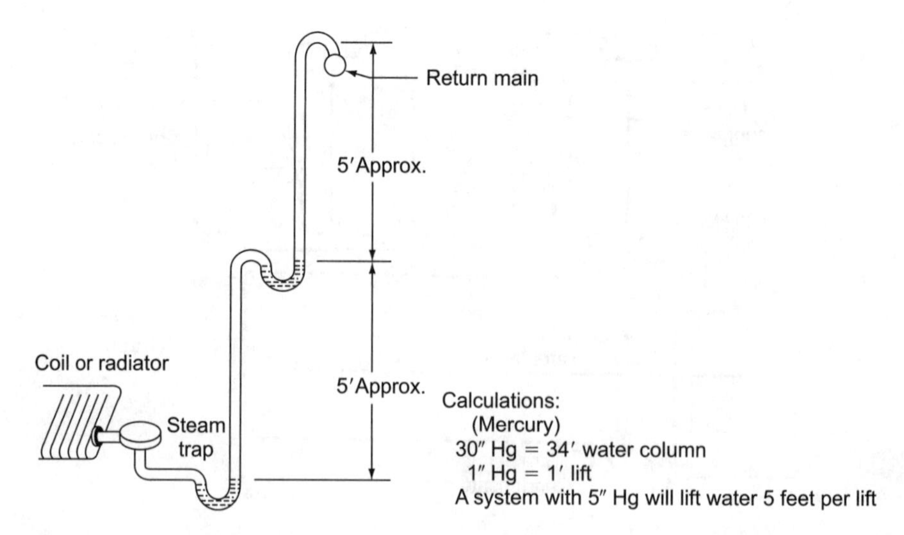

Figure 18–11
A condensate steam return from radiators below the main.

System designers use extensive tables showing the properties of saturated steam at each temperature reading. Conditions representative of heating steam pressures (0-15 psig) generally indicate that the heat release of a pound of steam is about 1000 BTU. This value is used as the heat content of steam for quick calculations.

REVIEW QUESTIONS

1. Name at least five types of terminal units.
2. Name the three main components of heating systems.
3. Can radiator valves be throttled in a one-pipe system?
4. What is the benefit of a Hartford loop piping arrangement?
5. Must steam heating fluids be pumped through a system in order to function?

CHAPTER

19

Steam Equipment and Piping

LEARNING OBJECTIVES

The student will:

- Describe basic concepts of satisfactory steam heating piping.
- Identify the components of various types of heating systems.
- Identify the types of steam traps and where they can best operate in a steam piping system.
- Discuss the steam supplied terminals available for systems.

HEATING SYSTEM COMPONENTS

There are a number of components necessary for a steam heating system to operate safely and effectively. In this chapter, we examine these components and their uses in the system.

Radiator Valve

The radiator valve must have a large opening and large vertical movement (lift) closure mechanism. The larger opening and bodies are necessary to accommodate steam entrance and condensate drainage to the one-pipe steam system. The most common radiator valve has a right angle pattern, but straight patterns are also available.

Valves in a one-pipe system are not meant to be partially closed or used for throttling steam to the unit. Partially closing the valve will increase the incoming steam velocity and allow the steam to push back (blow back) the condensate.

Radiator valves use either packing or packless bonnets. The packless bonnet uses a diaphragm to contain the steam inside the valve. The operating handle pushes down on the diaphragm, which in turn pushes on the valve closure plug. If the diaphragm cracks or breaks, a substantial leak results.

Air Vent Valve

Air vent valves are installed on radiators in one-pipe systems. These are available with adjustable or non-adjustable orifices, and for vacuum or non-vacuum systems. One-pipe systems must have air vents in order to allow the steam to enter the radiator (terminal). If the air could not escape through the air valve, the system would become air-bound, meaning steam could not enter the radiator.

Air vents are cylindrical devices with a lightweight metal float inside. The float contains a mixture of water and alcohol. It is connected to a valve mechanism venting through the top. When water enters the valve, the float lifts and closes the vent port. When the valve is heated by steam, the alcohol boils and evaporates, resulting in a lighter float rising to close the valve. Figure 19–1 shows a typical steam radiator vent.

Main vent valves are required at the end of the return main. These are also available in vacuum and non-vacuum types. These main vents commonly have a capacity of up to 15 times more than radiator vents. Air vents on the mains that

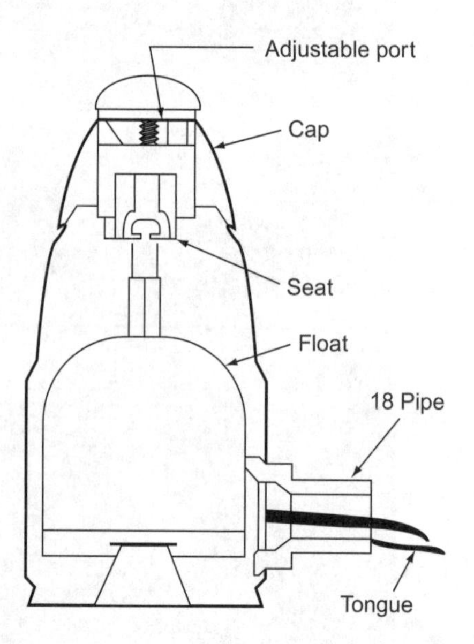

Figure 19–1
An adjustable steam radiator vent.

are not functioning will cause noticeable effects. For example, when air vents are not present or inoperable on mains, the boiler may short cycle. This is caused by the boiler firing when it is trying to compress the air. The pressure controls may shut off the boiler firing. Noise and water hammer are also indications of inoperable air vents.

Traps

Radiator traps are necessary for two-pipe systems. They allow the condensate to drain from the unit, which allows steam to enter. The traps also prevent steam from entering the designated condensate return line. The traps are generally thermostatic and usually have a right-angle bend. This trap contains a bellows, which expands to close the outlet port when steam temperatures are present. When the bellows element cools to 180°F, the outlet valve opens and condensate or air can flow to the return line. There are a number of different types of traps that may be utilized in a steam system. A brief explanation of each type follows.

Float Traps

Float traps close the outlet valve when no condensate is present and open it when condensate collects in the trap body. Float traps are best suited for radiator and convector applications. Figure 19–2 shows a float trap.

Float and Thermostatic Traps

These traps combine the operations of float traps and thermostatic traps. These are the most commonly used traps for unit heaters, air handler heating coils, and main drips. Float and thermostatic (F&T) traps continuously discharge condensate, at varying intensities at different times. They are the preferred choice for exchangers, air handler coils, and condensate drip locations on mains. Figure 19–3 shows an F&T trap.

A typical placement of a bucket or F&T trap in a piping system is illustrated in Figure 19–4.

Inverted Bucket Traps

This trap operates with a bucket-shaped moving element, mounted upside down, that acts like a float when steam or air enters the trap. Air passes through an opening in the bucket, so it is eventually released from the unit. If the trap fills with condensate, the bucket sinks and the exit valve is opened, allowing the condensate to flow to the return line. These traps are used on higher-pressure systems (up to 250 psi) where minimum air venting ability is acceptable. Figure 19–5 shows an inverted bucket trap.

In the Field

Service and repair technicians should remember that the piping system itself acts as a terminal—in a sense, a radiator. For that reason, traps are required to address condensation accumulation.

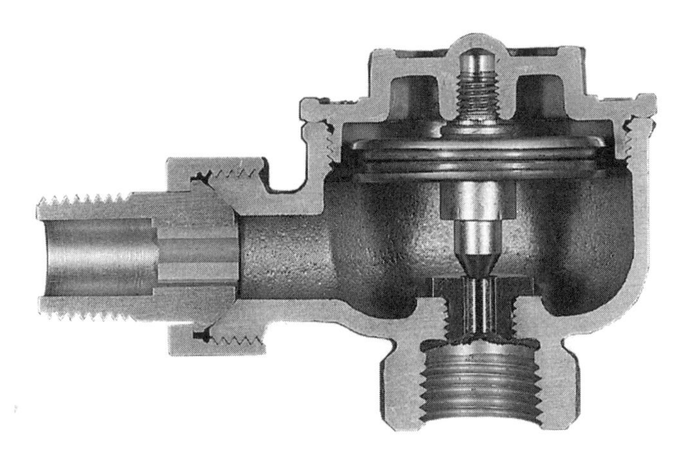

Figure 19–2
A thermostatic steam trap. Courtesy of Watts Water Technologies, Inc.

Figure 19–3
A steam float and thermostatic (F&T) trap. Courtesy of Watts Water Technologies, Inc.

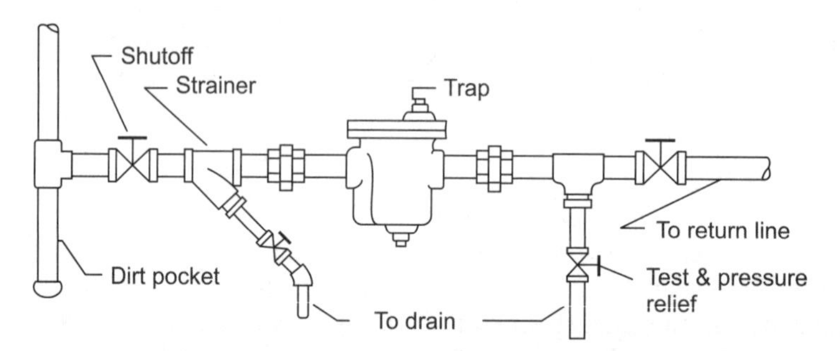

Figure 19–4
Placement of a bucket trap in a steam piping system.

Thermodynamic Trap

This trap closes based on the venturi effect of steam rushing past a disk through an opening. Initially, the disk is pushed off the seat by the inlet pressure and flow of condensate against the disk. During equilibrium operation, some flash steam pressurizes the space above the disk, causing the disk to move part way to the seat. When all the condensate is discharged, steam pressure accumulates above the disk, and the high velocity through the disk-seat area causes reduced pressure, which snaps the disk closed. The disk stays closed until condensate forms and the process repeats. Thermodynamic traps are used on higher-pressure systems (from 75 psi to 600 psi), as they require significant pressure drops between the unit condensing the steam and the condensate system return. Figure 19–6 shows a thermodynamic steam trap.

Figure 19–5
An inverted steam bucket trap. Courtesy of Watts Water Technologies, Inc.

Figure 19–6
A thermodynamic steam trap. Courtesy of Watts Water Technologies, Inc.

CONDENSATE PUMP AND VACUUM PUMP SYSTEMS

Condensate Pumps

Condensate pumps are used whenever a return line must be installed at an elevation below the boiler water line, or if the supply pressure is too high for gravity return to the boiler. A receiver tank (vented to atmospheric pressure) receives the returning condensate. When sufficient condensate has accumulated in the tank, a float switch activates a pump to return condensate to the boiler.

An alternate arrangement gives the tank a float valve connected to the water system (with approved backflow preventer). The pump starts up whenever the water level in the boiler falls to a preset elevation. This tank/pump/float-valve arrangement is called a boiler feed water system.

Vacuum Pumps

A vacuum pump system is similar to the condensate pump, except that the pump is capable of pumping air as well as water. A vacuum pump can reduce the return system pressure below atmospheric pressure, thereby speeding up circulation and reducing the steam temperature, which improves system efficiency and system comfort.

Vacuum pumps operate on various principles and tend to be considerably more expensive than condensate pumps. Many vacuum systems can operate satisfactorily with a condensate pump only. This option should be considered if a replacement is necessary.

AUTOMATIC BOILER LOW-WATER CUTOFF AND WATER FEEDERS

Operating a boiler without enough liquid is detrimental to the boiler and can be unsafe. Mechanical devices called low water cutoffs are used to address these concerns. They are float-activated devices that add water to the boiler when the boiler water level drops to an unsafe level. The simplest low water cutoff device simply disconnects the burner if the boiler water falls lower than a set level. These controls may also contain a switch that operates an alarm. Unsafe levels could result from leakage or large condensate collections unavailable to the boiler. Figure 19–7 shows an installation on a large boiler. In these devices, the float is linked to a valve mechanism in the water supply pipe. These devices are extremely valuable on solid fuel boilers, since the fire cannot be shut off immediately by the operation of a switch.

Electronic feeders provide waiting times that help stabilize the system while waiting for normal condensate to return. This prevents the boiler from flooding, a condition in which abnormal amounts of water prevent steam from being creating in the boiler container area, resulting in the boiler becoming inoperable.

Combination Water Feeder and Low Water Fuel Cutoff

This device combines a mechanical water feeder with a float-operated switch that cuts off the burner if the boiler water is below a safe level. The water-makeup feeder operates at a boiler level about 2 inches below the normal water line, and the cutoff switch operates at a level $\frac{1}{2}$ inch to 1 inch below the feeder operating level. Figure 19–8 shows one type of this combination system.

Many cutoff switches are double throw switches. If the low water cutoff switch operates, the switch makes an alarm contact that rings a bell, lights a lamp, or sounds another type of alarm when the low-water circuit opens.

Electric Water Feeder

An electric water feeder is a boiler makeup valve that is operated by a solenoid. The electric power to this device is controlled by a boiler water level controller. These valves are easily fouled, and are somewhat less reliable than mechanical makeup

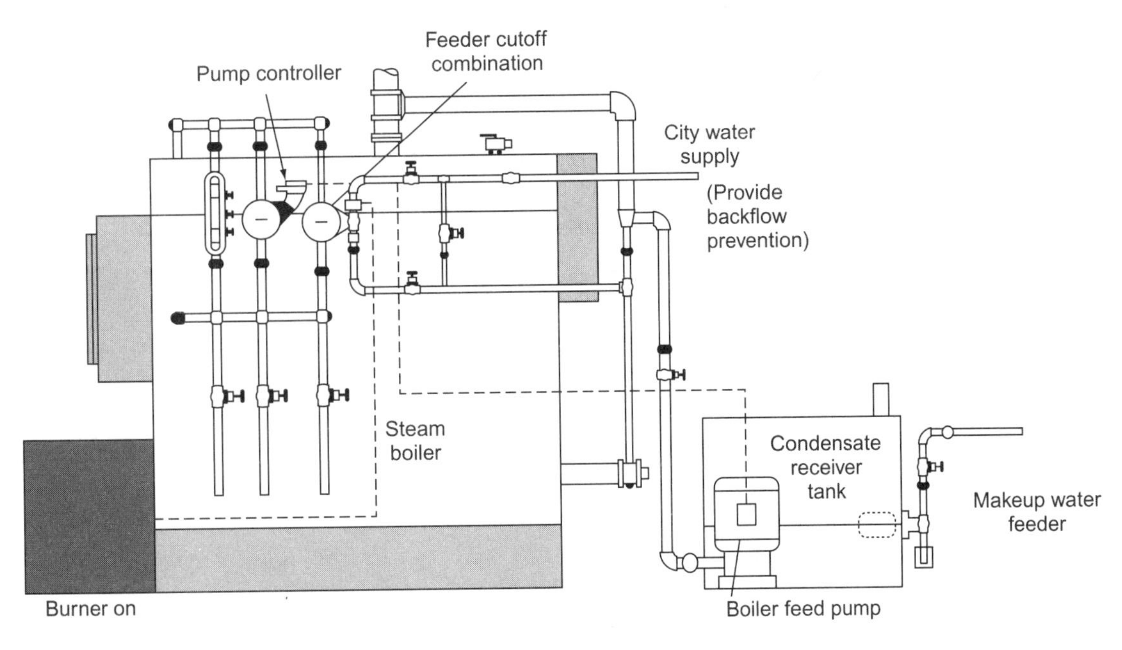

Figure 19–7
A typical steam boiler piping system.

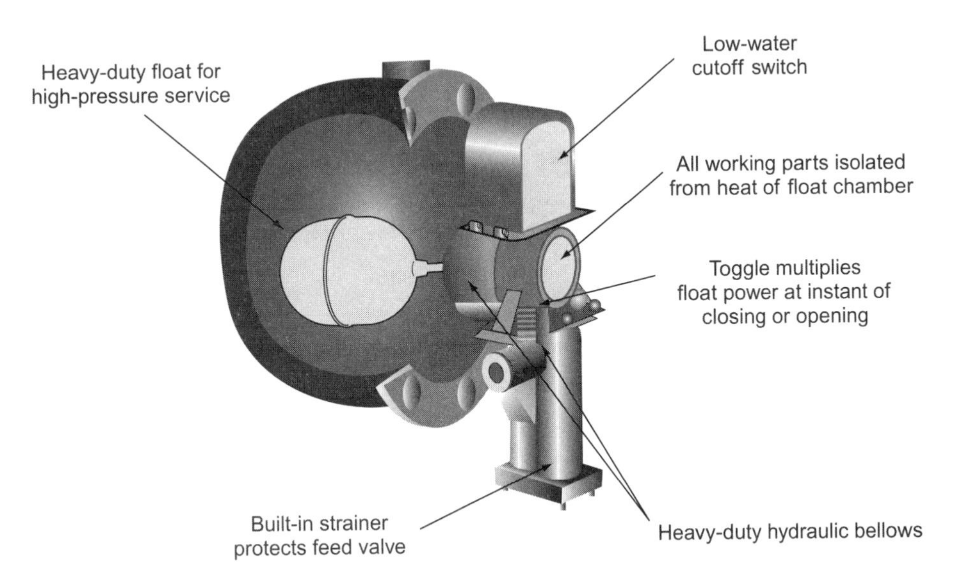

Figure 19–8
A combination water feeder and low water cutoff device for a steam boiler.

valve assemblies. The makeup feeders are equipped with a strainer to protect the valve seat and plunger. Figure 19–9 shows an electric water feeder.

Pump Control and Low Water Cutoff

This is a combination feed pump operating switch and low-water fuel cutoff. As boiler water drops, the feed pump is turned on. If the water level continues to drop, the fuel cutoff switch turns off the burner after another $\frac{1}{2}$ inch to 1 inch drop. Figure 19–10 shows the combination pump control and low-water cutoff.

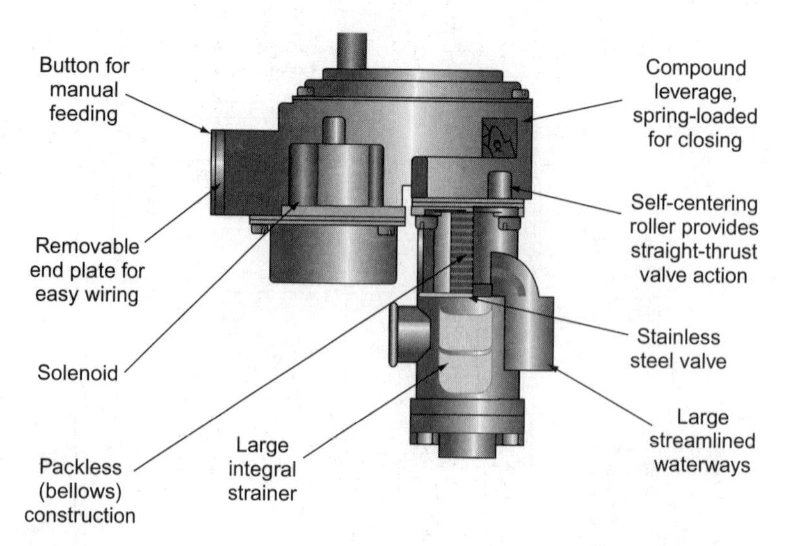

Figure 19–9
A steam electric water feeder valve.

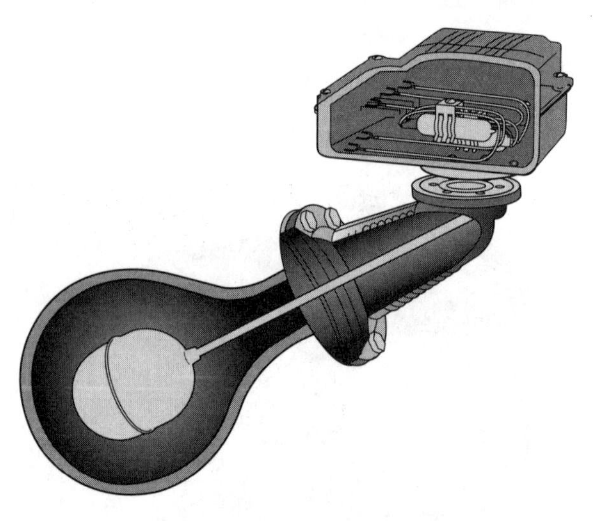

Figure 19–10
A steam pump control and low water cutoff.

Figure 19–11
A steam safety relief valve.

PRESSURE REDUCING VALVES

High-pressure boilers with some low-pressure load areas within the system require pressure reducing valves to supply the low-pressure loads. Downstream pressure is applied against the spring setting of the control valve. This balance is how the reducing valve operates. A bypass valve is usually installed to carry steam around the reducing valve, in case it must be removed or isolated for servicing.

SAFETY VALVES

Steam safety valves are similar in operation to the water heater relief valves you are familiar with. They are spring-loaded in the closed position. At a certain pressure applied at the inlet, the valve opens to allow a maximum outflow, relieving excessive pressures. They are equipped with a test lever that lifts the closure mechanism off the valve seat. The pressure setting is usually sealed at the factory, so they are not adjustable in the field. The valves have a steam volume rating, expressed in pounds per hour, as well as the pressure rating. Boiler codes require safety valve capacity to equal or exceed the input rating of the boiler. Figure 19–11 shows a safety valve.

The boiler codes permit multiple safety valve installations to protect a boiler. This practice is a good idea. Multiple valves may be less expensive than one large valve, and in the event of a safety valve failure, it is much less expensive to replace the one that has failed. For 15 psi applications, valves up to 2 inches can usually be combined to be less costly than valves $2\frac{1}{2}$ inches and larger. Valve construction is usually bronze in sizes up to 3 inches and pressures to 200 psi. For larger sizes or higher pressures, construction would be cast iron or cast steel.

When checking the condition of a boiler, always operate the test lever to ensure that the valve is not stuck in the closed position.

In the Field

You must exercise common sense when you evaluate the boiler size and consider where the discharge is piped. For example, if the valve does not reseat, can the boiler continue to operate and meet the structure's needs? Will the discharge injure others nearby or create excessive temperatures in the surrounding location, temperatures which were not in effect at the time of the original boiler installation?

PIPING DETAILS

There are many ways a heating system can be piped. Some designs are better than others, because the system is designed not only for good performance but also for ease in servicing. A few of the basics will be considered in the following sections, which reflect good design to serve both purposes. Previously we have discussed one- and two-pipe systems with some variations. Systems will have many different variations.

Hartford Loop

The Hartford loop, named after the Hartford Insurance Company in 1919, is a piping arrangement at the boiler designed to equalize pressure between the boiler supply and return. This arrangement prevents the boiler pressure from pushing boiler water into the return system, which would significantly lower the boiler water amounts. Firing the boiler with insufficient water would permanently damage the boiler. Worse, a red-hot dry-fired boiler receiving condensate water back down from a supply or fill would allow the water to flash into steam, possibly causing an explosion. The return to the equalizer is at an elevation about 2 inches below the normal water line. This effectively protects the boiler from dry firing.

In the Field

Several trade journals and magazines contain articles by well-known experts and could intimidate the average reader. It is wise to remember that their expertise results from a passion for their profession and from years of experience. However, you have the gift of ready access to technical information on the Internet and additional information such as code commentaries and association training manuals (PHCC), advantages that these experts did not have when they were starting out.

Header Details

Cast iron boilers may require more than one connection to the boiler. While this is common in newer systems, older systems' headers and uptake sizes should conform to the boiler manufacturer's recommendations. Always follow the manufacturer's instructions. Undersizing this piping will lead to excessive noise, unstable water lines, and reduced system capacity.

It is good practice to provide swing joints in the connection(s) to the boiler tapings and in the connection to the steam main. Such joints provide the flexibility required to minimize stresses in the boiler itself and in the piping system. The system connection is made *after* all boiler connections and *before* the connection to the equalizer ell. The equalizer line should reduce below the water line, ideally with a reducing tee where the Hartford loop ties in.

Another consideration is the take-off from the main, out to and possibly up to a radiator terminal. Note the take-off in Figure 19–12, with a tee and rolling 45. This combination allows for pitch on the horizontal run and keeps the incoming condensate in a one-pipe system from interfering with the branch's steam supply ability.

In the Field

The Hartford loop is a perfect example of a correctly designed system with the required attention to design detail. Steam system operations are simple, yet the original designs and installations required a great deal of skill and shrewd understanding.

Low-Water Cutoff and Automatic Feeder Piping

The low water cutoff connections to the boiler must not contain any valves. This piping must be at least as large as the low water cutoff body tapings. The piping should include nipples with caps in crosses (not elbows) so that the interior of these pipes can be inspected when the boiler is drained. The feeder piping should

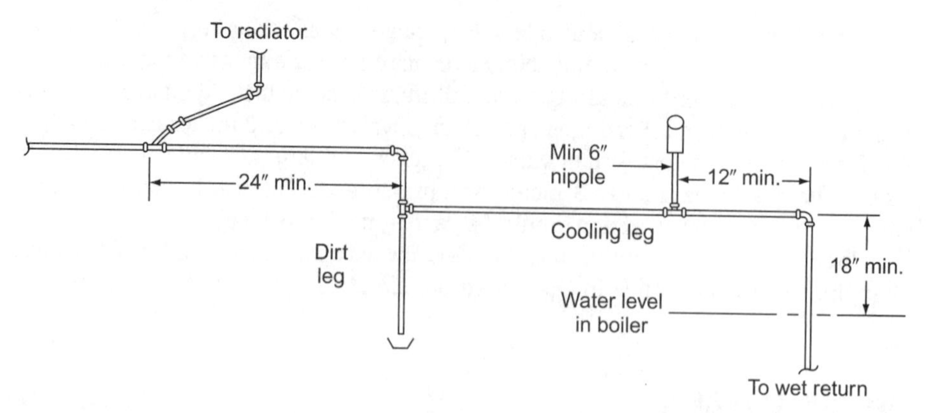

Figure 19–12
The illustration shows the end of a main piping run with a wet return.

include bypass and isolation valves, check valves, and unions located to aid servicing. In addition, the makeup water pipe should connect to the boiler return line in a gravity system. If the water in your area is scaling, increase the pipe by at least two sizes where the makeup connects to the return.

Main End Piping

A typical steam main end with an air vent is shown in Figure 19–12. The main end drops to the wet return. No trap is needed, as the wet return isolates each main end from all the others. If the wet return is not accessible, a valve, strainer, and F&T steam trap are required at the end of the cooling leg, and the outlet of the trap connects to the dry return main.

Note the use of dirt legs (collection legs) in all the figures in this unit. The dirt leg is a straight piece of pipe below a tee. The straight piece is capped on the bottom. Any solids that come down the line fall into the capped nipple, and water flows out the side branch of the tee. Dirt legs are generally provided at all low points where condensate collects, at the main ends, base of risers, and near valves and regulators. These collectors also serve to prevent water hammer against delicate trap parts, because any fast-moving water simply slams into the water in the dirt leg.

Wye Strainers

These strainers are installed to intercept foreign materials before they can enter any critical components with close-fitting parts. Thus, traps, regulating valves, and automatic control valves are protected with strainers.

TERMINALS

We have considered typical terminals, such as radiators, in this chapter and the previous chapter. While they are the norm, other terminals must be discussed as well.

Air Handler Steam Coil

The piping is similar to a unit heater, except that a vacuum breaker is needed on the inlet piping. Either a commercial vacuum breaker or a check valve may be used. The reason for this difference is that an air handler coil can have cold outdoor air blowing across it, which can create a vacuum in the coil. Such a vacuum will tend to hold condensate in the coil, where, in extreme conditions, it could freeze and break the tubes.

Note that a vacuum breaker installed on the return line does not protect the coil! The vacuum is broken in the return line, but the condensate is still held in the coil. Thus, the vacuum breaker must be installed on the supply piping if it is to protect the coil. Figure 19–13 shows a typical installation.

Unit Heater, Two-pipe

The piping is taken from above the center of the steam main, through a gate valve and union, to the heater. The F&T trap is supplied through a dirt leg, strainer, and gate valve. The return gate valve is not needed in small systems. Figure 19–14 shows a typical two-pipe unit heater.

The supply and return piping for high-pressure systems is similar to that described above, except that an inverted bucket trap is used. Condensate may be lifted to an overhead return main if convenient. Note that if the condensate is lifted to the return main, a check valve should be used at the outlet of the trap, and the connection should be made to the top of the return main. A word of caution when using a unit heater on a high-pressure steam system: be sure that the unit heater core is rated for the steam pressure to be used.

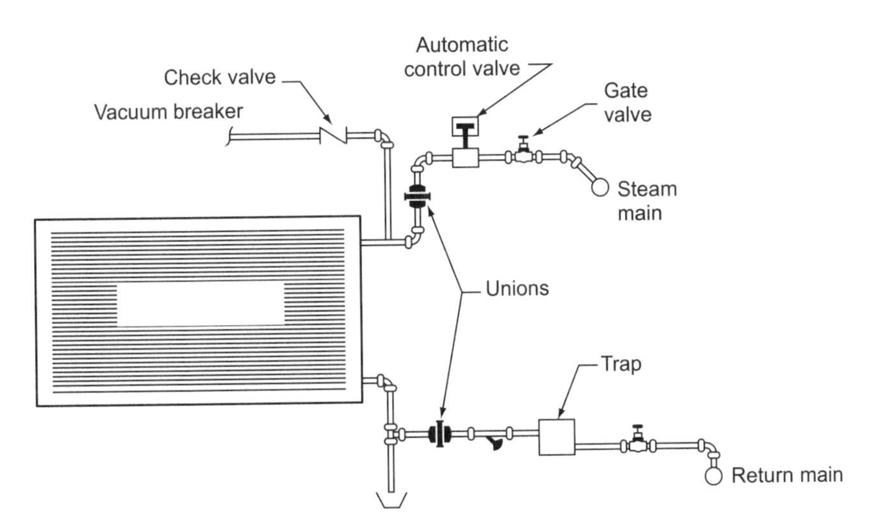

Figure 19–13
A typical steam air handler coil piping.

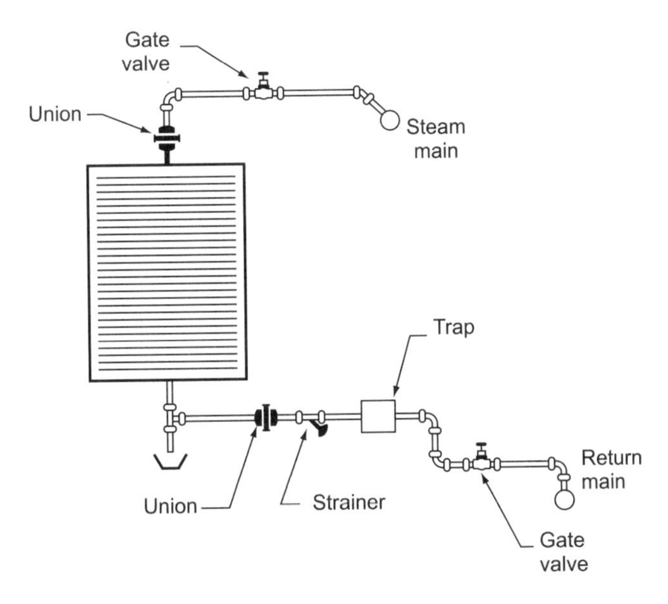

Figure 19–14
A typical steam two-pipe unit heater system installation.

REVIEW QUESTIONS

1. What is the purpose of dirt legs in steam piping systems?
2. What is the purpose of a double throw switch on many low-water cutoffs?
3. What is the purpose of radiator air vents at terminals on one-pipe systems?
4. Name at least three of the steam traps used in one- and two-pipe steam systems.
5. Why are angle radiator valves with large openings and large lift required for one-pipe steam systems?

CHAPTER 20

Hot Water Heating and Hot Water Specialties

LEARNING OBJECTIVES

The student will:

- Describe the basic principles of hot water heating systems.
- Describe the main accessories and specialties used in hot water heating systems.

HOT WATER HEATING

Hot water heating systems are similar to steam heating systems. In some cases, the same pieces of equipment are used. Each system has a fuel system, a method to convert the fuel to heat energy, a water supply system, a closed loop heat distribution system, and a control system to monitor the operation. Depending on availability, oil, natural gas, LP gas, or electricity are used as the fuel source. Because of increasing fuel costs, some companies are starting to promote the use of solar heating systems to help supplement the primary fuel source.

The principal difference between hot water and steam systems is that hot water heating systems run at temperatures and pressures that are below the saturation point of water. Typical hydronic systems operate at pressures less than 30 psig, which means they could have water temperatures as high as 274°F before steam is generated. In most cases, the water temperature will be controlled somewhere between 160°F and 170°F.

Like steam boilers, the hydronic boiler is a closed container that has a fire side (or combustion chamber) and a water side. The fire side contains all components necessary to hold the fuel and air mixture at the proper ratio and ignite it. It also contains the necessary sensors to monitor the flow of combustion gases and the presence of the flame. The water side has sensors to monitor the presence of the water as well as its pressure and temperature. All of these sensors interface with the burner controller, which controls the firing of the boiler. The surface that separates the fire side from the water side is called the heating surface. It was discovered long ago that increasing the amount of surface area on the heating surface causes the boiler efficiency to increase. The goal is to keep the combustion gases in contact with the heating surface long enough to transfer the greatest possible amount of heat. In an electric boiler, the heating elements are submerged into the water side, so there is no combustion chamber or flue system.

In the Field

Be extremely careful when working around large commercial electric boilers: they typically have very high voltage. The system must be de-energized and locked out before any troubleshooting can occur.

HOT WATER SPECIALTIES

Hot water specialties, or boiler trim, refer to special control devices that enable the boiler to function properly. These devices are usually a mixture of electrical and mechanical components that monitor system conditions. Some perform a certain task, such as closing a fuel gas valve or opening to relieve pressure. Others send a signal to another device, such as a burner controller, to signal another operation. Some devices work as a backup to protect the system in case the primary control fails. Figure 20–1 shows the basic location of some of the major components.

DEFINITIONS

Air Bound

Being air bound is a condition in which so much air has collected in the system, or part of the system, that the flow of water is stopped. This not only inhibits the flow through the system, but also reduces the amount of heat transfer.

Air Eliminator Fitting

This device separates entrained air from the moving water and directs the air to the compression tank or to an air vent. While these devices are not essential to the operation of a hot water system, they are very helpful in maintaining satisfactory operation throughout the heating season without the need for periodic manual venting of the system.

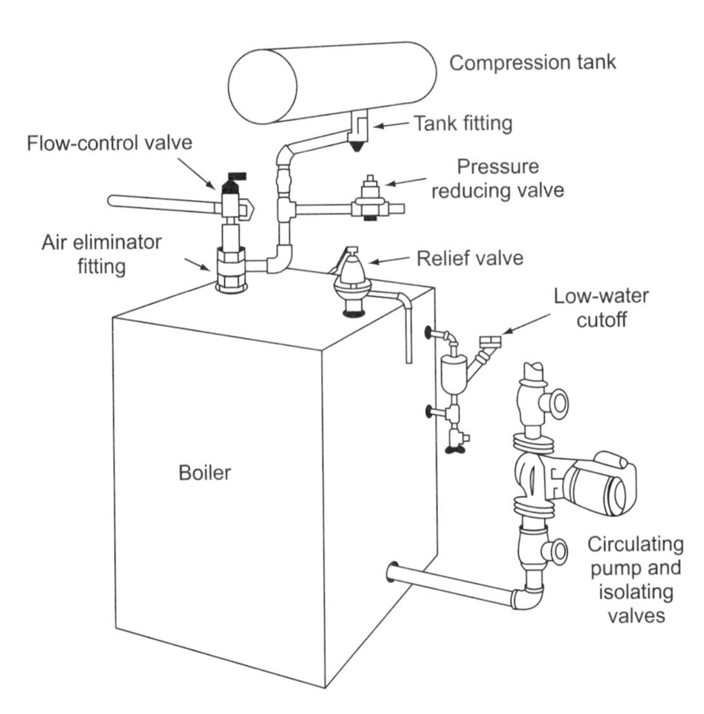

Figure 20–1
This typical boiler setup shows the location of most of the hot water specialties.

Air enters the system as dissolved gas in the make-up water, as residual unvented air from the original system fill, or as other gases formed from chemical reactions in the system. One form of this fitting uses a dip tube down into the boiler to affect the separation (Figure 20–2), while another form uses a series of baffles to achieve the separation (Figure 20–3).

Recent designs of boilers by several manufacturers include construction details that accomplish the separation of air at the boiler outlet. Boilers with this specially designed outlet have no need for these special fittings.

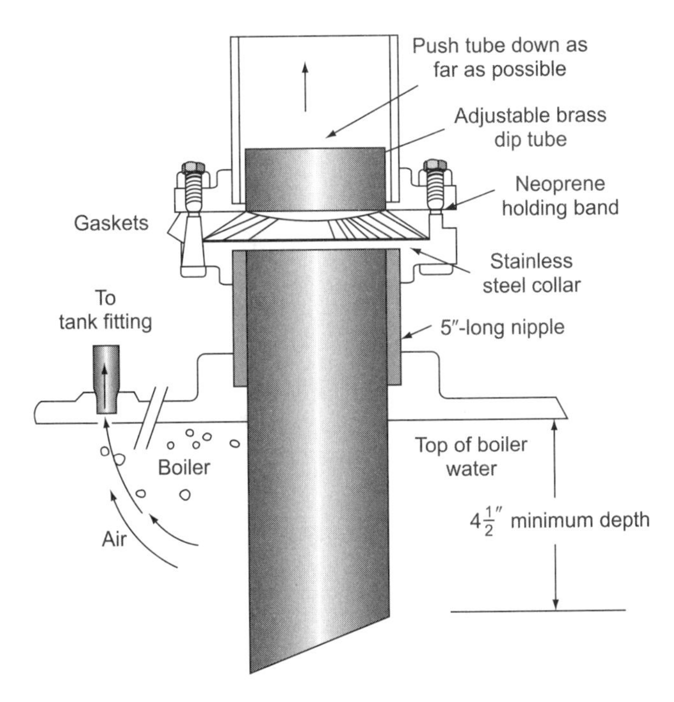

Figure 20–2
This type of air elimination fitting helps prevent air in the boiler from entering the piping distribution.

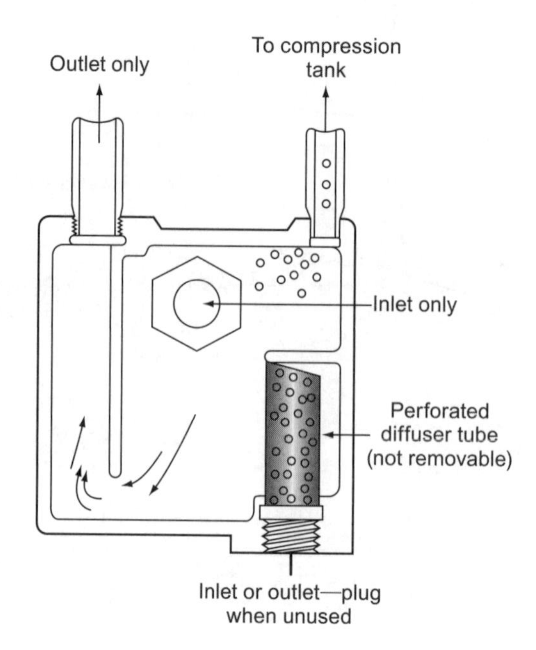

Figure 20–3
This air elimination fitting uses an internal baffle to trap air and discharge it.

Air Vents

Air vents (Figure 20-4) are made to be easily opened manually, or to open automatically. They are installed at all system high points so that collected gases may be removed. In a well-designed system, these vents need to be used only on the initial fill, to eliminate the air contained in the system before the water is introduced into the piping.

Automatic air vents can be a problem. Some types do not hold a tight seal, and most will permit air to enter the system if the contained pressure goes to a vacuum. Manual air vents are less expensive and more reliable.

Aquastat

An aquastat is a temperature-sensing device that is used to measure the temperature of water in the system. Aquastats consist of a temperature probe and a set of electrical contacts. The temperature probe will sense a change in the system water temperature and cause the electrical contacts to open or close.

Figure 20–4
Automatic air vent. Courtesy of Bell & Gossett.

Balancing Valve
A balancing valve is a valve designed to add flow resistance to one circuit in a water system, diverting more flow toward another portion.

Baseboard Heat
Baseboard heat uses a liquid-to-air heat exchanger to transfer heat to a space. The heating terminal is usually made of cast iron, steel, or copper and is located at the baseboard in a room.

Boiler, Hot Water
A hot water boiler, or hydronic boiler, is a pressure vessel in which water is heated, without boiling, through the addition of heat from an energy source.

Branch
A branch is a pipeline which connects the hot water piping circuit to the terminals.

BTUH
BTUH, sometimes notated as BTU/hr, is an expression of power, read "British thermal units per hour."

Circuit
The circuit refers to the network of piping (water flow path) that extends from the boiler supply opening to the boiler return opening.

Circulator or Circulating Pump
The circulator, or circulating pump, is a motor-driven device that moves water through the system. The pump is usually a centrifugal pump that is capable of moving high volumes of fluid at relatively low pressures. The pump should be selected based on how much volume of water is needed to move through the system at the required pressure drop. Since the vast majority of these systems are closed loops, the height of the system does not have much effect. Circulating pumps for this type of application can either be mounted directly in line with the piping or mounted on a concrete pad on the floor. See Chapter 26 for more details.

Compression Tank
The compression tank is a sealed tank that contains the air and water (ideally about half and half) that cushion water volume changes in the system. These volume changes occur because the water expands as the temperature increases and contracts as the temperature decreases. Older types of compression tanks were equipped with a special tank fitting that facilitated the flow of water to and from the tank. The air fitting allowed air to flow only into the tank, not out. Newer designs contain a rubber diaphragm that separates the air cushion from the water side. This design prevents the air from being absorbed into the system.

Direct Return
A direct return is a two-pipe system in which the first unit supplied is the first one connected to the return main.

Downfeed System
A downfeed system is a hydronic system with the supply main above the terminals.

Drain Cock
A drain cock is a valve located at the lowest point of the system to provide a drain for the system.

Expansion Tank
An expansion tank is a tank that operates at atmospheric pressure and is located at the highest point of the system. This tank allows for volume changes of water in the system due to changes in temperature.

Flow Control Valve

The flow control valve (Figure 20-5) consists of a check valve with a fairly heavy closure disk. The valve closes against reverse flow (for parallel pumping installations), and does not lift under the force of gravity alone, preventing gravity flow.

Flow Fittings

Flow fittings are used in one-pipe, or mono/low, systems to develop the force to divert water flow into, or aspirate flow out of, terminal heating units. See Figure 20-6.

Forced Convection

Forced convection systems use a pump to circulate the heat transfer medium.

High Limit Control

A high limit control is a safety control that turns off the burner when the boiler reaches a set temperature.

Figure 20–5
Flow control valve. Courtesy of Bell & Gossett.

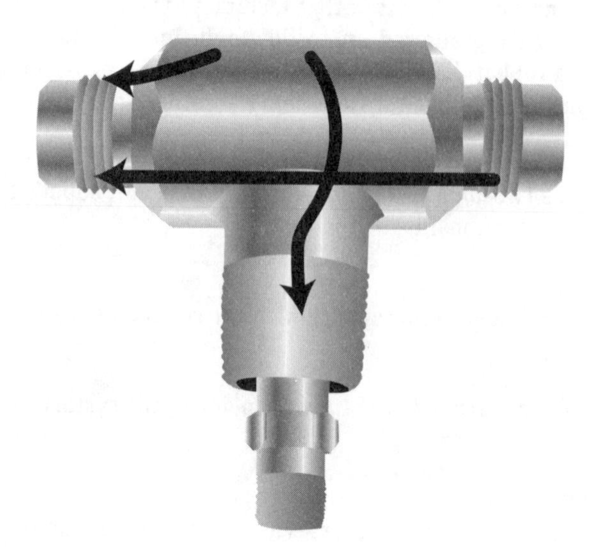

Figure 20–6
Flow fitting.

Low-Water Fuel Cutoff

The low water fuel cutoff (Figure 20-7) is piped directly to the side of the boiler and contains a switch to turn off the burner if the water level drops below a certain line. The ASME boiler code (adopted in most jurisdictions) requires a low-water cutoff on hot water boilers of 400 MBH input and greater.

Make-up Water Line

The make-up water line is the connection from the domestic water system to the boiler or hot water system. It is used to fill the system or replenish it when needed to control the water level.

Multiple Zone

A multiple zone system permits heating in parts of a building independently of other sections.

One-pipe Fitting

A one-pipe fitting is a specially designed tee fitting for use in a one-pipe hot water heating system.

One-pipe System

A one-pipe system is a piping system in which a single main is used as both supply and return to the radiation.

Outdoor Reset

Outdoor reset is the concept of controlling the circulating water temperature based on the outdoor temperature. As the outdoor temperature decreases, the building will lose more heat, and the boiler water temperature should increase to compensate. The converse is true as well; an increase in outdoor temperature calls for a reduction in boiler water temperature. This can be controlled by a thermostat, as shown in Figure 20–8, or by a programmable schedule built into a building automation system.

Packaged Boiler

A packaged boiler is a preassembled boiler with major operating components, including the burner, controls, and various other required devices, which are assembled at the factory.

Figure 20–7
Low-water fuel cutoff.

Figure 20–8
Outdoor reset thermostat.

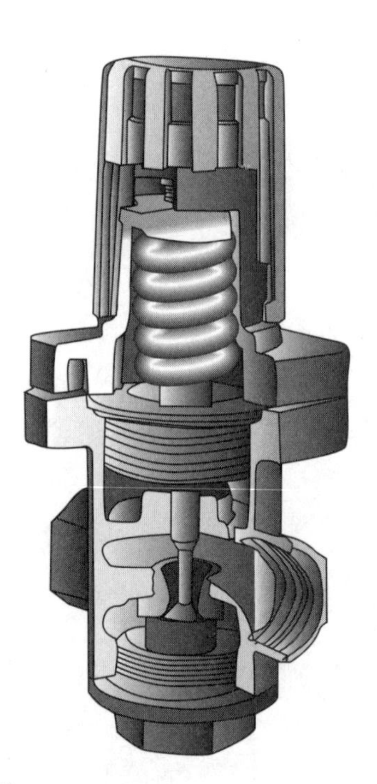

Figure 20–9
Pressure reducing valve.

Pressure Reducing Valve

Pressure reducing valves (Figure 20-9) are usually set to maintain 12 psig, but they are adjustable over a wide range of values to permit connection from the water supply to the heating system, while maintaining the lower heating system pressure at all times. These valves may not operate, however, after long periods of disuse, so always install a bypass hand valve around these units, with both valves downstream of the backflow preventer.

Radiator Valve

A radiator valve is a valve installed on the terminal to control flow through the terminal.

Relief Valve

The relief valve shown in Figure 20–10 is a device that is made to open when the pressure of contained fluid exceeds the setting built into the valve. Because of their unique design, they will immediately pop open and then close quickly after a suitable pressure drop has occurred. The usual relief valve setting for hot water systems is 30 psig, but a large range of settings and capacities are available for any condition likely to be encountered. As with steam safety valves, it is good practice to combine smaller valves (up to 2 inches) to handle the required capacity for a boiler. This is *not* the typical T&P valve found on a residential water heater. It will only open based on pressure, not temperature. They should be tested annually.

Return Piping

Return piping is the part of the system that carries water from the terminals back to the boiler.

Reverse Return

A reverse return system is a two-pipe arrangement in which the first unit connected to the supply is the last one on the return.

Supply Piping

Supply piping is the part of the system that carries water from the boiler to the terminals.

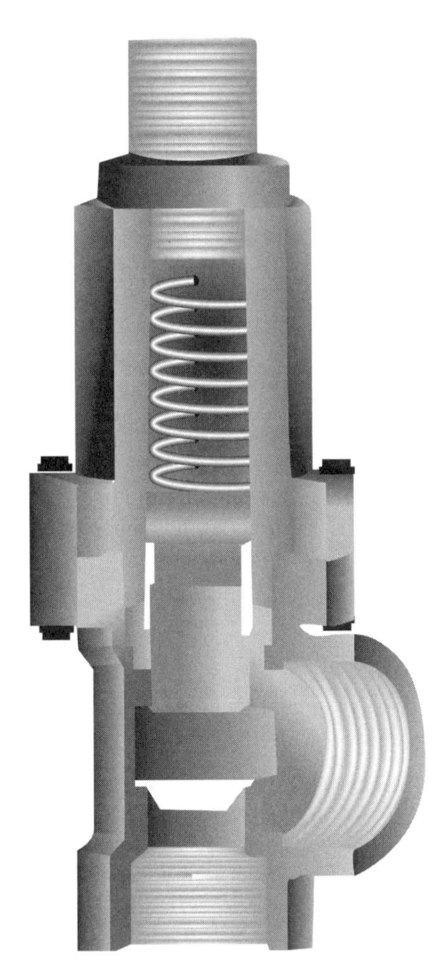

Figure 20–10
Pressure relief valve.

Figure 20–11
This pneumatic zone control valve can limit the amount of water delivered to a circuit.

Terminal Unit

The terminal unit delivers heat to the room. It may be an air handler coil, convector, radiator, baseboard, radiant heater, unit heater, or similar device.

Two-pipe System

In a two-pipe system, supply and return are in separate piping systems.

Unit Heater

A unit heater is a device with a hot water coil, casing, and fan arranged to force air through the coil to heat a building space.

Upfeed System

An upfeed system is a system in which the supply main is below the terminals.

Zone Valve

A zone valve is a power-operated valve that is controlled by the action of a thermostat or thermostatic device. See Figure 20–11.

REVIEW QUESTIONS

1. If the water level gets too low, what will shut off the flow of water?
2. As water in a system heats up and cools down, what accommodates the changes which take place in terms of volume?
3. Do compression tanks and expansion tanks operate at the same pressures?
4. In order for proper flow to occur, what needs to be purged from the system?
5. Name the concept that requires an increase in system temperature when the outdoor air temperature drops.

CHAPTER 21

Forced Hot Water Systems

LEARNING OBJECTIVES

The student will:

- Contrast the different types of forced hot water heating systems.
- Discuss the different heating capacities for various pipe sizes.

NATURAL OR GRAVITY CIRCULATION

Before there were reliable pumps to circulate water, hot water systems relied on natural circulation. As water was heated in the boiler, it would become less dense. This change in density would cause the hotter water to rise. As the hot water rose, the cooler, denser water would move in to take its place. While this type of circulation was inexpensive, it was very difficult to control and had very limited circulation distances. It would be very rare to find this type of system still in operation.

FORCED CIRCULATION

Forced hot water systems use a circulating pump to eliminate the dependency on gravity flow effects to move water through the system. A pump (or circulator) can produce much greater flows than gravity effects. This means much smaller pipe sizes are sufficient to deliver the required heat, and it also means that heating terminals can be located below the boiler or at a significant horizontal distance from the boiler.

SERIES LOOP SYSTEMS

Series loop systems (Figure 21–1) use baseboard radiation units connected in series with one another to form a loop. This loop starts and ends at the boiler. Since the tubing in the baseboard is, in effect, the pipe main, each branch must be kept to the maximum total load that is appropriate to the tubing size used. Table 21–1 shows generally accepted maximum heating limits based on the pipe size of the valve.

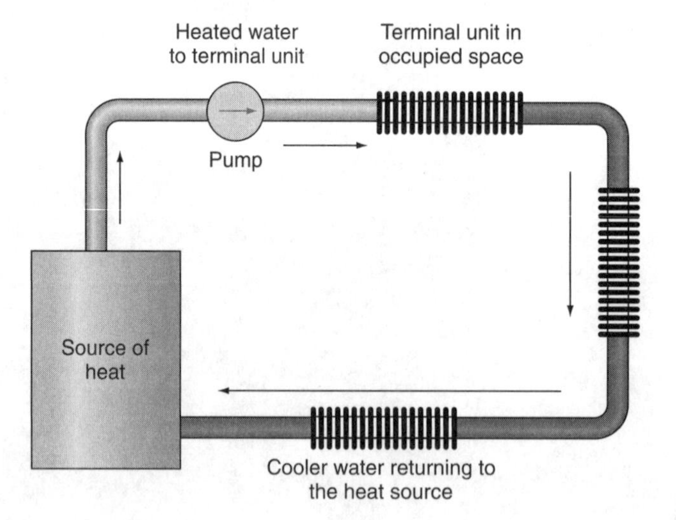

Figure 21–1
All terminal units are connected in series with each other in series loop systems.

Table 21-1 Heating Capacity For Various Pipe Sizes

Pipe Size (Inches)	Maximum Series Load (BTUH)
$\frac{3}{4}$	35,000
1	70,000
$1\frac{1}{4}$	140,000

One-pipe Systems

One-pipe systems (Figure 21–2) were developed in the 1940s to economize on piping installation. These systems work fairly well if all the terminals are similar in flow and pressure drop characteristics, but it is inherently a poor system.

The single main must be full-sized all the way around the building and uses special monoflow tees, so there is very little cost savings when compared to a two-pipe arrangement. In addition, the system offers no flexibility to solve problems, should any arise. Because the main pipe is so large compared to the branch runs, special diverter tee fittings are used to divert flow to heating terminals connected to a single pipe main. Diverter tees are commonly called monoflow tees; they have a cone inside one of the run ports. This cone causes enough flow resistance to divert a larger than normal portion of the flow into the branch circuit.

Flow Rate Through a One-pipe System

The quantity of water needed to flow through a one-pipe system can be calculated rather easily by using the following formula. Suppose a boiler is required to supply 200,000 BTUH with a 20°F temperature drop between the boiler discharge and the boiler supply. The flow rate in gallons per minute can be calculated as follows:

$$\text{Water flow (gpm)} = \frac{Q_T}{500 \times \Delta T}$$

$$\text{Water flow (gpm)} = \frac{200{,}000 \text{ BTUH}}{500 \times 20°\text{F}}$$

$$\text{Water flow (gpm)} = 20 \text{ gpm}$$

Where: QT = total heat required (BTUH)

ΔT = drop in water temperature between building supply and building return

Two-pipe Systems

Two-pipe systems are hydronic piping systems that have a supply and a return main that are separated by the heating or cooling terminals. These systems are the most flexible in design and have more tolerance for variations in demands. A given

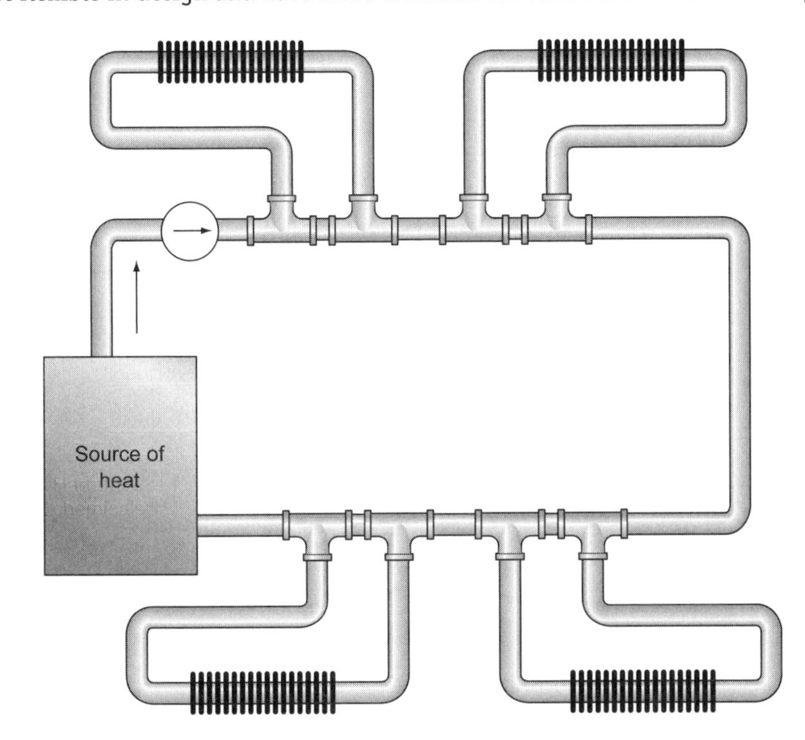

Figure 21–2
A typical one-pipe system.

system may contain any mix of terminal types; however, it is necessary to have similar terminals on any given zone. If one section of the system receives more flow than necessary, balancing valves must be employed to adjust the pressure drop across any terminal to maintain circuit flows and pressure drops in the zone. Two-pipe systems, in general, may be reverse return or direct return.

Two-pipe Direct Return System

In the two-pipe direct return system, all of the heating terminals are in parallel with each other. As shown in Figure 21–3, they are all supplied by a hot water supply line and all discharge into a common hot water return line.

In this arrangement, the first heating terminal supplied will have the shortest supply line and the shortest return line. Additional units will have progressively longer supply and return lines. Since the resistance to flow will increase with the addition of piping, the majority of flow will be to the first unit, then the second, then the third, and so on.

In order to balance the flow to each unit, balancing valves can be added to the beginning of each branch. A balancing valve is a manually operated valve that increases the resistance to flow as it is closed. Once balancing valves have been set and the system is balanced, there should be no need for further adjustments.

By increasing the flow resistance (closing the balancing valve) to the first branch, more flow will be available to the rest of the system. As the branches get further away from the boiler, the balancing valves will be closed less and less. Theoretically, there should be no reason to have a balancing valve in the last run.

Two-pipe Reverse Return System

The two-pipe reverse return system is similar to the two-pipe direct return system, except that the return piping is different. As an attempt to balance the flow resistance, the return piping is rearranged so that the first heating terminal supplied by the supply main will be the last one connected to the return main (Figure 21–4). In this arrangement, the total length of piping to each unit will be very close to equal, so fewer balancing valves should be required. Unfortunately, the reverse return system usually results in a longer total circuit length and is seldom the most efficient way to accomplish satisfactory balancing.

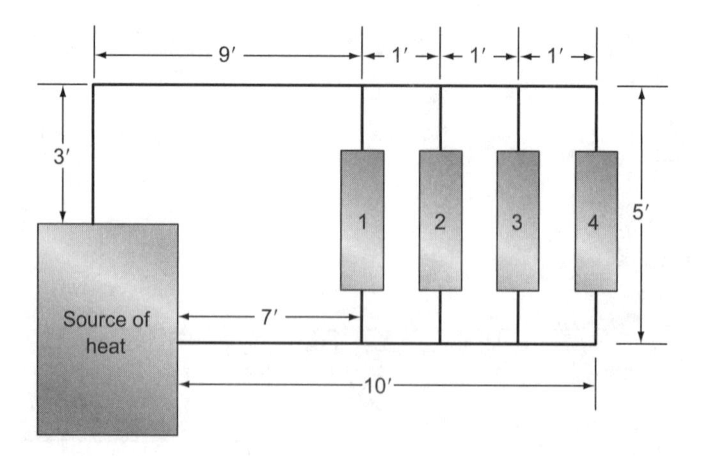

Figure 21–3
Notice that the first unit has much shorter piping length than the last unit when using direct return.

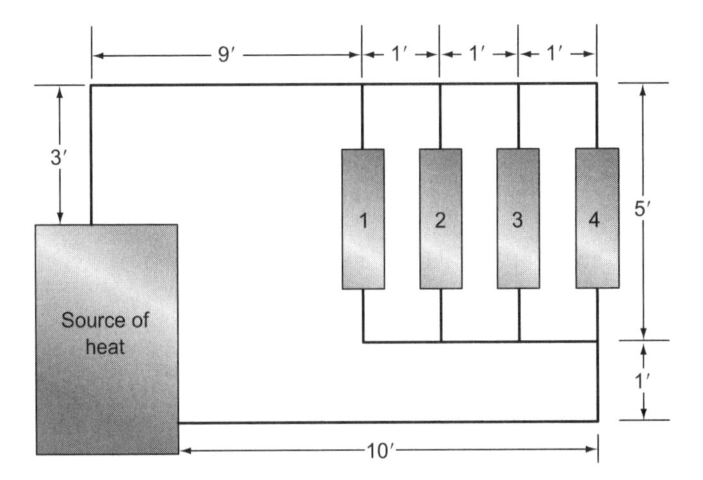

Figure 21–4
With reverse return piping, the piping length to each unit is more nearly equal.

PRIMARY-SECONDARY PUMPED LOOP SYSTEMS

Primary-secondary pumped loop systems have two separate loops. The primary loop is a relatively short loop with a circulating pump that draws water out of the boiler and then returns it back to the boiler. The secondary loop is a branch off the primary loop and contains its own circulating pump. If multiple heating units are required in the secondary loop, they may be arranged in any of the previously mentioned arrangements.

The short section of piping in the primary loop between the secondary loop supply and the secondary return is known as the common piping. It is better to keep this piping section as short as possible; most boiler manufacturers limit this to less than 2 feet. This way, the flow through the secondary loop is more easily controlled by the operation of the secondary loop pump.

ZONING

Zoning is a method of creating a separate loop for each area of a building. By breaking the building into smaller areas, the heat supplied to each area can be adjusted as the load fluctuates. For example, some large buildings with large southern window exposure may not have much demand for heat in those areas, but spaces internal to the building or on the other side may still require heat. In this case, each zone would have its flow controlled by a zone valve. Zone control valves are automatically controlled valves that receive either an electrical signal or a pneumatic pressure in response to a signal from a zone thermostat. The usual arrangement is to interconnect the zone valves in such a way that when all are closed (i.e., no heat is needed anywhere), the boiler burner and circulator are turned off.

INSTALLATION OF FORCED HOT WATER SYSTEMS

Requirements for forced hot water systems include the following considerations:

1. Slope all piping upward in the direction of flow, and provide accessible manual vent devices at high points. In this way, trouble-free circulation is ensured.
2. Flow control valves should be installed so that they prevent gravity flow when the circulator is off. Gravity flow patterns will usually be different

from those produced by the circulator; therefore, gravity flow will result in uneven heating input to the building spaces.

3. A compression tank should be installed to provide air to minimize pressure variations as water volume changes due to changes in system temperature.
4. The minimum operating pressure must be high enough to maintain water at the highest terminals, and low enough to avoid relief valve discharge at highest operating temperatures.
5. Air vents should be installed at each terminal and at system high points. Manual air vents are preferred, as automatic air vents can permit air to enter the system if the pressure goes lower than normal.

PIPING DETAILS

There are preferred piping details and considerations for hot water systems. Typical of these are the following:

1. Keep piping system and branches as simple and short as possible.
2. Consider expansion and contraction in piping design, hanger selection, and hanger location.
3. Slope piping upward in the direction of flow, and install vents at high points.
4. Reducer fittings in supply piping should be eccentric type (unless air venting can be ensured by other means).
5. If a horizontal pipe must be offset vertically around an obstruction. Be sure to locate air vents at high points where the flow drops down.

SYSTEM DIVISION

When dividing the system into two or more circuits, important aspects include the following:

- Avoid bull-headed tees.
- Install balancing valves on the branches, especially if the branches are short.
- Consider the use of zone valves on each branch to improve system performance.

Unit Heater Piping

It is preferable to arrange the flow to travel upward through the heater to achieve natural venting. Install an air vent at the high point of the piping.

REVIEW QUESTIONS

1. ______ pipe forced hot water systems are less adaptable to change and adjustment than ______ pipe systems.
2. Can zoning give better control over the distribution of heat and water flow?
3. Describe the ideal length of the primary loop.
4. What kind of valves are necessary in two-pipe direct return systems?
5. Reverse return systems require less pipe than ______ systems.

CHAPTER 22

Hydronic Heating and Controls

LEARNING OBJECTIVES

The student will:

- Describe the types of units used to distribute heat to a building.
- Describe the controls necessary for different types of hydronic heating.

RADIANT HEATERS

In earlier chapters, we discussed how hot water is heated and how special controls are used to control the boilers and keep them safe. We have also discussed different methods of distributing the hot water through the building. In this chapter, we will discuss how the different terminal units dissipate heat to their surroundings. We will also discuss some of the controls used to control the amount of heat that is transferred.

Radiator Units

Most people have probably seen the large cast iron radiators that were popular terminal heating units in early construction. Because these units were made of cast iron, they were slow to heat up and slow to cool down. One advantage of this was that the heat was gradually supplied to the space, so that it was not very noticeable. The main disadvantage was that they were slow to respond to load changes. Their mass did allow for quiet operation. Figure 22-1 shows a typical cast iron radiator.

Radiators work by a combination of radiation and convection. Hot water moves from the distribution piping into the radiator, which in turn warms up the radiator. The radiator then radiates heat out into the space. As the radiator gives off heat, the surrounding air warms and becomes less dense. The warm air rises and the cooler ambient air comes in to take its place. This causes air currents to move up the wall and transfer heat into the space. Typically, water supplied to radiators is at temperatures less than 180°F.

The heat output of this style of radiator is determined by these three factors:

1. The temperature of the surrounding air, which is usually assumed to be 70°F.

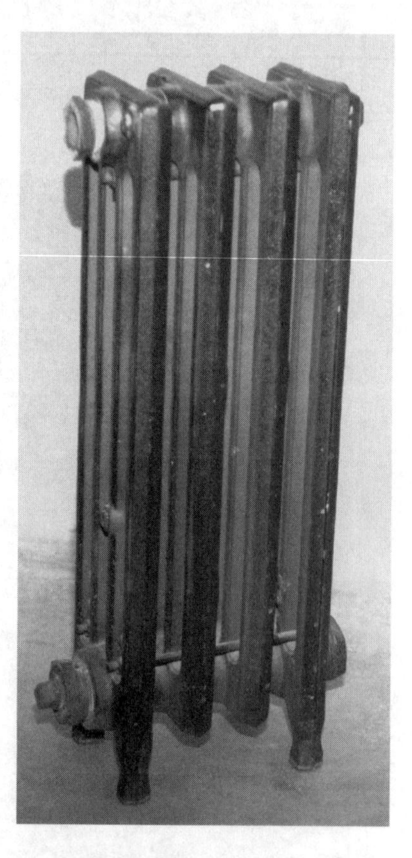

Figure 22–1
Cast iron radiators give a good steady flow of heat to the conditioned space. Courtesy of Bill Johnson.

2. The temperature of the radiator's surface, which is directly related to the temperature of the water or steam inside the radiator. Understandably, the hotter the radiator gets, the more heat it will radiate.
3. The surface area of the radiator. The larger the radiator, the more heat it can radiate. A radiator's relative radiating surface area is measured in terms of square feet of "Equivalent Direct Radiation," or "EDR."

Finned-tube Baseboard Units

In order to have a more rapid response to heating load changes, finned-tube baseboard units were constructed. These units typically consist of a section of copper tubing with aluminum fins attached along its length. Even though these units are built with materials that conduct heat at a much faster rate, they still require a certain amount of radiating surface area. Designers came up with the idea of making them longer and much lower in profile. This means they take up less wall area while still distributing heat into the space. To conceal and protect the finned tube, the assembly is placed within a sheet metal casing. The metal housing is also equipped with dampers so that the amount of air moving across it can be controlled. They are generally available in sections 2 ft to 8 ft in length.

Finned-tube baseboard units are typically rated by the amount of BTUH that each linear foot can dissipate. Manufacturers usually publish charts that list the BTU/hr/ft rating of each model, based on an expected design temperature difference. The system designer can then use the expected ambient temperature and the expected hot water temperature to calculate the required length of the heater needed.

Radiator Controls

Whether a cast iron or finned-tube radiator is used, the only parameter that can be controlled individually at the terminal unit is the rate of flow of the water. As mentioned earlier, the temperature of the boiler water can be controlled by an outdoor reset schedule, but this would affect the entire building. In many cases, a thermostatic radiator valve is used. When this valve senses an increase in the ambient temperature of the space it throttles to close, which restricts the flow to the radiator. Ideally, this valve will reach a set point that will balance the heat loss of the room and remain in that position. This will give a continuous flow of heat to the building space and prevent the boiler from cycling on and off.

Fan Coil Units

Fan coil units are similar in construction to an air conditioning evaporator or condensing coil. The coil consists of several horizontal tubes that allow water to circulate back and forth across an air stream. The horizontal tubes are connected together with thin fins, usually made of aluminum. The aluminum fins give the coil the necessary surface area to dissipate heat into the building air. A fan is mounted to blow air across the coil. The fan and coil arrangements can be enclosed in a metal housing that contains ductwork to mix in and filter the outdoor air as well. See Figure 22-2. They can also be designed for use in large shop areas; these units will be capable of throwing the air much farther. See Figure 22-3.

Fan Coil Controls

In the case of fan coil units, there are more choices of parameters to control. One option is to leave the fan on at a continuous speed and vary the temperature of the coil by controlling the flow rate of the water. This can be accomplished with a properly sized control valve like the one shown in Figure 22-4. This type of valve can be remotely controlled by a pneumatic or electric thermostat or by a computer-based building automation control system. In the case of the remote thermostat, the thermostat can be mounted on the wall or in the return air ductwork. The idea is for the thermostat to control the flow of water to the fan coil and thus to control the discharge air.

Figure 22–2
This fan coil unit is controlled by a computerized building automation system.

Figure 22–3
This fan coil unit has no flow control valve for the hydronic water. The two-speed fan is cycled on and off by a thermostat.

Another option is to leave the flow of water constant and cycle the fan on and off to control the room temperature. This option works well, except that on days of moderate temperatures, the air within the space can become stale due to the lack of circulation.

Figure 22–4
This is a pneumatically controlled two-way control valve.

Radiant Heating

Radiant hydronic heating is considerably different from the hydronic convective heating that we have been discussing until now. In the case of fan coil units, the heat from the hot water was passed to the building air. The air, in turn, heated the occupants and the building. In radiant heating, the idea is to heat the building shell to a temperature around 85°F. The building structure will then radiate heat to the occupants within.

Radiant heating systems usually consist of long runs of plastic tubing embedded within the floors, walls, and in some cases, the ceiling. The most common type of tubing is PEX-AL-PEX, two layers of cross-linked polyethylene with a layer of aluminum between them. The layer of aluminum is an oxygen barrier to prevent the migration of oxygen into the system. This is important because oxygen will cause the metallic components in the system to corrode.

Radiant Floor Designs

The most common location for radiant heating systems is in the floor. Depending on the time and location, the installation of the floor heating system can change. In new construction, the piping will be embedded in a concrete slab. The slab can be on grade: the first floor for single-story buildings, or in the basement for buildings with basements. In either case, a layer of insulation and a vapor barrier are installed between the slab and the ground. This prevents the majority of heat loss to the ground. As shown in Figure 22–5, the tubing is secured to wire mesh and then both are embedded in concrete.

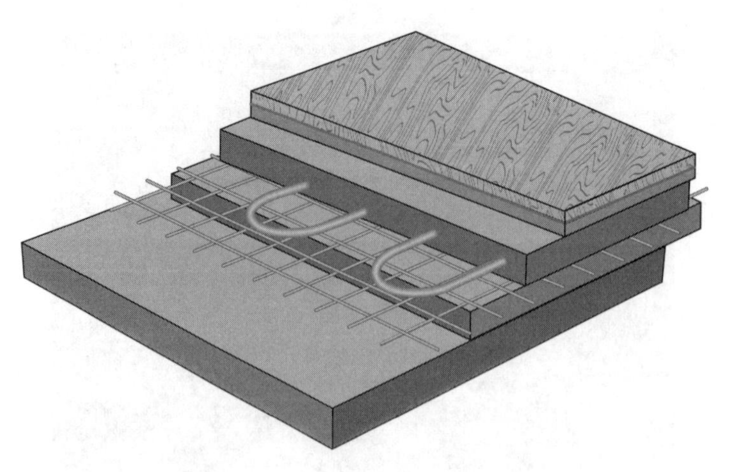

Figure 22–5
Cutaway of a slab on grade radiant piping system.

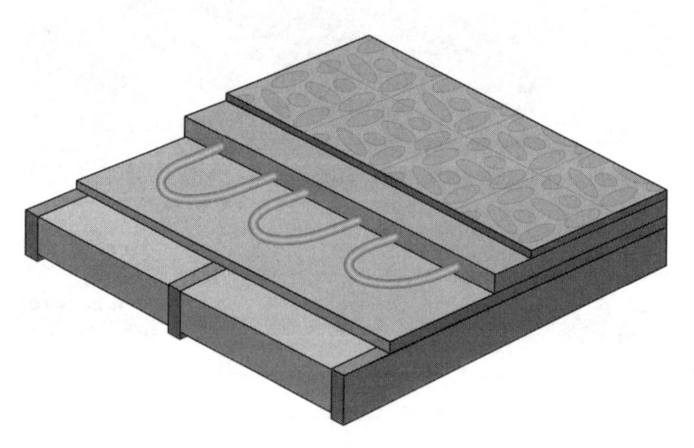

Figure 22–6
Cutaway of a thin-slab radiant piping system.

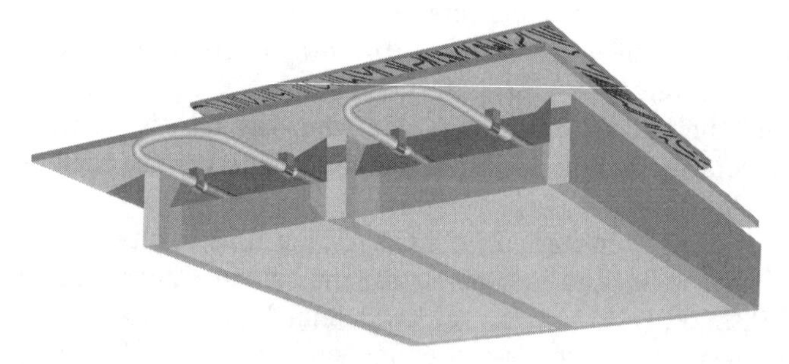

Figure 22–7
Cutaway of an under-mount radiant piping system.

On floors between stories or on pre-existing framed floors, a thinner slab of concrete is poured over plywood subflooring. In this case, the tubing is stapled directly to the subflooring and then concrete is poured to a height of at least $\frac{3}{4}$ inch above the tubing. As shown in Figure 22–6, the space between joists is heavily insulated if the space below is unconditioned.

When the flooring in a pre-existing building can not be covered or removed, a third option can be used. As shown in Figure 22–7, the supply tubing is stapled into the subflooring from below. Again, insulation is installed below the tubing. This is probably the least efficient of the three choices.

In any of the three installation options, the pattern in which the tubing is to be laid must be chosen. Conventional theory says that the hottest water should be

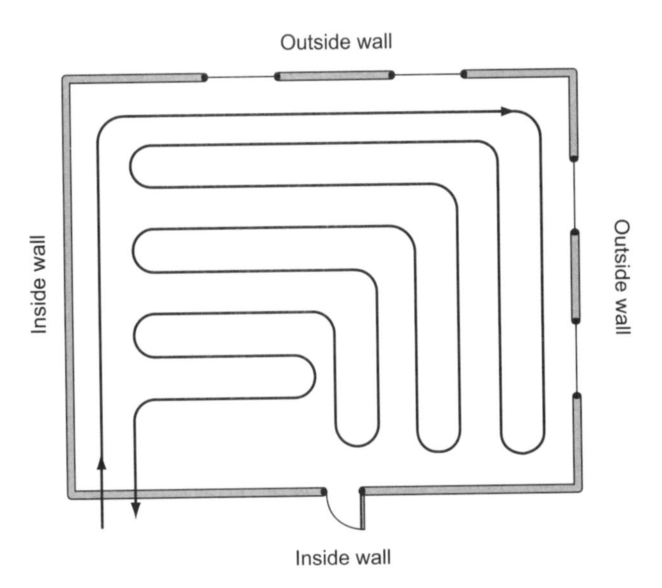

Figure 22–8
The water temperature will decrease as it flows inward toward the area of lower heat loss in this serpentine pattern.

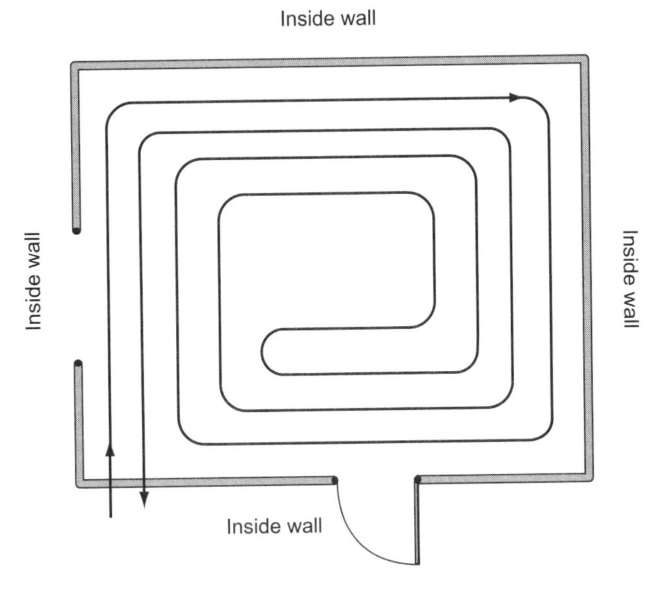

Figure 22–9
This pattern provides the most even distribution of heat.

sent to the biggest heat loss, which is usually an outside wall. Figure 22–8 shows two variations of a serpentine pattern, in which the water coming from the boiler is directed toward the outside wall(s). When there are no outside walls, the counterflow spiral pattern gives the most evenly distributed heat transfer (see Figure 22–9). The spacing between the tubing runs depends on the slab thickness, water temperature, and tubing size.

Radiant Floor Controls

The primary method of controlling the temperature in a radiant floor system is to control the rate of flow of water to each zone. As with fan coil units, a thermostat can be used to monitor the temperature of each conditioned space. The thermostat will send an electrical signal back to the control valve which will open and close to control the flow of water in that loop. Figure 22–10 shows a typical radiant floor system manifold with control valves directly mounted to the distribution manifold.

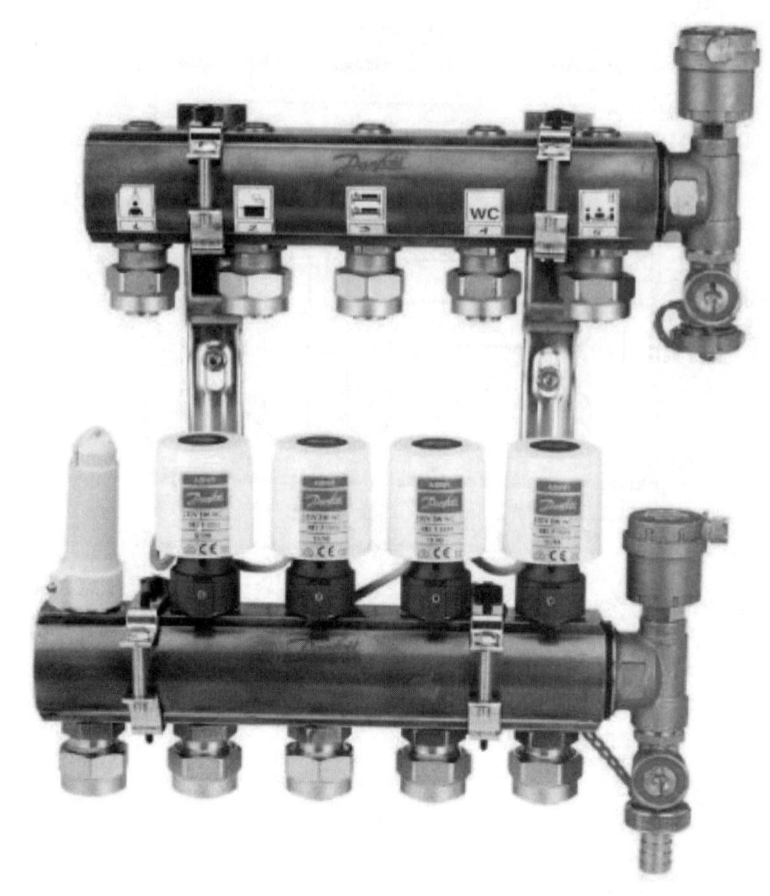

Figure 22–10
Distribution manifold for a radiant floor heating system. Courtesy of Danfoss.

REVIEW QUESTIONS

1. Explain how cast iron radiators handle sudden heating load changes.
2. Does each radiator control the temperature of the water flowing through it?
3. Is copper tubing the preferred material for radiant floors?
4. Explain how fan coil units operate, pull in, and circulate air.
5. Can radiant floor heating systems be installed in old construction?

CHAPTER 23

Forced Air and Humidification

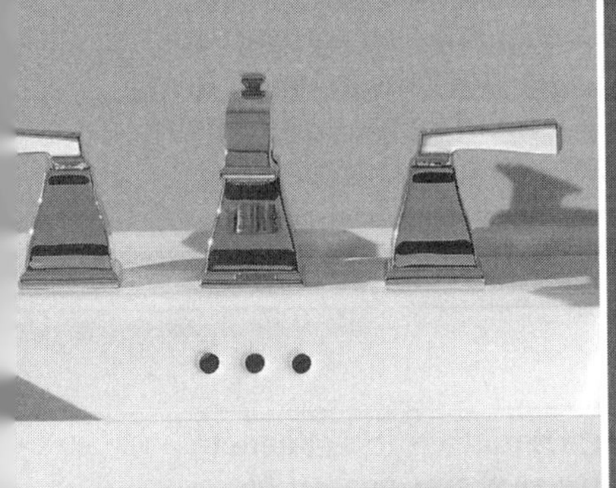

LEARNING OBJECTIVES

The student will:

- Describe forced air systems.
- Discuss basic principles of humidification.

FURNACES

The principles of forced air heating systems must be understood by anyone attempting to service or make any adjustments to these systems. In many states, individuals who are engaged in the installation or service/repair of heating systems must be licensed to perform these services. Furnace manufacturers offer efficient, safe, and reliable products for today's market. The following are the five different configurations offered:

1. *Upflow/highboy* draws return air in at the bottom of the unit and supplies air out of the top.
2. *Downflow/counterflow* draws return air in at the top of the unit and supplies air out of the bottom of the unit.
3. *Horizontal* has airflow entering on one side and exiting out of the other. This airflow direction is determined by the manufacturer. This model is unique to the Carolinas because of crawlspace construction; this model furnace could also be installed in attics.
4. *Lowboy* models were popular in areas that could build basements and use the unit in that location. Lowboy units are lower in height than an upflow type furnace, but these two units operate in the same way.
5. *Package/rooftop* are self-contained units. These units are designed for outdoor installations in buildings in which interior space is not available.

The type of fuel for the furnace will typically depend on what is available. Where it is available, natural gas can be supplied through a series of gas pipe lines from the local supplier directly to the point of use. In areas where natural gas is not available, LP and fuel oil must be supplied by on-site storage tanks. In some cases, such as hospitals, LP or fuel oil may be used as a backup to the primary heat source. Electricity may be used as a source of heat or as a backup for use with a heat pump. The heat pump is available in all of the same configurations as the other units and uses a reverse cycle refrigeration system for the source of heat.

Furnace Ratings

Furnaces are rated in terms of MBH (thousands of BTUH). Manufacturers adhere to guidelines set by government agencies or private groups that test and set these standards. The American Gas Association, AGA, and Underwriters Laboratories are examples of these agencies.

Presently, federal guidelines require manufacturers to produce fossil-fuel burning units that have an efficiency rating of at least 80 percent. AFUE stands for Annual Fuel Utilization Efficiency. It is a measure of what percentage of the fuel's heat content is converted into usable heat. For example, a 90 percent AFUE means that for every BTU worth of gas used, the system will provide 0.9 BTU of heat. The higher the AFUE of the unit, the more efficient it will be.

Ductwork

Ductwork is used to deliver conditioned air to the space (supply duct) and return the air (return duct) to be recirculated and reconditioned through the unit. The furnace is often referred to as the air handler. The air is filtered either at the air handler or with a duct-mounted return air filter grill. Air filtering is very important because it helps keep indoor air quality higher; it also keeps the system cleaner, which helps with heat transfer and system efficiency.

The duct system has many configurations, all consisting of a main duct, branch ducts, and various designs of supply registers and return grilles. Supply registers and return grilles may be floor-, ceiling-, or wall-mounted. The designer has a wide range of options.

The ductwork may be constructed of a variety of materials; sheet metal is still the most commonly used. When the need to insulate ductwork came about, designers turned to ductboard, a stiff material that they could use to make the duct and at the same time satisfy the insulating requirement. Because of high moisture content in most crawl spaces, ductboard should never be used in a crawlspace. Another popular product is flexible duct, which is a thin plastic tube with a helical wire support. Both the plastic tube and support wire are insulated with fiberglass and a thin aluminum foil outer shell. It comes in 25-foot lengths and should be pulled straight and tight.

Furnace Venting

Furnaces are referred to as condensing or noncondensing furnaces. Manufacturers of older models of furnaces realized that a lot of the heat produced by the burning fuels went up the chimney, wasted to the outdoors. If these flue gases could be retained or the flow rate slowed through the heat exchanger, then more of the burning process would be used to heat the conditioned space and not wasted. One of the results of this was a reduction in flue gas temperatures. This caused the flue gases to reach their dew point temperatures, which resulted in condensation forming in the chimney or vent system. Figure 23–1 shows a condensing furnace.

Condensing furnaces rated at an efficiency of 90 percent or higher have a flue gas temperature of approximately 105°F. The reduction in flue gas temperatures causes the gases to reach a dew point temperature, forming condensation in the vent system. A condensate drain must be added to the vent system. The vent system is more complex than in the past and manufacturers' guidelines regulating the size and length of the vent and the termination point must be followed. The lower vent temperature allows the use of PVC pipe and also makes it possible to terminate the vent horizontally. It is important for the technician to place the vent exit away from openings in the building to prevent flue gases from entering those openings.

Noncondensing furnaces rated at 80 percent efficiency have a vent temperature of approximately 550°F. The resulting flue gas temperatures are well above their dew point temperatures. These furnaces are referred to as fan-assisted flue gas furnaces. Manufacturers' venting tables must be followed when installing vent systems for these furnaces. When these furnaces were originally introduced, it was thought that they could be vented and terminated horizontally. This failed, mostly because of the vent materials used, and the practice was stopped. Vent tables also must be consulted when venting in common with other appliances. For further information concerning appliance venting, refer back to *Plumbing 301*.

Furnace Controls

Furnace controls have evolved from the simple on-off switch used as a thermostat, with little or no means of heat anticipation, to the high-efficiency programmable thermostats that we use today. Missing from current furnace controls are the thermocouples, stack switches, and the hard-to-adjust fan and limit switches. Replacing these controls are computerized control boards. When the thermostat sends a signal to the board that there is a call for heat, the board does a precheck of the furnace control system to make sure all controls are in the correct position. Listed below is a step-by-step startup and shutdown procedure of a more current gas furnace.

1. Thermostat sends signal to control board.
2. Control board completes pre-startup check, completes circuit to vent fan motor.
3. Vent fan pressurizes vent system, causes vent pressure switch to make and does a pre-purge of the heat exchanger and vent system. Also checks condensate drain.
4. Pressure switch signals control board to try for ignition. Control board starts electric spark and timed gas flow to burner section. Flame proofs send

Figure 23–1
This gas-fired 90 percent efficient condensing furnace is equipped with a fresh air intake so no indoor air is lost to combustion.

a signal back to control board. This process is call flame rectification. The flame touches the metal of the burner and grounds the circuit. Over-temperature switches and a flame rod provide safe operation of this process.
5. Furnace blower starts after a timed sequence.
6. Furnace runs until thermostat is satisfied.
7. Control board opens circuit to burner section. Vent fan runs to complete a post-purge of the heat exchanger and vent system. Furnace blower remains on for a preset time.

The control board of many of today's furnaces is equipped with lights that are designed to light up in a particular sequence that alerts the service technician to a problem.

The more current oil furnace uses a control that replaces the stack switch. This control senses light instead of heat. When the thermostat sends a signal to the burner section for heat, a timed sequence begins. A motor starts that drives a fuel pump and combustion air fan. An ignition transformer that provides a spark is energized. A photoelectric cell or cad cell that is sensitive to light begins to look for the flame. Once the flame is established, a blower motor starts, and warm air is delivered to the conditioned space. The standard oil burner pump pressure has increased from 100 psig to over 200 psig on some burners. The oil furnace is a dependable, efficient unit. These conditions depend on the original start-up of the furnace. If the installation technician fails to make the proper startup checks and adjustments to the oil furnace, the result is a dirty flame that produces soot. This soot becomes an insulator, blocking the heat exchanger passages and producing an inefficient unit.

The checks and adjustments are:

1. Check for correct burner nozzle.
2. Check oil pump pressure.

3. Check electrode alignment.
4. Check vent temperature.
5. Check smoke.
6. Check CO_2.
7. Check burner efficiency.

Figure 23–2 shows a flue gas analyzer that can be used to make adjustments to the burner.

ELECTRIC FURNACES

The electric furnace has improved heater elements that offer longer life than the earlier models. The electric furnace can claim higher efficiency because all of the heat that it produces goes directly into the conditioned space. With an electric furnace, there is no need for a chimney or combustion air intake because there are no burned fuel byproducts.

BUILDING HUMIDITY

Maintaining a comfortable humidity level in a conditioned space during the heating season is a difficult task. The colder the outdoor air, the lower the dew point temperature of the air will be. At these conditions, the air is saturated and has plenty of moisture, but when this air is heated, the humidity level drops and the conditioned space becomes uncomfortable.

Moisture can be added to the conditioned space by the use of humidifiers. These devices add water vapor to the conditioned space. The units can be placed directly in the area to be humidified or can be added to the duct system or mounted directly to the furnace.

Humidifiers can be controlled by a humidistat, a control that reacts to humidity change, much in the same way a thermostat reacts to heat change in a conditioned space.

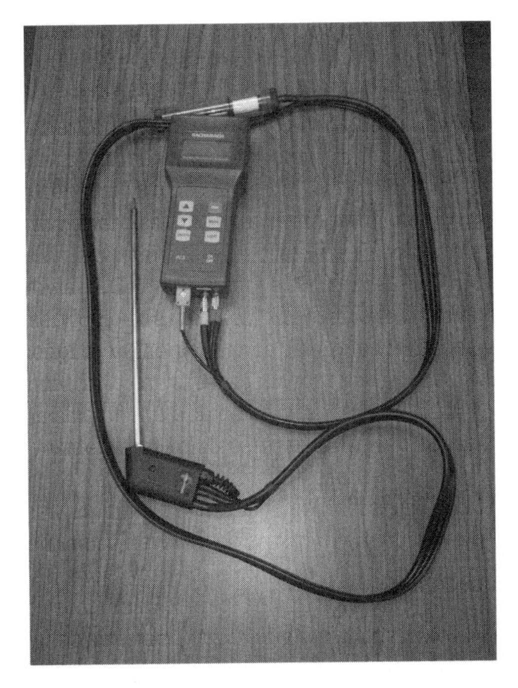

Figure 23–2
This flue gas analyzer can be used to make adjustments to the burner.

Humidifiers work well when the water supplied to them is mineral-free. Minerals cause problems when they separate from the water and clog up the media that is used to transfer the water to the air. The nozzle of the humidifier spray becomes blocked and the water is no longer a mist going into the conditioned space. Lime in the water will show up as a white dust on polished furniture.

Increasing the humidity level in the conditioned space can cause problems, with condensation forming on window sills and any other areas where outdoor air might be able to infiltrate.

The amount of water that will need to be added to the conditioned space can be determined by the use of the psychrometric chart or several charts available from the manufacturers of humidifiers or the American Society of Heating, Refrigeration, and Air Conditioning Engineers (ASHRAE).

Example of Sizing Procedure

Example

Estimate the air changes for the house. A tightly sealed, well-insulated house will have $\frac{1}{2}$ air change per hour. An average house will have $1\frac{1}{2}$ air changes per hour.

The formula for required moisture is:

$$H = \frac{V(W_2 - W_1)R}{8.345}$$

where

H = water required in gallons per hour

V = building volume in cubic feet divided by 1000

W_2 = pounds of water per 1000 cubic feet of air at final conditions (indoors) (Table 23-1)

W_1 = pounds of water per 1000 cubic feet of air at initial conditions (outdoors) (Table 23-1)

R = air changes per hour

8.345 = weight of water (pounds per gallon)

The house in the following example has 8 foot ceilings and an area of 1500 square feet. The desired condition indoors is 68°F with 35 percent relative humidity. How much moisture must be added if the outdoor conditions are 0°F with 50 percent relative humidity?

$$V = (volume)/1000$$

$$Volume = 8 \times (1500) = 12000$$

$$V = \frac{12000}{1000} = 12$$

$$H = \frac{12(.0381 - 0.029)1.5}{8.345} = 0.76$$

$$H = 0.76 \text{ gal/hr or approximately 18 gallons per day}$$

Psychrometric Chart

Determining the amount of moisture that must be added to the conditioned space can also be accomplished by referring to the psychrometric chart. See Figure 23-3. The psychrometric chart and its use were introduced in *Plumbing 301*. To use the chart, follow the steps below.

1. Measure the volume of the conditioned space in cubic feet.
2. Determine the dry bulb temperature and relative humidity required for the conditioned space.

Table 23-1 Humidity Temperature Chart. Courtesy of Plumbing–Heating–Cooling–Contractors—National Association.

Temp.	% Relative Humidity												
°F	20	25	30	35	40	45	50	55	60	70	80	90	100
−10	.008	.010	.014	.014	.016	.018	.020	.022	.024	.028	.032	.036	.040
0	.012	.015	.018	.021	.024	.027	.029	.032	.035	.041	.047	.052	.059
10	.020	.025	.029	.034	.039	.044	.049	.054	.059	.069	.078	.088	.098
20	.031	.039	.047	.055	.063	.071	.079	.087	.095	.110	.126	.142	.161
30	.051	.064	.077	.089	.103	.116	.128	.141	.154	.180	.203	.232	.258
40	.077	.097	.117	.135	.155	.174	.193	.213	.233	.272	.310	.358	.390
50	.123	.141	.171	.198	.227	.259	.284	.312	.341	.399	.456	.521	.574
60	.162	.205	.246	.288	.329	.370	.412	.453	.495	.579	.633	.747	.831
62	.176	.220	.264	.308	.353	.398	.442	.486	.532	.622	.711	.801	.893
64	.189	.236	.284	.332	.379	.427	.475	.524	.572	.668	.764	.863	.960
66	.203	.254	.306	.356	.407	.458	.510	.561	.613	.717	.821	.926	1.030
68	.217	.272	.326	.381	.436	.491	.547	.602	.657	.768	.880	.991	1.106
70	.232	.291	.349	.408	.467	.527	.585	.645	.704	.824	.944	1.063	1.186
72	.249	.312	.375	.438	.501	.564	.628	.691	.755	.884	1.013	1.139	1.273
74	.267	.334	.401	.469	.536	.605	.671	.740	.809	.947	1.083	1.224	1.364
76	.285	.357	.429	.502	.574	.647	.720	.792	.886	1.012	1.160	1.310	1.461
78	.305	.382	.458	.535	.614	.692	.769	.848	.925	1.082	1.241	1.400	1.564
80	.326	.408	.490	.574	.656	.740	.823	.906	.991	1.160	1.330	1.500	1.676
82	.348	.436	.534	.612	.701	.790	.881	.970	1.060	1.240	1.422	1.609	1.792
84	.371	.464	.558	.654	.749	.844	.939	1.033	1.129	1.322	1.520	1.710	1.916
86	.389	.496	.597	.698	.799	.901	1.003	1.105	1.212	1.418	1.627	1.833	2.048
88	.422	.523	.635	.745	.851	.960	1.068	1.178	1.288	1.510	1.731	1.958	2.189
90	.450	.563	.678	.793	.908	1.024	1.141	1.258	1.375	1.613	1.853	2.095	2.338

- Pounds per 1000 cubic feet of air (based on standard conditions)
- Moisture Content* (Courtesy of ASHRAE)

3. Plot these conditions on the psychrometric chart. In this example we will use 70 degrees dry bulb and 40 percent relative humidity. At these conditions, we can read on the chart that a pound of air in the conditioned space occupies 13.5 cubic feet per pound and that this air can hold 46 grains of moisture per pound.
4. Once we plot the desired indoor conditions, we now have to refer to the ASHRAE guide or another recognized text to determine the average heating season outdoor conditions. For our example we will use 30 degrees dry bulb and 20 percent relative humidity. The chart shows us that the outdoor air contains 6 grains of moisture per pound of dry air.
5. The next step is to measure the internal volume of the conditioned space. For our example, we measure the interior height width and length. We will use a dwelling of 1200 square feet, with a ceiling height of 8 feet. This gives us a total of 9600 cubic feet. If we refer back to the psychrometric chart, we find that each pound of this air occupies 13.5 cubic feet. This gives us a total of 711 pounds of dry air.
6. The difference in moisture content of the outdoor air and the air in the conditioned space is 40 grains of moisture per pound. If we multiply this

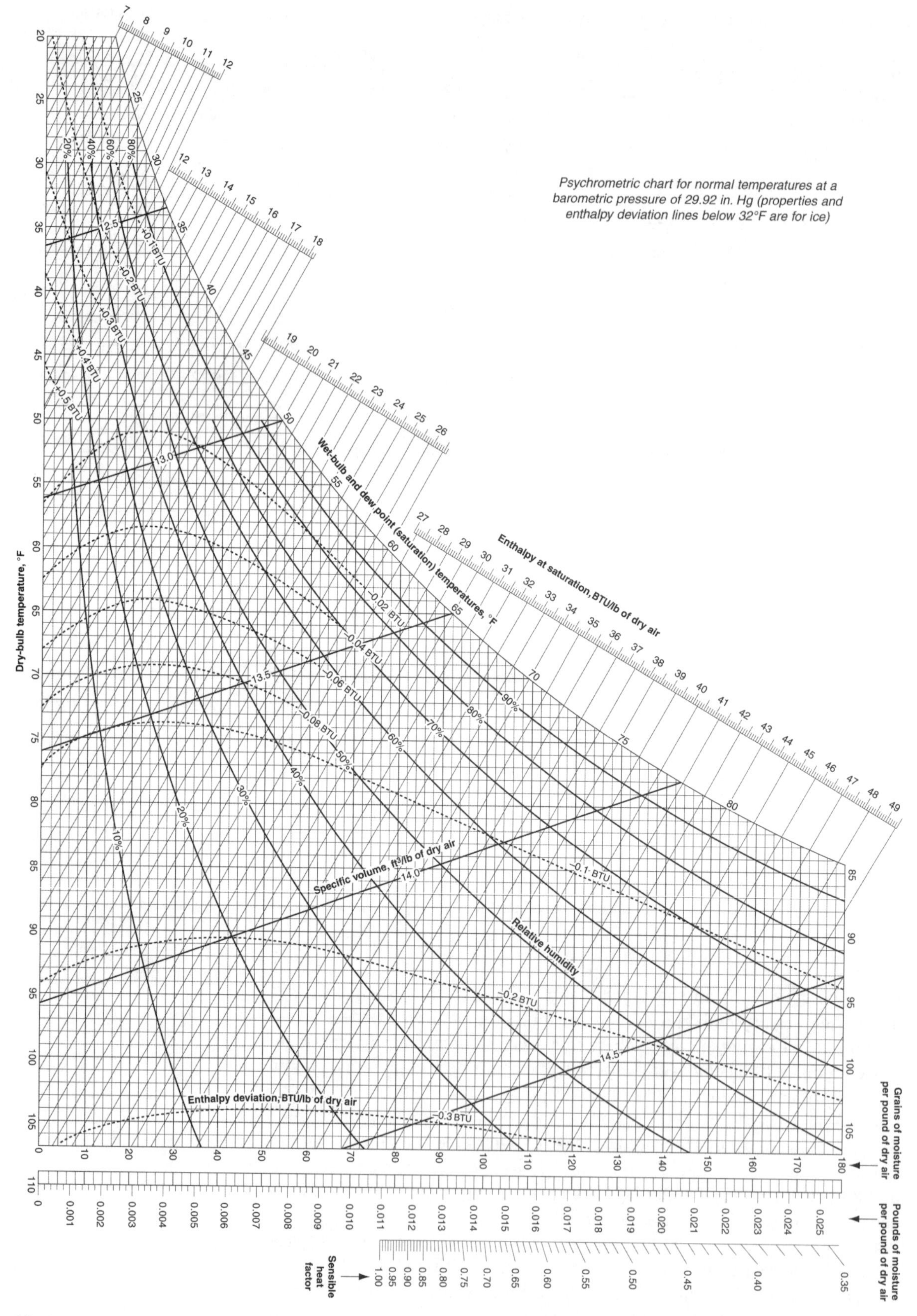

Figure 23–3
Psychrometric Chart. Courtesy of RSES.

difference times the number of pounds of dry air, we can determine the amount of moisture that the humidifier will have to add to the conditioned space. 711 pounds of dry air times 40 grains of moisture per pound equals 28440 grains of moisture.

7. Humidifiers are rated in gallons per hour. There are 7000 grains in one pound of water and water weighs 8.34 pounds per gallon. If we divide 28440 grains by 7000, we determine the number of pounds of water needed. 28440 divided by 7000 equals 4.06 pounds of moisture,
8. To determine the gallons of moisture needed, divide the 4.06 pounds by 8.34 to determine the gallons of moisture required.

REVIEW QUESTIONS

1. What is the minimum efficiency of a fossil fuel–burning furnace?
2. What is formed in the heat exchanger of a high-efficiency furnace that requires a drain?
3. What control replaced the stack switch in the more current oil furnace?
4. What is used to provide humidity in the conditioned space in the heating season?
5. Name a chart that can be used to determine the amount of moisture in a conditioned space.

CHAPTER 24

Solar Systems and Conservation Methods

LEARNING OBJECTIVES

The student will:

- Describe how solar heating systems can be applied to hydronic systems.
- Describe basic methods and devices for conservation in heating systems.

SOLAR ENERGY AS HEATING INPUT

In Chapter 22, the principal source of heat for hydronic heating systems was the burning of fossil fuel. In this chapter, we will discuss solar energy as a possible heating input. While it is a free source of heat, every possible step to improve efficiency should be taken.

Since a hydronic solar system is simply another form of a hydronic heating system, all the considerations of piping, air removal, and liquid flow that we have studied must be observed. Figure 24–1 shows the piping and equipment arrangement typical of a hydronic solar space heating system.

The collector array receives solar radiation and converts it to thermal energy, which is stored in the form of heated water, usually in a large, well-insulated tank. The collector (called a flat-plate collector) is normally mounted on the roof of the building where it will receive sunlight for the majority of the day. The collector is mounted facing south at approximately a 30° angle to receive the maximum amount of solar radiation. More advanced systems have a motorized tracking system that adjusts the angle of the collector to follow the sun. An insulated tank is traditionally located at a convenient place in the building. Depending on the amount of heated water needed, the tank may be an insulated tank that resembles an electric water heater or a custom-built tank constructed on-site. Where space in the building is limited, the tank can be buried outside, well below the frost line. Purchased tanks like the one shown in Figure 24–2 have an internal heat exchanger and some form of backup heat source.

To increase efficiency, it is important to insulate all of the piping and valves. Full port isolation valves should be added to aid in the service or repair of any of the components.

A circulating pump is used to develop the necessary fluid flow for the desired heat transfer. These pumps are normally small, fractional-horsepower pumps. The pump is started and stopped by a controller that monitors both the collector temperature and the storage tank. When the controller senses that the collector is hotter than the storage tank, it will turn on the pump. The pump will continue to run as long as the storage tank is cooler than the collector.

In the Field

Since the sun cannot be turned off, any closed loop portion must have a temperature and pressure relief valve to maintain safe conditions.

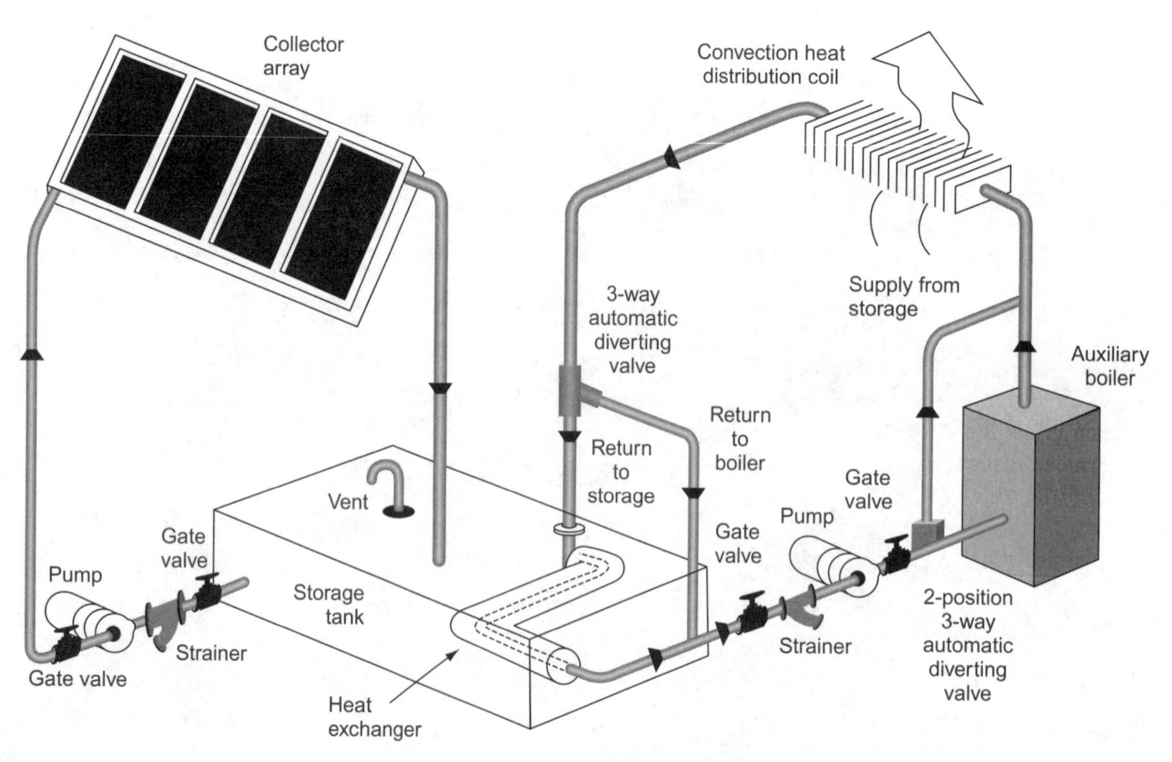

Figure 24–1
Solar hydronic heating system.

Figure 24–2
Commercially available storage tank with backup heat source.

In Figure 24–1, the storage tank is vented, allowing the water in the collector to drain back whenever the power fails or during the night. This feature makes sure that the collector water does not freeze or cool down during non-operating times.

A heat exchanger can be located in the storage tank whenever any of the following conditions warrant its use:

- Building loop pressure is too high for the collector array.
- A drain back system is used.
- Potable water system contains hard water.
- A mixture of antifreeze is used to protect the collector from freezing.

The heat exchanger joins the collection and distribution parts of the system together. To protect the potable water system, a double wall exchanger is used whenever an antifreeze mixture is needed.

The distribution system is composed of piping, valves, circulator, and terminal units that deliver the heat energy to the space to be heated. The system shown in Figure 24–1 includes an auxiliary boiler. In smaller systems, it may simply be a typical residential water heater. The boiler heats the water to the final required temperature. Thus, the solar equipment can provide the heating as long as possible, with aid from the boiler when needed. This arrangement provides the best choice of capital cost and operating savings. A solar system sized to provide the full heating load on the coldest day will be oversized on any day of lesser load. The best approach is to obtain 50 to 60 percent of the required capacity from the solar portion, and the balance from the boiler.

The mix of sizes does not mean that the operating energy will be in the same ratio. This ratio will be closer to 75 percent to 85 percent for the solar. No matter what the design temperature for a region is, extremely low temperatures occur only for very few hours per heating season.

Figure 24–3 shows a different way to use solar energy. The collector system is the same as the one described above, but the terminal unit in the distribution system is a water source heat pump. The fluid from the heat exchanger in the storage tank is piped to a separate heat exchanger. This second heat exchanger allows the heated water to give off its heat energy to the cooler refrigerant of the heat pump.

Traditional heat pumps extract heat from the outdoor environment during the heating cycle. This process works extremely well until the outdoor temperatures drop below 35°F. With the solar system acting as substitute heat source, the heat pump can operate even under the most extreme design conditions. Solar system efficiency and effectiveness increase rapidly as fluid temperature goes down, so this scheme makes solar payback much better.

Figure 24–4 shows another variation of how solar heat can help supplement an existing heating system. The heated fluid from the heat exchanger is pumped through a heating coil placed in the return ductwork to a furnace. Instead of the hydronic boiler auxiliary in Figure 24–1, the furnace is the supplemental heating system. (Only the blower is required to distribute the heat from the solar coil.)

In this type of system, a two-stage thermostat located within the condition space is used to control the system. When the building begins to cool down, the thermostat turns on the circulating pump, which circulates the building loop water through the heat exchanger and the heating coil in the ductwork. It also turns on the indoor blower to circulate the indoor air. Under light heating loads, this will be enough to heat the building, so the fossil fuel heater never turns on. If there is more heating demand than the solar heating system can supply, the building temperature will continue to drop. The additional temperature drop will activate the second-stage heat, which in this case is the furnace. Now both the solar heating system and the furnace will work to heat the building until the indoor thermostat is satisfied.

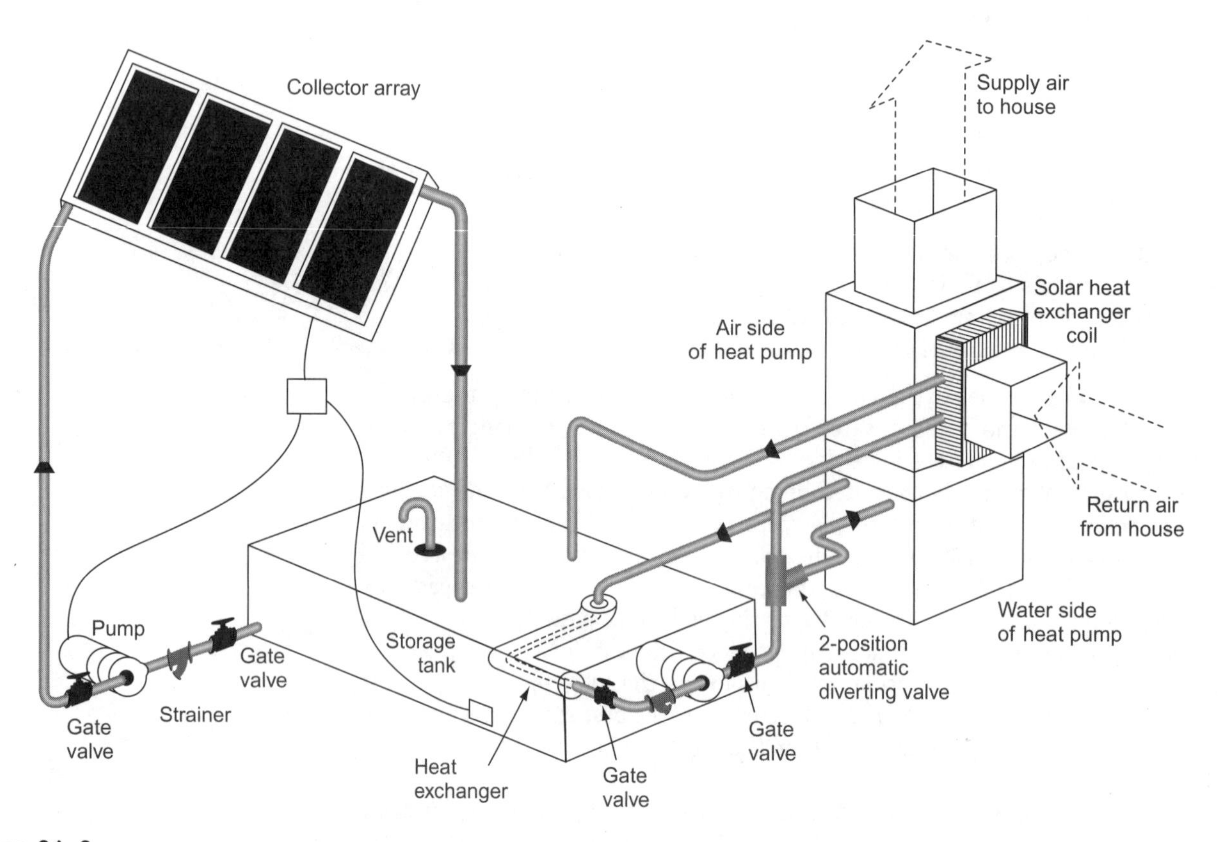

Figure 24–3
This solar heating system acts as a heat source for a water source heat pump.

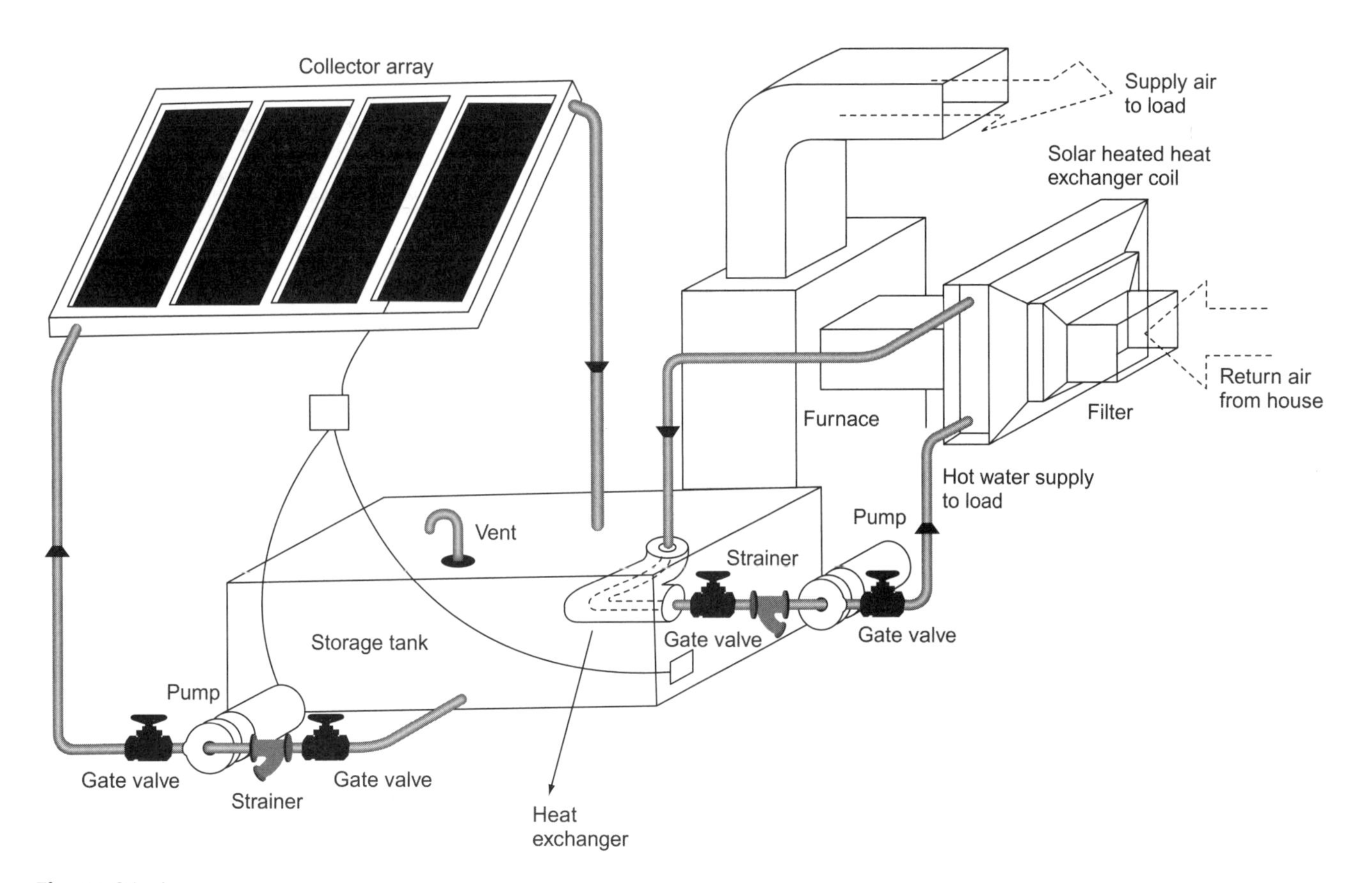

Figure 24–4
This solar heating system supplies heat for a duct mounted water coil.

Installation and Maintenance Tips

Pertinent items to consider for solar installations include the following:

- Slope all piping to aid venting and draining, and provide openings for both.
- Piping should be insulated.
- Over-pressure protection must be provided.
- Appropriate piping materials and methods must be used for the pressures and temperatures to be encountered.
- Collectors must be solidly attached to the structure, with waterproof penetrations of roof or walls. Wind loads should be considered when selecting the supporting system.
- Pressure-test collectors before elevating to the point of installation.
- Observe proper safety practices when working on the roof. Rubber-soled shoes should be worn.
- Work on asphalt shingles only in moderate temperatures. These shingles may break if very cold, or deform if they are very hot.
- Keep the collectors covered until the fluid is in place and the system is ready to operate.
- The system should be tested and flushed after completion.

The heat transfer fluid's characteristics determine the level of backflow protection of any connection to the potable system, and may also restrict material choices for system components. Some fluids may break down and become aggressive materials if they are overheated for an extended time.

Review the characteristics and requirements of the system for any other details that may call for special design or material selection.

CONSERVATION METHODS

In this section, we will discuss ways to reduce energy consumption in the operation of air conditioning systems.

Combustion Equipment

Under normal working conditions, fossil fuel-burning appliances located within the living space get their air for combustion from the indoor conditioned living space. This conditioned air is drawn in by the indoor draft blower and discharged with the flue gases to the outdoors. When this air is withdrawn from the conditioned living space, additional air is brought into the building to keep the building at a neutral pressure. This additional air is usually in the form of unconditioned air through infiltration. Infiltration air comes through cracks around windows, doors, attics, and crawlspaces. In cases where the building is extremely airtight, this lack of makeup air can cause the fuel burning appliance to operate poorly or, under extreme conditions, even cause a fire. To avoid this, consider using outdoor air for combustion if the burner can be adapted to this mode of operation. In this method, a separate pipe connects the flue intake to the outdoors. Now all combustion air comes from the outdoors and is exhausted to the outdoors; no indoor air is used. This will reduce building infiltration and therefore reduce heat loss to the conditioned space.

Consider flue dampers. Reports are mixed as to their effectiveness and safety, but they do offer the chance to save energy. Check with a knowledgeable person on current experience with these devices in your geographic area and proceed accordingly.

Examine and adjust the burner and heat exchanger to optimum efficiency. It is best to use a flue gas analyzer for this test. Establish a procedure to recheck this efficiency on a regular basis (at least once per heating season). If major equipment replacements are necessary, investigate upgrading the equipment with high-efficiency products. Keep heating terminals clean and unobstructed for the most effective heat delivery.

Air Distribution Systems

One of the most effective steps to improve the efficiency of the building is to properly seal and insulate the air conditioning duct systems. It has been estimated by the U.S. Department of Energy that air distribution systems in typical construction are only about 60 percent to 75 percent efficient. Leaks in the supply side of the duct system that allow air to leak outside the building envelope will cause the building to become negatively pressurized. This negative pressure will cause additional infiltration air to be drawn into the space. The converse is true about leaks on the return side; it will make the building positively pressurized by bringing unconditioned attic or crawlspace air into the building. As part of the Energy Star rating system, ductwork must be tested to determine the amount of leakage to outside the building envelope.

It is also a good idea to install balancing dampers on each branch run, to aid in balancing the airflow. Balancing the airflow can help prevent some rooms from being cooler than others. Inexperienced technicians may install a branch run off the end of the trunk line; this should *never* be done. The vast majority of the air will be blown into the end run, with very little available for the other branches.

System Considerations

Lubricate circulator and blower bearings and drive motors. Adjust the blower belt to the proper tension and alignment.

Set-back thermostats produce significant energy savings when a family (or business) operates on a predictable schedule most of the time. Be careful reducing the off-schedule temperatures when a heat pump with electric strip heat is being used for heat. A large jump in programmed temperatures will make the electric strip heat come on in order to warm the building. All energy savings will be diminished.

Balance the heating system (hydronic or forced-air) to minimize overheating in parts of a building. When dealing with hydronic floor heating systems, install zone controls that permit keeping unused areas cooler than the rest of the space. Do not get in a habit of closing supply registers on an air conditioning system; this will cause a reduction in airflow across the evaporator (cooling) coil. This reduction in airflow can cause the coil to ice up and thereby allow liquid refrigerant back into the compressor, eventually leading to a compressor failure.

Insulate pipe or ducts that pass through unheated areas and make sure to seal any openings made through the building envelope. These are areas that can cause large heat losses from the building. Figure 24–5 shows a common mistake: there is no insulation behind this fiberglass shower stall. In this case only the thin fiberglass wall and the exterior sheathing offer any insulation from the cold outside.

Building Actions

Do the obvious things: add or improve insulation, improve fit of weather-stripping around doors and windows, and add double or triple glazing. Check for the possibility of air-trap vestibules at frequently used outdoor entrances and exits. For additional tips and training concerning energy conservation or "green building," visit the United States Green Building Council's Web site at http://www.usgbc.org.

Figure 24–5
This newly constructed home has no insulation between the fiberglass shower stall and the outside wall.

REVIEW QUESTIONS

1. What is the function of a solar collector?
2. A circulating pump moves heated water from the collector to an ______.
3. A differential thermostat prevents the circulator from operating when the collector is ______ than the storage tank.
4. What percentage of the heating demand for a building should a solar heating system be sized to supply?
5. How can leaks in an air conditioning ductwork system affect the efficiency of a system?

CHAPTER

25

Hydraulic and Pump Theory

LEARNING OBJECTIVES

The student will:

- Discuss the basic principles of hydraulics and relevant terminology.
- Define pump terminology and theory.

HYDRAULIC THEORY

Hydraulics is the study of liquids at rest and in motion. Pneumatics is the study of gases. Both the principles of hydraulics and pneumatics can be used to describe some of the conditions seen in a plumbing system.

Liquids and gases are both fluids. Because they are fluids, they will take the shape of any container that holds them. When a fluid is contained within a pressurized container, the pressure will be exerted equally in all directions. Depending on the temperature and pressure, a substance may change from liquid to gas or vice versa, or possibly into a solid. The following definitions are used to describe some of the important characteristics of liquids and gases and also the key characteristics of pump performance.

Definitions

Absolute Pressure

Absolute pressure is the gauge pressure plus the pressure of the atmosphere, normally 14.696 psi at sea level at 70°F. Absolute pressure is indicated as psia.

Atmospheric Pressure

Atmospheric pressure is the weight of the atmosphere's gases pressing down on the earth. Normal atmospheric pressure at sea level at 70°F is 14.7 pounds per square inch, which is written as 14.7 psia. A conventional pressure gauge lying on a table at sea level at 70°F will read "0," because both the inside and outside of the gauge are exposed to the same atmospheric pressure.

Compressibility

Compressibility is the ability of a fluid to occupy less volume as its pressure increases. Since the volume of gas decreases when more pressure is applied, it is considered compressible. Since the very high pressures needed to compress a liquid are well above those seen in normal applications, liquids are considered to be non-compressible.

Density

Density is an important characteristic of liquids and gases that describes its weight per unit of volume. Density is expressed in terms of pounds per gallon (or pounds per cubic foot) for liquids, or pounds per cubic foot for gases. Since ice will float in water, we know that ice is less dense than water. This is also apparent to anyone who has filled ice trays with water, the ice will always expand to a larger volume than the water.

Dynamic Discharge Head

Dynamic discharge head is the static head pressure, read with a gauge at the discharge of the pump outlet, plus the velocity head (at the gauge location), minus the friction head due to flow in the pipe.

Dynamic Suction Lift and Dynamic Suction Head

Dynamic suction lift and dynamic suction head are suction lift and suction head minus the velocity head. In many plumbing applications, the fluid velocity is relatively low so velocity head is sometime considered negligible.

Pressure

Pressure is simply the weight of liquid (or gas) distributed over a unit area. Since the weight of the fluid is influenced by gravity, the greater the vertical distance below the fluid surface to the point of measurement, the higher the pressure. It is usually expressed in pounds per square inch gauge (psig), which means it is the pressure read on a gauge connected at that point.

Specific Gravity

Specific gravity is the weight of a substance compared to the weight of an equal volume of water for liquids or air for gases. A liquid with a specific gravity of 1.5 is 50 percent heavier than water. Therefore, when combined with water, the liquid in question would sink. If a gas, such as natural gas, has a specific gravity of 0.60, it is 60 percent as dense as air. This means that the less dense gas would rise to the top of the container.

Vacuum

Vacuum is any pressure less than atmospheric pressure. Vacuum conditions may occur readily in plumbing systems.

Viscosity

Viscosity is a measure of a fluid's resistance to flow. A fluid with a high viscosity will have a high resistance to flow while a fluid with a low viscosity will flow easily. For most fluids, the viscosity will decrease as the temperature of the fluid increases.

Head

Head is a term used to describe the amount of energy contained in a unit of fluid at any given location in a piping system. The energy contained in the fluid may be the result of the fluid being elevated above some point of reference, kinetic energy in a flowing fluid, or the fluid being under pressure.

The basic consideration of head as applied to pumps is that the pump should be able to lift the fluid a certain number of feet above its current fluid level because the pump is adding energy to the fluid.

Elevation and Pressure Head

When a pressure is given in psi it is referred to as a pressure head. This psi reading can easily be converted into an elevation head pressure (in feet) by the following formula:

$H = 2.3 \times (P)$

$P =$ pressure in psi at the base of a water column

$H =$ height of the water column in feet (ft wc)

Example 1

What would be the elevation head pressure equivalent to a pressure gauge reading of 40 psi?

$H = 2.3 \times (40 \text{ psig})$

$H = 92$ ft wc

An elevation head pressure (in feet) can be converted to an equivalent pressure head (in psig) by simply rearranging the previous formula.

$P = H/2.3$

Since most people are more comfortable with multiplication, dividing by 2.3 can be replaced with the numerical equivalent, multiplying by 0.434.

Example 2

What would be the pressure head (psig) equivalent to a column of water 100 feet tall?

$P = 0.434 \times 100$ ft

$P = 43.4$ psig

Friction Head

Friction head is the pressure loss (usually stated in feet) due to the resistance of flow of a fluid through a piping segment or system. Dimension "A" in Figure 25–1 is the pressure loss due to flow from test point 1 to test point 2. Dimension "C" in Figures 25–1 and 25–2 is the friction loss in the pump suction pipe.

As a fluid travels through a pipe, the outermost fluid will tend to cling to the inside wall of the pipe; the rougher the inside the pipe is, the more the fluid will tend to stick. When flow is relatively slow, the fluid against the inside pipe wall will travel at a slower pace than the fluid in the middle of the pipe. This causes layers of fluid to slide on each other as they move through the pipe, which results in the least amount of friction head loss. This type of flow is called laminar flow. As the flow increases, the outer layer of fluid gets pulled along more rapidly and causes a swirling motion. This turbulent flow is less orderly, so there is more friction head loss. This type of flow is called turbulent flow. This is similar to a large group of people trying to leave a room. If everyone lined up in a single line, the participants could smoothly walk out the door. If everyone ran for the door at the same time, there would be a jam up at the door and very few people could leave.

It should also be pointed out that even in laminar flow, there is a small loss of pressure due to flow. This pressure loss increases as the length of pipe and the number of fittings increases. It also increases when the inside diameter of the pipe is too small for the flow rate required.

Suction Lift

Suction lift is the distance that a liquid must be raised from the surface of the supply to the inlet of the pump plus the friction head of the suction pipe—this applies when the pump is above the source. See Dimension "B" in Figure 25–1.

Suction Head

Suction head is the elevation of the surface of the source above the pump inlet minus the friction head of the suction pipe—this applies when the pump is below the source. See Dimension "E" in Figure 25–2.

Static suction lift is the vertical distance from the center of the pump down to the surface of the liquid. See Dimension "D" in Figure 25–1.

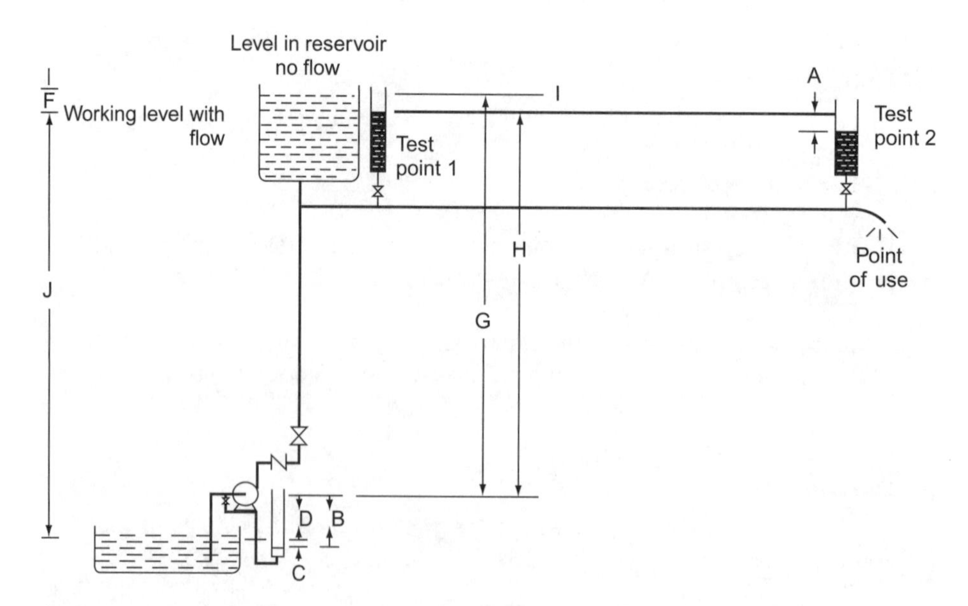

Figure 25–1
This illustration shows the location of typical pressures associated with pumping systems.

Static Suction Head

Static suction head is the vertical distance from the center of the pump up to the surface of the liquid. See Dimensions "E" plus "C" in Figure 25-2.

Static Discharge Head

Static discharge head is the vertical elevation from the center of the pump up to the point of free discharge when there is no flow in the system. See Dimension "G" in Figure 25-1.

Suction Head

Suction head is the energy necessary to draw water into the pump inlet, measured in absolute pressure.

Total Head (Pump)

Total head is the distance from the surface of the supply to the "no-flow" level. See Dimensions "J" plus "F" in Figure 25-1.

Total Head (System)

The head required to move water in the system. See Dimension "J" in Figure 25-1.

Velocity Head

Velocity head is determined by the equation

$$H = \frac{V^2}{2g}$$

where H = velocity head in feet
V = velocity of flow in feet per second (ft/sec)
g = acceleration due to gravity: 32 ft/sec/sec

Note: Because most plumbing practices limit the velocity of the fluid to less than 8 ft/sec, velocity head is negligible and may be ignored.

Example 3

What would be the velocity head loss of water flowing through a pipe at a velocity of 6.5 ft/sec?

$$H = \frac{(6.5 \text{ ft/sec})^2}{2 \times 32}$$

$$H = \frac{42.25}{64} = 0.66 \text{ ft}$$

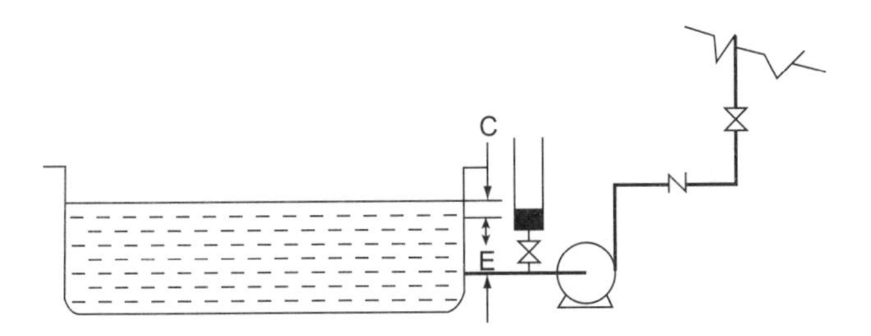

Figure 25-2
Here the fluid level is above the intake of the pump. This is sometimes called a flooded intake. A shutoff valve should be installed in the intake line to facilitate servicing the pump.

APPLICATIONS OF PUMPS

Pumps are used in plumbing and heating systems to move liquids from one place to another, or to maintain pressure on a plumbing or heating system to ensure adequate flow of liquids when required. Thus, potable water system pumps maintain required system pressure for the water delivery required by system parameters. Sewage pumps are used to elevate storm or waste water from building subdrainage systems. Pumps are also used in hydronic heating and cooling systems to move liquids from the boiler or chiller to the air conditioning terminals within the building. Pumps are used in fire-protection systems to develop flow to sprinkler heads.

The remainder of this chapter will discuss the details of pump theory, construction, and application.

PUMP THEORY

Lift

The static suction lift, "D" in Figure 25–1, is limited to the atmospheric pressure on the surface at the source. In theory, this value could be as much as 33.9 ft at sea level (atmospheric pressure = 14.7 psi), but actually, the maximum value will be about 25 ft at sea level, and proportionally less as elevation increases.

It is desirable to keep the suction lift as small as possible, to prevent the pump from cavitating. Cavitation is a process where a gas bubble is introduced into the suction side of the pump. This gas bubble can be in the form of air that was introduced into the fluid by splashing, by a low fluid level near the suction intake of the pump, or by the boiling off of the liquid itself. All the previous discussion in this chapter has been for water at 70°F maximum temperature. As suction lift increases, or water temperature increases, a point is reached where the vapor pressure of the fluid equals the suction pressure. At this point, the liquid boils, and all lifting ceases!

If the suction lift is sufficiently small, the liquid will remain liquid, and liquid transfer will continue. Thus, the suction lift should be kept as small as possible, and the fluid temperature as cool as possible.

The cavitation problem occurs when the gas bubble travels from the low-pressure side of the pump to the high-pressure side. On the low-pressure side, the gas bubble increases in size, but once pressure is applied it violently collapses. When the bubble collapses, it forms a shock wave that generates heat and can

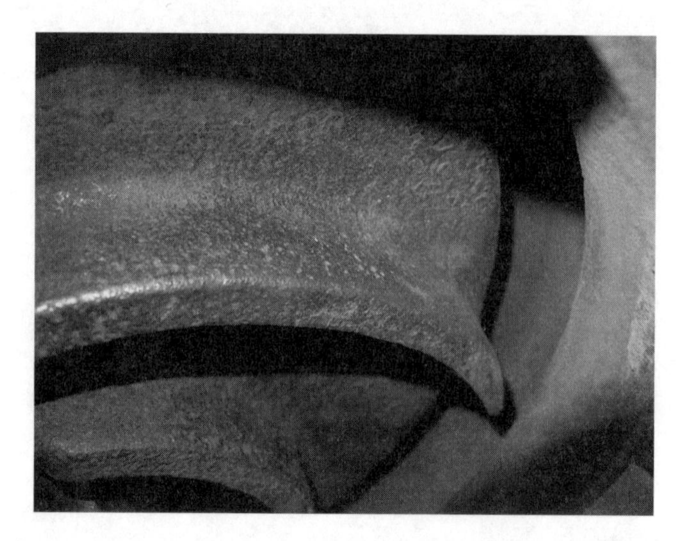

Figure 25–3
Pump damage (pitting) due to cavitation. Cavitating pumps will run hotter and louder than normal.

Table 25-1 Suction Lift Limitations With Altitude

Altitude Above Sea Level	Standard Atmospheric Pressure (Lbs/Sq.in)	Barometer Reading (Inches of Mercury)	Equivalent Head of Water (Feet)	Maximum Practical Dynamic Suction Lift (Feet)
0	14.7	29.929	33.95	25
1000	142	28.8	32.7	23.8
2000	13.6	27.7	31.6	22.7
3000	13.1	26.7	302	21.3
4000	12.6	25.7	29.1	21
5000	12.1	24.7	27.9	19
6000	11.7	23.8	27	18
7000	112	22.9	25.9	17
8000	10.8	22.1	24.9	16

actually remove pieces of metal from the pump components. Figure 25-3 shows the results of cavitation on the interior of a pump.

Some pump types cannot produce a significant amount of vacuum. For these types, suction lift must be kept to an absolute minimum. Table 25-1 indicates how the amount of vacuum, or suction head, decreases as the elevation increases; this in turn sets a practical limit to the amount of suction lift that a pump can produce.

If the lift required is greater than what can be developed by a pump located above the liquid to be pumped, a deep well, submersible, or jet pump must be used. These pump types are described in Chapter 26.

REVIEW QUESTIONS

1. What is cavitation?
2. What two factors determine whether a substance is a liquid, solid, or a gas?
3. What is the difference between gauge pressure and absolute pressure?
4. What is the velocity head pressure of a liquid that has a velocity of 8 ft/sec?
5. What happens to the amount of suction lift available for a pump as the elevation above sea level decreases?

CHAPTER 26

Pump Types, Pump Uses, and Piping Design

LEARNING OBJECTIVES

The student will:

- Contrast the different pump designs.
- Discuss the uses of pumps in plumbing and heating systems.
- Discuss basic piping design concepts.

PUMPS

Pumps raise liquids either by pushing the fluid or by reducing the pressure in the suction pipe so that the atmospheric pressure on the free surface of the reservoir raises the liquid into the pump. Most pumps lift water by a combination of both of these effects. As described in the previous chapter, the suction lift of the best lift pumps is limited to about 75 percent of the head because of atmospheric pressure. Remember the head equivalent for water of atmospheric pressure at sea level (atmospheric pressure = 14.7 psi):

$$\text{H (in feet)} = (14.7 \text{ psi})\left(\frac{1 \text{ ft WC}}{0.434 \text{ psi}}\right) = 33.9 \text{ ft}$$

$$\text{H (practical)} = 33.9 \text{ ft} \times .75 = 25 \text{ ft}$$

The following sections describe some of the pump types that are utilized in hydronic systems.

Positive and Non-positive Displacement Pumps

Pumps can be classified into two categories, positive displacement and non-positive displacement pumps. Positive displacement pumps typically have a tight internal mechanical seal that separates the low pressure side from the high pressure side of the pump. As it progresses through its pumping cycle, it encloses a certain volume of liquid that is moved through the pump on each cycle of operation. Use caution when working with these pumps because they can generate very high pressures if they continue to run. They should always include a relief valve of some type. Non-positive displacement pumps work similarly, but there is more space between moving internal pumping parts, allowing for a certain amount of slippage. Non-positive displacement pumps cannot generate as high a discharge pressure. When their discharge piping is blocked they will generate heat, due to internal slippage.

Another pair of subcategories is fixed displacement and variable displacement. Fixed displacement pumps move the same amount of liquid with each revolution of the power source (usually a motor) as long as there is no internal slippage of the pump. Variable displacement pumps are capable of changing the amount of liquid they move per revolution. In many cases, the discharge pressure will control the amount of liquid pumped. For example, if the discharge pressure increases, the internal mechanisms of the pump shift to lower the amount of liquid being pumped. This will help control the pressure and flow rate of the system.

Reciprocating Pumps

Reciprocating pumps consist of one or more cylinders and pistons, similar to a automobile engine or air compressor. The piston is connected to a crankshaft; when the crankshaft turns, the piston moves back and forth in the cylinder. A check valve on the inlet prevents the fluid that was pulled in on the intake stroke from backing out on the pressure stroke. Similarly a check valve on the outlet prevents backflow from the discharge region on the suction stroke, so the pump takes in liquid from the supply on the suction stroke and moves the liquid to the discharge chamber on the pressure stroke. If either check valve is defective, the pump will not work properly.

Rotary Pumps

Rotary or spiral rotor pumps move a fixed volume of liquid through the pump as the rotating parts are turned. A rotary gear pump can trap the fluid between the teeth of each gear. As one gear turns it turns the other gear, which pumps the fluid.

Turbine Pumps

Turbine pumps incorporate a series of small blades or fins on the perimeter of a rotating disk. The disk is placed in a close-fitting housing, and the many small vanes induce a pressure rise. Multistage turbine pumps are used to develop larger pressure increases.

Centrifugal Pumps

Centrifugal pumps employ a rotating impeller within a carefully shaped enclosure called a volute. The water is drawn in at the center of the impeller and discharged at the perimeter. The pressure rise comes from the centrifugal force imparted to the water by the spinning impeller. The best volute designs ensure that most of the energy imparted to the water is converted into useful discharge head. Various centrifugal pump configurations are available:

Split-case

This pump can be repaired without disturbing the piping.

Vertical Drive

This is similar to the pump design of a turbine pump. The vertical drive pump has an extended shaft turning an impeller located below the liquid level.

Submersible

The submersible pump houses the motor in a watertight enclosure so that the entire unit may be placed in the water to be pumped. The suction pressure of this pump is the net depth of the pump elevation below the free water level in the well. Figure 26–1 shows a typical submersible pump used in a residential well.

Figure 26–1
A typical submersible pump has the motor and pump in a one-piece watertight assembly.

Conventional Coupled Shaft

Conventional coupled shaft pump and motor sets with mechanical seals or packing glands are used for general heating and cooling work.

Non-priming

The non-priming pump must be primed before operation. This can best be accomplished by locating the pump below the surface of the water supply.

Self-priming

Self-priming pumps are capable of removing the air from the suction pipe and lifting the liquid into the pump inlet.

Centrifugal Pump Impellers

Many variations on pump impellers are used for specific applications. The open impeller is least efficient, but it can move liquids containing larger solids without becoming locked up. The semi-open design will handle some solids, with some improvement in efficiency. The closed design is best suited for liquids only.

Jet Pumps

Jet pumps combine a centrifugal pump with a venturi effect. Some of the discharge of the pump is returned to the pump inlet, where it is introduced in such a way as to reduce the pressure at that point. This lifts water from well below the surface level of the water. Jet pumps are capable of lifting water from greater depths than conventional centrifugal pumps. See Figure 26–2.

PUMP USES AND PIPING DESIGN

Of the many pump types available, certain pumps fit different applications better than others. After considering the following list of requirements and characteristics, the selection of the appropriate pump type will be narrowed to a few choices. If in doubt, pump manufacturers are a great source of assistance in selecting the best one for your application.

Intended Use

Examples include:

- Water distribution
- Waste service (sanitary, clear water, or storm)
- Hot or chilled water circulation
- Feed water to steam boilers
- Other applications

Special Conditions

Typical conditions that would require special consideration include:

- Does the liquid contain solids?
- Is the temperature unusual (high or low)?
- Is the viscosity high or low?
- Is the liquid chemically aggressive?

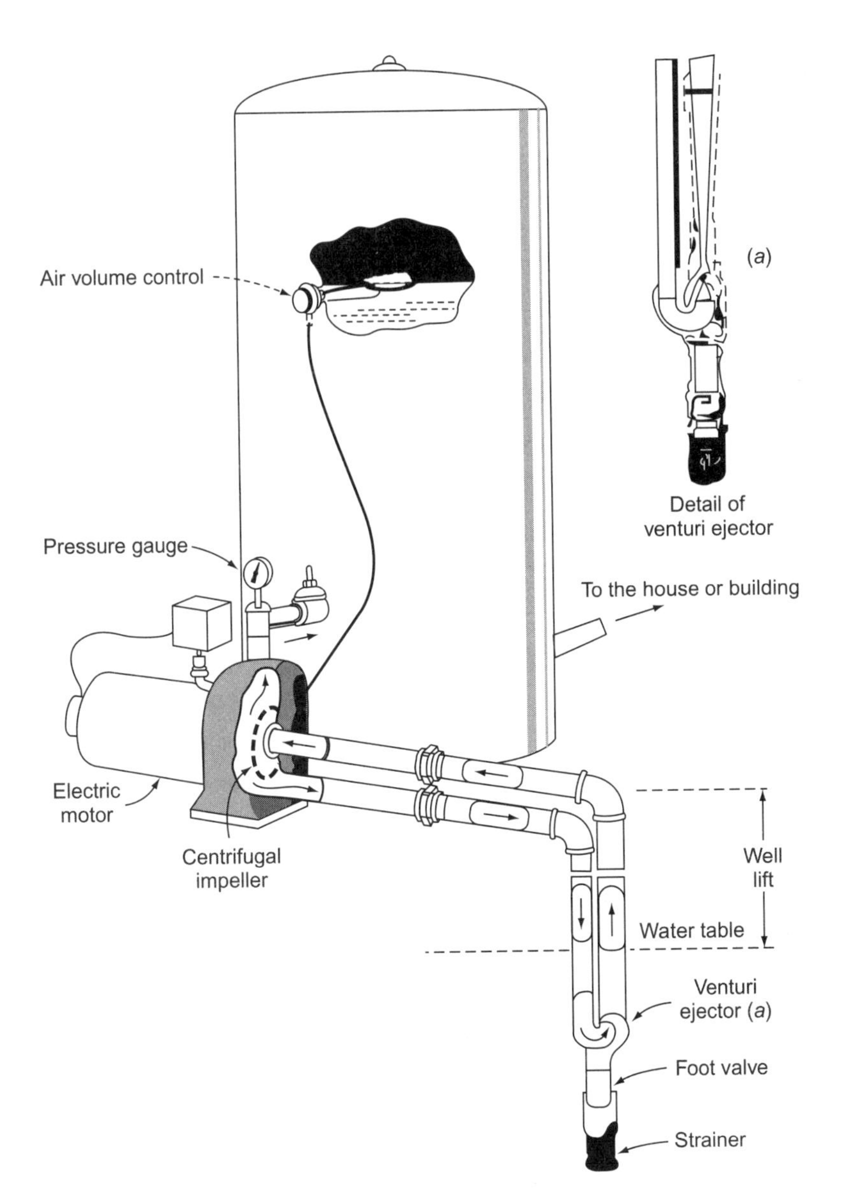

Figure 26–2
Typical jet pump installation.

Installation Requirements

Normal system defining parameters include the following:

- Is a self-priming pump required?
- What is the pressure rise required?
- What type of piping is required?
- How much space is required for service and repair?

Energy Source

- Electric
- Pneumatic
- Engine-driven

Cost

- What are the cost constraints and limitations?
- First cost compared to lifetime cost?

After reviewing these topics, the pump can be selected. Table 26–1 lists the choices of materials used for pump construction based on the acidity of the liquid being pumped. Always consult the pump manufacturer in extreme cases; special seals and bearings may be required.

Table 26–2 lists pump choices based on application or use.

Using these tables as guides, select the pump from recommendations of the manufacturer and your own experience, or the experience of others.

Table 26-1 Materials Used In Pump Bodies

pH Value	Conditions	Pump Construction
0–4	Acid	Alloy steels (stainless) or engineered plastics
4–6		All bronze
7	Neutral	Standard (irons)
6–8		Bronze or iron fitted
8–10		All iron
10–14	Alkaline	Alloys of bronzes or steels

Table 26-2 Required Pump Characteristics. Courtesy of Plumbing–Heating–Cooling–Contractors—National Association.

Type of Pump	Practical Suction Lift (ft)	Usual Well-Pumping Depth (ft)	Usual Pressure Head (ft)	Advantages	Disadvantages	General Remarks
			Positive Displacement (Reciprocating)			
Shallow	22–25	Shallow	100–200	Positive action.	Pulsating discharge.	Best suited for capacities of 5–25 gpm against moderate to high heads.
Deep	22–25	Up to 600	Up to 600 above cylinder (deep well)	Discharge against variable heads. Pumps water containing sand and silt. Especially adaptable to low capacity and high lifts.	Subject to vibration and noise. Maintenance cost may be high. May cause destructive pressure if operated against closed valve.	Adaptable to hand operation. Can be installed in very small diameter wells (2" casing). Pump must be set directly over well (deep well only).
			Positive Displacement (Rotary)			
Gear (oil and shallow well pumps)	22	22	50–250	Positive action. Discharge constant under variable heads. Efficient operation.	Subject to rapid wear if water contains sand or silt. Wear of gears reduces efficiency.	
Helical rotary (deep well)	Usually submerged	50–500	100–500	Same as shallow rotary. Only one moving pump device in well.	Same as shallow well rotary except no gear wear.	Flexible drive coupling can help reduce vibration. Best adapted for low capacity and high heads. Strainers recommended.

Table 26-2 Required Pump Characteristics Courtesy of Courtesy of Plumbing–Heating–Cooling–Contractors—National Association. (Continued)

Type of Pump	Practical Suction Lift (ft)	Usual Well-Pumping Depth (ft)	Usual Pressure Head (ft)	Advantages	Disadvantages	General Remarks
Non-positive Displacement Pumps (Centrifugal)						
Straight centrifugal (single stage for shallow well)	20 max.	10–20	100–150	Smooth, even flow. Pumps water containing sand and silt. Pressure on system is even and free from shock. Low starting torque. Usually reliable and good service life.	Loses prime. Efficiency depends on operating under design heads and speed.	Very efficient pump for capacities above 60 gpm and heads up to about 150 ft.
Regenerative vane turbine (single impeller shallow wells)	28 max.	28	100–200	Same as straight except not suitable for pumping water containing sand or silt. Self-priming.	Same as straight except maintains prime easily.	Reduction in pressure with increased flow not as straight centrifugal.
Vertical line shaft turbine (multi-stage deep well)	Impeller submerged	50–300	100–800	Same as shallow well turbine. All electrical components are accessible above ground.	Efficiency depends on operating under design head and speed. Requires straight well large enough for turbine bowls and housing. Lubrication and alignment of shaft is critical. Abrasion from sand.	
Submersible turbine (multi-stage deep wells, etc.)	Pump and motor submerged	50–400	50–400	Same as shallow well turbine. Easy to frost-proof installation. Short pump shaft to motor. Quiet operation. Well straightness is not critical.	Repair to motor or pump requires pulling from well. Sealing of electrical equipment from water vapor critical. Abrasion from sand.	3500 RPM models, while popular because of smaller diameters or greater capacities, are more vulnerable to wear and failure from sand and other abrasives.
Non-Positive Displacement Pumps (Jet, private water supply)						
Shallow well	15–20 below ejector	Up to 15–20 below ejector	80–150	High capacity at low heads. Simple in operation. Does not have to be installed over the well. No moving parts in the well.	Capacity reduces as lift increases. Air in suction or return line will stop pumping.	
Deep well	15–20 below ejector	25–120; 200 max.	80–150	Same as shallow well jet. Well straightness not critical.	Same as shallow well jet. Lower efficiency, especially at greater lifts.	The amount of water returned to ejector increases with lift: 50% of total water pumped at 50 ft lift and 75% at 100 ft lift.

PIPING DESIGN

There are two general categories of water piping systems: those that are completely closed, and those that are open to the atmosphere at some point.

Closed Systems

Most heat transfer systems, such as arrangements for hot water heating or chilled water cooling, fall under the category of closed systems. To calculate pressure drops due to flow in these systems, you do not have to consider elevation, because all elevation changes cancel out as you travel around a closed system.

Open Systems

Arrangements that are open to the atmosphere include potable water supply systems, cooling tower systems, and lawn sprinkler and fire sprinkler systems. Pressure drop calculations must take elevation changes into account, because the changes do not cancel out in systems that are open to atmosphere.

Pressure Drop/Flow Rate Curves For Piping

Figure 26–3 shows the pressure drop versus flow for copper tube type L. The *x*-axis represents the pressure drop in psi for each 100 foot of tubing. The *y*-axis represents the flow rate of the fluid being pumped in gallons per minute. The graph

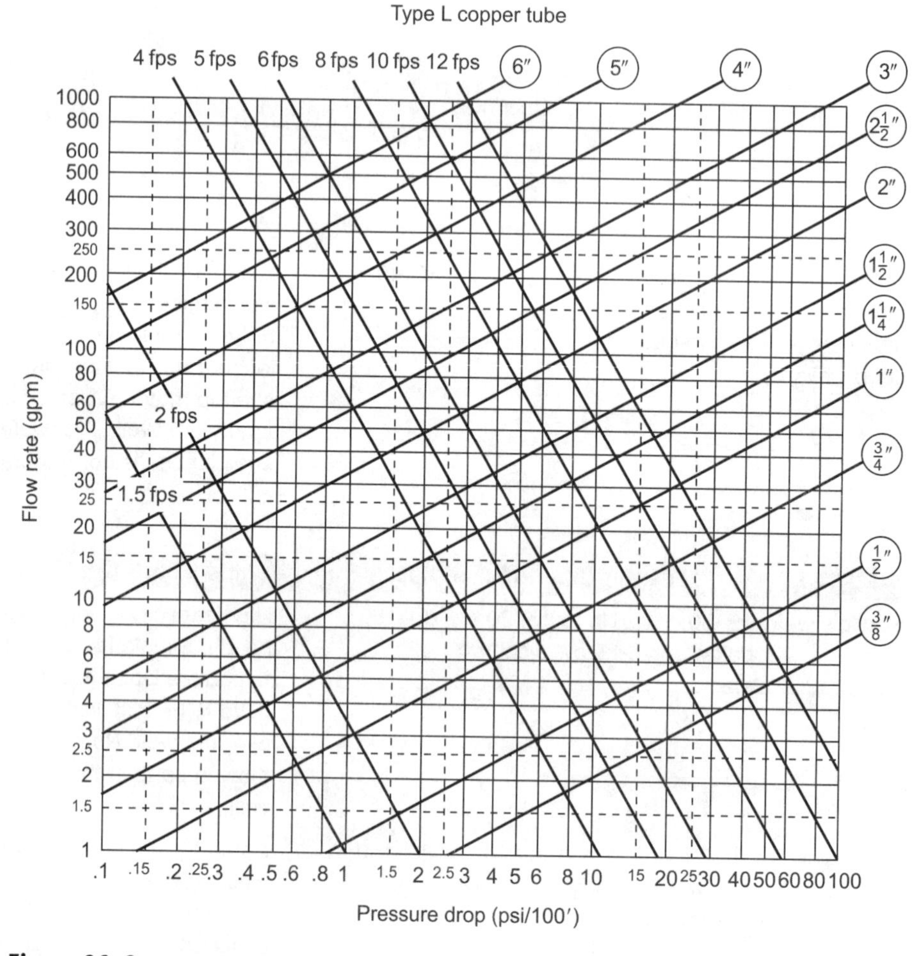

Figure 26–3
Piping flow for a hydronic heating system.

has lines that slope diagonally downward from right to left for tubing sizes $\frac{3}{8}$ inch to 6 inches. The graph also has velocity flow lines that slope diagonally downward from left to right, 1.5 fps to 12 fps. This is a typical format for this chart. Others are available for plastic pipe, steel pipe (of various inside wall roughness conditions), and pipes made of other materials.

The pressure drop is given in psi. Other presentations of this chart may use feet of head. Be careful to recognize these units!

The same information is available in other forms also. Some manufacturers provide circular slide rules, and the data are also presented in tables in many reference works.

A few examples will show the methods.

Example 1

Using Figure 26-3, determine the pipe size required to have approximately 10 gpm at a velocity of 4 fps

Answer: 1 inch

Example 2

Using Figure 26-3, determine the pressure drop in 200 feet of 1 $\frac{1}{2}$ inch pipe, when the flow rate is approximately 30 gpm of flow.

Answer :

$$\left(\frac{3.0\text{ psi}}{100\text{ ft}}\right)(200\text{ ft}) = 6.0\text{ psi}$$

Example 3

Figure 26-4 shows the piping system for a hydronic heating system with the flow indicated for each portion of the installation and the sections marked for listing in Table 26-3. (Note that Figure 26-3 presents pressure drop in psi, so the table includes a conversion column of psi to feet of pressure drop.)

The pipe sizes are selected based on a moderate pressure drop per 100 ft. The usual maximum for small systems is 4 ft of pressure drop per 100 ft of pipe length. The fitting and valve allowance is taken as 50 percent of the piping pressure drop (i.e., not including boiler, flow valve, strainer, or radiation).

The sum of the pressure drops is the pressure rise the pump must maintain when it is moving design flow (30 gpm in this example). What frequently happens is that the head calculated is higher than what is available from standard, off-the-shelf pumps. When this occurs, it is desirable to check the pressure drop table to see if pipe sizes can be increased, or radiation changed, so that the system pressure drop will be equal to, or less than, what can be obtained from a stock pump. If these choices are not feasible, a nonstandard pump will be needed to match the requirements as shown in the table.

Note that radiation paths C to E and D to G are parallel to D to F, and require slightly less pressure drop at design condition. Therefore, we used the highest pressure loss path to calculate the required pump characteristics.

Table 26-3 Values for Figure 26-4

Item	Flow (gpm)	Length (ft)	Selected Size (inches)	Pressure Drop Per 100' (psi)	Pressure Drop Per 100' (ft) (psi × 2.3)	Pressure Drop (ft) (pd/100) × Pipe Length
Boiler	30	—	—			2.0
AB	30	5	2	0.9	2.1	0.1
Flow valve	30	—	2			2.3
BC	30	10	2	0.9	2.1	0.2
CD	25	15	2	0.7	1.6	0.3
Radiation (D to F)	10	—	1			8.0
FG	15	15	1.5	1.2	2.8	0.4
GH	30	43	2	0.9	2.1	0.9
IJ	30	5	2	0.9	2.1	0.1
Strainer	30	—	2			1.0
Fitting and Valve Allowance	(0.5)(AB + BC + CD + FG + GH + IJ) (0.5)(0.1 + 0.2 + 0.3 + 0.4 + 0.9 + 0.1)					1.0
Total Pressure loss:						16.3

Example 4

Consider Figure 26-5. A cooling tower on the roof cools the water from a chiller condenser. Table 26-4 itemizes the pressure pattern.

Note that the tower has a net pressure drop of 8 ft, since the discharge pipe terminates 8 ft above the level of water in the tower sump, and the system is open to atmosphere at this point.

The total pressure requirement and flow must be supplied by the pump. These systems are much less likely to use a stock pump, but it is still best practice to check the pressure requirements to see if any component can be modified or changed to improve the pressure loss pattern. The pipe sizes should be selected for a maximum of 4 ft of pressure loss per 100 ft of pipe. Table 26-4 includes a column of pressure drop in psi which is converted to feet for consistent units.

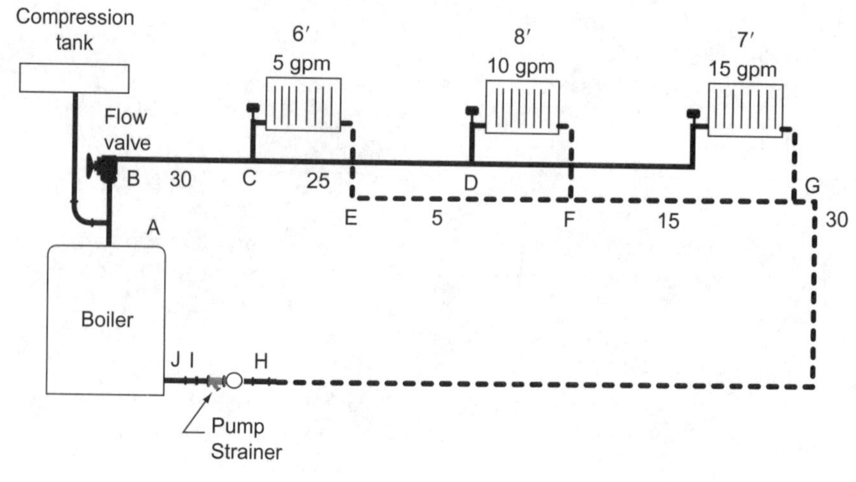

Figure 26-4
Piping flow for a hydronic heating system.

Example 5

Figure 26–6 shows a cold potable water supply system. These systems have been described and analyzed in earlier chapters of this series of texts. For potable water systems, the pressure loss per 100 ft of pipe is usually selected to be 8 to 10 psi per 100 ft, or 4 fps to 8 fps (depending on pipe material) flow velocity, whichever results in lower flow. The design flow for each segment of the potable system is determined by the correlation of water supply fixture units (WSFU) and gpm. Once the design flow in gpm is selected, similar design charts and basic information are used as described above.

If the bathroom group is 20 ft above the pneumatic tank, and if the minimum pressure at the bathroom must be 15 psi at 5 gpm flow, the pressure table is as follows.

Table 26-4 Values for Figure 26-5

Item	Flow (gpm)	Length (ft)	Selected Pipe Size (inches)	Pressure Drop Per 100′ (psi)	Pressure Drop Per 100′ (ft)	Pressure Drop (ft)
Discharge Pipe (from pump to condenser)	100	22	3	1.1	2.6	0.6
Strainer			3			2.0
Condenser	100					20.0
Pipe to Tower	100	45	3	1.1	2.6	1.2
Net Elevation of Tower						8.0
Suction to Pump	100	17	3	1.1	2.6	0.4
Fitting & Valve Allowance			(0.5)(Pipe Friction) (0.5) (0.6 + 1.2 + 0.4)			1.1
Total Pressure Loss:						33.3

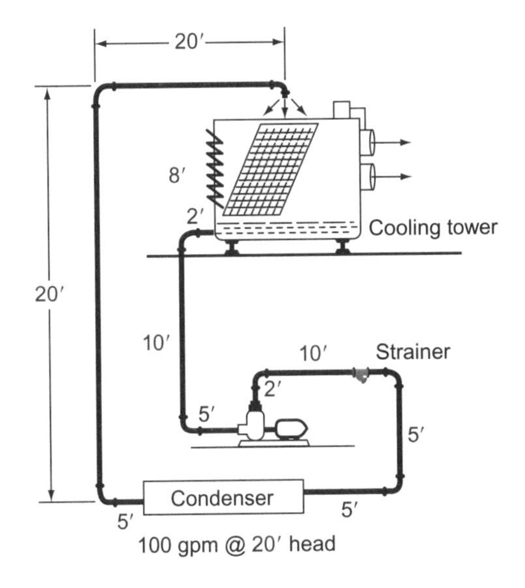

Figure 26–5
Piping flow for a hydronic cooling system.

The minimum pressure at the pneumatic tank must be equal to the total calculated above, and the pump must be able to supply at least design flow at a pressure of about 20 psi above the value calculated above. In this way, the pump will have acceptable on and off times, preventing short-cycling.

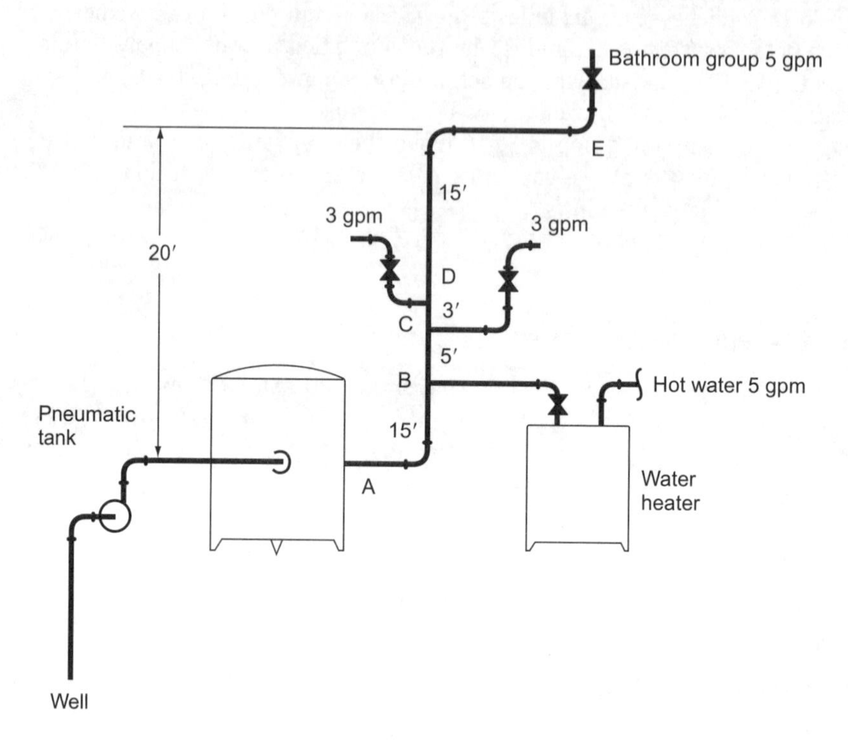

Figure 26–6
Potable water system.

Table 26-5 Values for Figure 26-4

Item	Flow (gpm)	Length (ft)	Selected Size (inches)	Velocity (fps)	Pressure Drop Per 100' (psi)	Pressure Drop (psi)
AB	16	15	1	6	6	$(\frac{6}{100})(15) = 0.9$
BC	11	5	1	4	3	$(\frac{3}{100})(5) = 0.15$
CD	8	3	0.75	5.5	6.5	$(\frac{6.5}{100})(3) = 0.2$
DE	5	15	0.75	3.5	2.8	$(\frac{2.8}{100})(15) = 0.42$
Elevation at E			20′ converted to psi			8.7
Minimum Required Pressure at E						15.0
Fitting Allowance			(0.5)(AB + BC + CD + DE) (0.5)(0.9 + 0.15 + 0.2 + 0.42)			0.84
Minimum Pressure (psi) Required at the Pneumatic Tank:						26.2

REVIEW QUESTIONS

1. In a closed hydronic system, do changes in the elevation of the piping flow path affect the hydraulic performance of the system?
2. The maximum design pressure loss in piping for water distribution systems is ______ psi per 100 ft, unless this results in excessive flow rates through the piping.
3. Jet pumps can lift water from lower levels than conventional ______ pumps.
4. Positive displacement pumps will reduce their flow rate automatically when what increases?
5. The maximum practical suction lift for a pump at sea level is how many feet of water column?

CHAPTER 27

Pump Performance Curves, Installation and Maintenance

LEARNING OBJECTIVES

The student will:

- Discuss the significance of pump performance curves.
- Describe the principle concerns of pump installation.
- Describe the concept of pump maintenance.

PUMP PERFORMANCE CURVES

Flow/Pressure Rise

Centrifugal pump operating capability can be shown by a family of curves drawn on graph paper. The curves show flow (usually in gpm) plotted against pressure rise (usually in feet) for cool water. A separate curved line can be drawn on a single chart to represent the pressure/flow relationship of each impeller diameter available for that pump. Additional groups of curves can be drawn for different available pump speeds. Typical pump speeds are 850 rpm, 1150 rpm, 1725 rpm, and 3450 rpm. Each of these speeds produces a family of pressure-flow curves with various impeller diameters.

Net Positive Suction Head

See Figure 27–1 for an example of pump performance at a given speed. The sloped vertical lines show the net positive suction head (NPSH) required at the pump inlet.

The net positive suction head is the absolute pressure required to maintain the inlet water in liquid state (rather than changing to vapor) at the temperature for which the chart is prepared (usually 60°F).

REQUIRED INPUT POWER

Figure 27–2 shows the horsepower lines for a particular pump with various impeller diameters. The motor for the pump must be adequate for the conditions encountered.

For example, if a pump with an $8\frac{1}{2}$ inch impeller is operating at point A in Figure 27–2, a 5 HP motor will carry the load. If, however, the actual operating head is lower than calculated, the pump and system characteristics may meet at

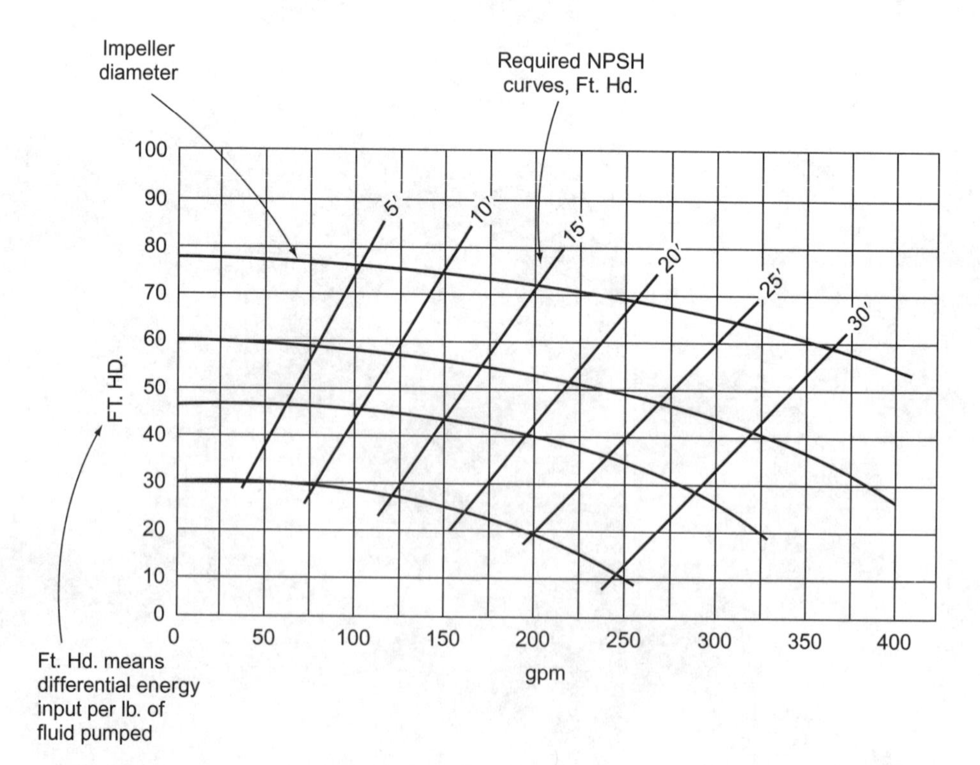

Figure 27–1
Typical centrifugal pump curve.

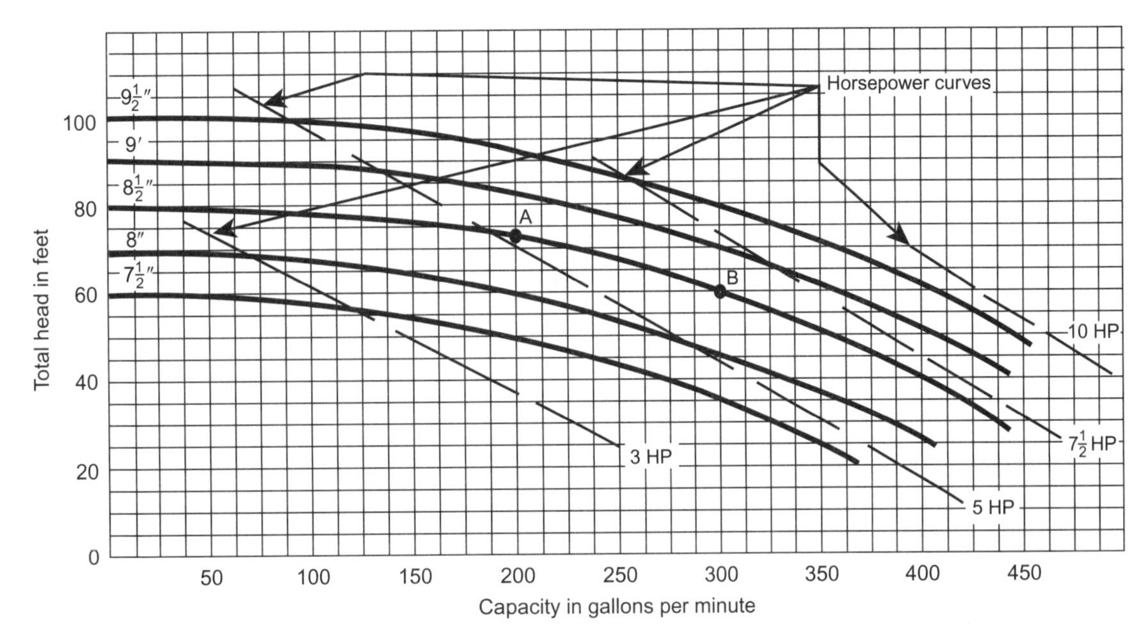

Figure 27–2
Horsepower curves have been added. Notice they slope downward from left to right.

point B. The required horsepower will be about $6\frac{1}{2}$ HP, and the 5 HP motor would be considerably overloaded.

Remember, if you reduce the discharge pressure of a centrifugal device (pump, blower, or fan) it will increase the amount it pumps, which will increase the power required.

Thus, when selecting the motor for a pump, you should select the size that equals or is above the required power all along the selected impeller diameter.

Pump Affinity Laws

There are some easy mathematical equations that can help you understand how a change in the pump's rpm can affect the volume of fluid being pumped, the head pressure, and the input power needed. In each of the following equations, the subscript 1 represents the beginning condition and the subscript 2 represents the ending condition.

$$Q_2 = Q_1\left(\frac{N_2}{N_1}\right)$$

$$H_2 = H_1\left(\frac{N_2}{N_1}\right)^2$$

$$P_2 = P_1\left(\frac{N_2}{N_1}\right)^3$$

Q = flow rate (gpm)
N = pump rpm
H = Head or pressure (psi or feet)
P = Power (watts or brake horsepower)

Example 1

A pump is running under the following conditions: $N = 1750$ rpm, $Q = 100$ gpm, $H = 30$ psi, and $P = 2$ hp. What will change if the rpm doubles to $N = 3500$ rpm?

$$Q_2 = 100 \text{ gpm}\left(\frac{3500}{1750}\right) = 200 \text{ gpm}$$

$$H_2 = 30\text{psi}\left(\frac{3500}{1750}\right)^2 = 30 \text{ psi} \times 4 = 120 \text{ psi}$$

$$P_2 = 2 \text{ hp}\left(\frac{3500}{1750}\right)^3 = 2 \text{ hp} \times 8 = 16 \text{ hp}$$

From this it is clear to see that doubling the pump rpm will double the flow rate, quadruple the pressure, and octuple the power requirement.

EFFICIENCY CURVES

Figure 27–3 shows the efficiency curves for the pump. For a pump that operates most of the time, efficiency is an important consideration. The pump should be selected to operate at the highest efficiency possible. For a pump that operates only intermittently, the greater first cost of the higher-efficiency pump may not be justified.

Figure 27–4 adds the NPSH curves to develop the full chart as presented. If the absolute pressure at the pump inlet is less than the value shown, some water will flash to steam and the pump is said to be cavitating. As described in Chapter 25, cavitation will cause increased noise, rapid impeller erosion, bearing wear, and erratic operation. Cavitation can be prevented by cooling the water, increasing inlet pressure, or using a larger pump (with a lower required NPSH).

PUMP SELECTION

It is best to select a pump that will operate in the middle third of the capacity-head curve. In this way, the pump will be at the most efficient range and have maximum tolerance for variations in the system pressure and flow requirements. Keep in mind that at the high-flow end (right side) of the pump operation, the pump will be noisy, susceptible to cavitation, and have no tolerance for increased flow.

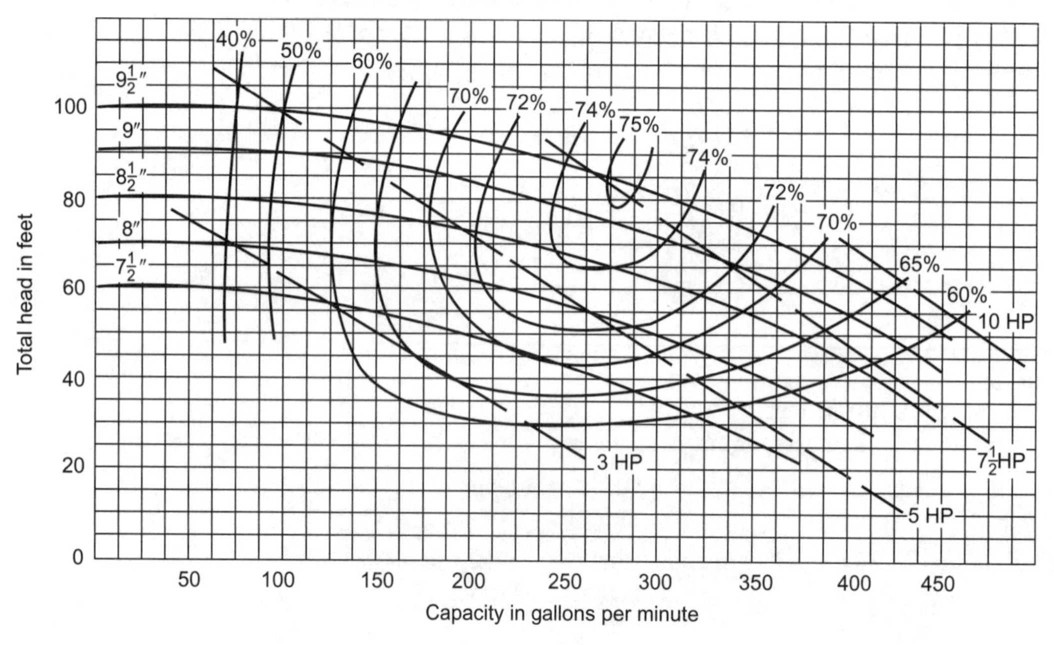

Figure 27–3
Efficiency curves are U-shaped.

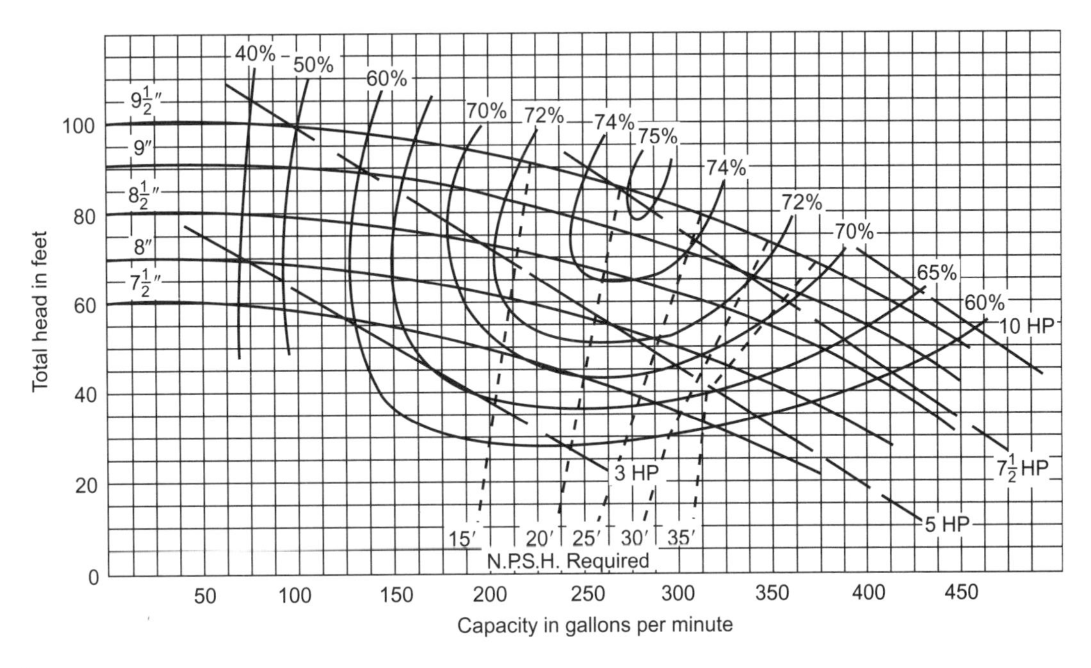

Figure 27–4
By adding all of the previous curves to one chart, a complete pump curve chart is formed.

Never extend the pump curves to select an operating point beyond the curves as shown by the manufacturer. Operation beyond the curves is "No Man's Land"—the pump will be noisy and so turbulent that the flow and head are completely unpredictable.

It is generally agreed that 1750 rpm pumps are more long-lived than 3450 rpm units, but some experience suggests that 3450 rpm pumps are just as reliable. The higher speed uses a much smaller, lighter pump (therefore, less costly) than 1750 rpm. The higher speed is usually noisier, however, with a higher-pitch sound.

System Curve

Figure 27–5 shows the concept of a system curve. This curve shows the flow/pressure drop relationship for a given closed system of piping and terminals. Since pressure drop is proportional to the square of the flow, the curve is a parabola starting at "0" flow and "0" pressure drop. This means that as the flow rate (gpm) increases, there will be a large increase in the operating head pressure.

To summarize, the following ideas determine the nature of the system curve:

1. In any piping configuration, with zero flow, there is zero system pressure drop.
2. With one unit flow, one unit pressure drop will result.
3. With two units flow, four units pressure drop will result.
4. With three units flow, nine units pressure drop will result.

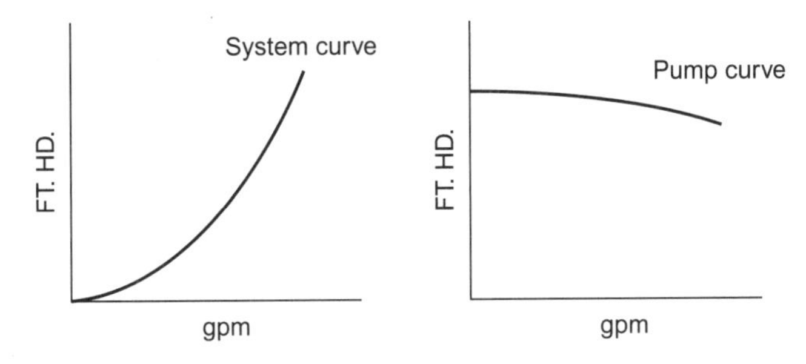

Figure 27–5
Piping system curve.

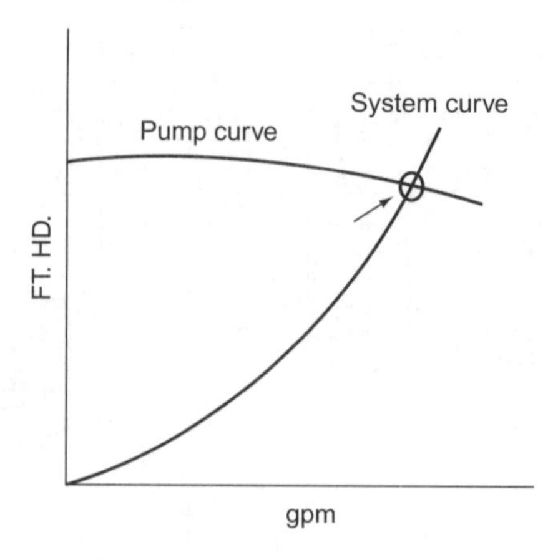

Figure 27–6

Operating Point

The system curve is a parabola going through (0, 0) on the chart, called the origin. Note the following facts:

- The system can only operate on the system curve.
- The pump can only operate on the pump curve.

Therefore, the only place that the two can operate together is at the intersection of the two curves. See Figure 27–6.

Example 2

A pump is needed to circulate water to transfer 2,000,000 BTUH with a 20°F temperature drop. Water has a specific heat of 1 and it weighs 8.33 lbs/gallon.

The required flow is:

$$\text{GPM} = \frac{\text{BTUH}}{(\text{weight of liquid})(\text{specific heat})(60\ \text{min/hr})(\text{change in temperature})}$$

$$\text{GPM} = \frac{2,000,000}{(8.33)(1)(60)(20)} = 200\ \text{gpm}$$

Calculate the pressure drop in the system and compute the total head required to develop the flow required by using the methods shown in Chapter 26.

PUMP SIZING

Example 3

Assume that the calculation in Example 2 shows that the required head is 30 ft when 200 gpm is circulated. To be safe, we select a pump with a 40 ft head at 200 gpm.

Figure 27–7 shows the effect of this oversizing.

- Point A is 200 gpm at 40 ft on the pump curve.
- Point B is the actual system point at 200 gpm at 30 ft head.

Point B and the origin determine a parabola which is shown as the system curve. The actual operating point will be at the intersection of the

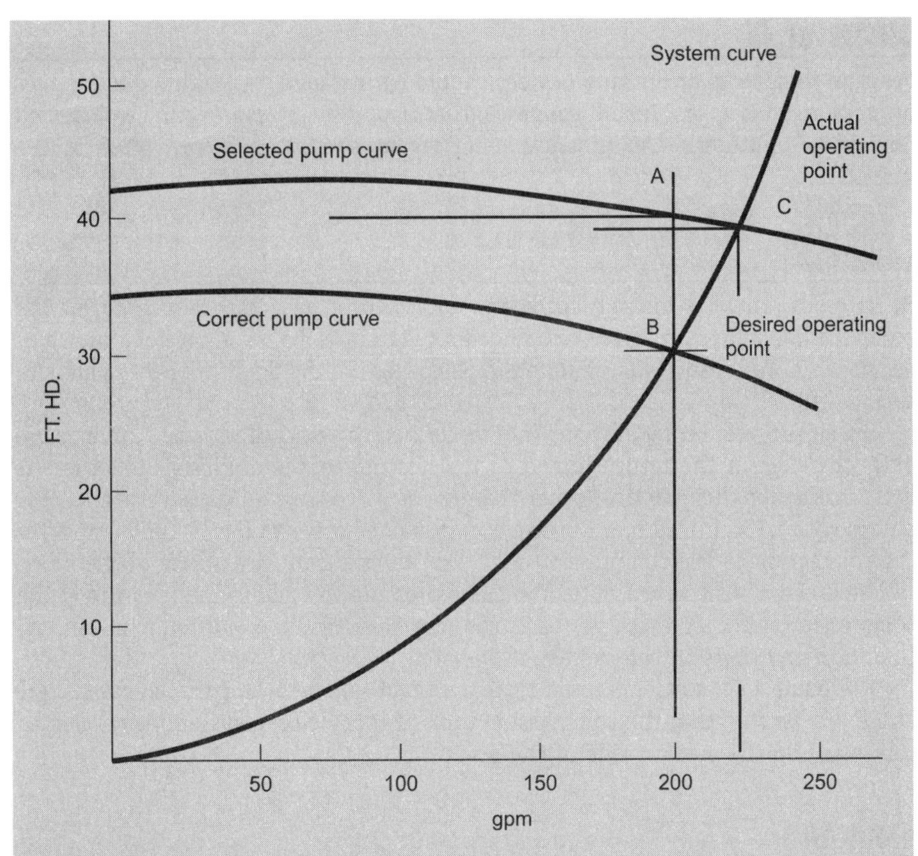

Figure 27–7
Diagram for Example 3 showing the effect of oversizing.

selected pump curve and the system curve, point C. In this example, the actual flow will be about 12 percent more than calculated, and the pressure about 25 percent more than required.

In most cases, this greater flow will not produce a problem, but sometimes it might; check this out for the particular job. Potential problems include more operating horsepower, greater operating cost, more noise, and greater erosion on system components.

PUMP INSTALLATIONS

To ensure long, trouble-free performance from a pump, the following general considerations for pump installations should be observed.

Electrical

Be sure the power supply provides what the motor requires. The voltage should be within 10 percent of the nameplate rating, while the wiring and starters should be rated for the current rating on the nameplate. Magnetic motor starters are required for both single- and three-phase motors. These starters protect the motor from overload or abnormal electrical conditions.

Smaller fractional-horsepower, single-phase motors can be controlled by simple relays or manual switches. These switches may be fused or nonfused. Most small single-phase and some three-phase motors contain an internal thermostat that opens if the motor gets too hot.

Steam or Air

Pressure regulators or limiting devices should be installed on turbine pump drives or pressurized ejectors. Install gauges so that important pressures can be observed. Emergency shutdown and pressure relief means should be incorporated in the installation.

Location

Be sure that there is enough space and access to inspect and service/repair the pump. Larger pumps may need extra head room to install a beam for hoisting major components. For open systems, place the pump close to the liquid supply to minimize suction lift.

For closed circulating systems (hot water heating or chilled water), pumps are often installed in the return piping to the heating source or chiller, but there is little reason for this practice, other than the need to save on piping material. For pumps with a total head of 25 ft or more, it is imperative to locate the pump after the connection to the compression tank. In this position, the intake of the pump will be under a positive pressure. The pump pressure will add to the pressure at the compression tank connection, and you are less likely to develop a vacuum condition anywhere in the system.

The pump in a solar hydronic system should pump toward the elevated heat absorbers. Be sure that the compression tank (if it is a closed system) is connected someplace on the suction side of the pump.

Mounting

Small pumps (usually fractional-horsepower) are mounted in line with the piping. This arrangement means that there is no extra piping or anything else required for the installation. Larger pumps are set on concrete pads and bolted down, with the piping brought to the pump. After the piping has been installed, the pump frame should be shimmed to be level in both directions.

If the pump is assembled on a channel base, pack the channel space with grout to increase the mass of the supporting systems and reduce vibration.

Pump Alignment

Aligning the pump, motor, and shafts can be completed in four easy steps using a straight edge and a set of feeler gauges. The importance of shaft alignment increases as the rpm and the horsepower of the unit increase. Many coupling and equipment manufacturers recommend tolerances based on the rpm and the diameter of the coupling. Table 27-1 shows some misalignment tolerances when using a rigid coupling. A rigid coupling only has two metal hubs with no rubber insert between them.

While it is more common to see couplings with rubber inserts, the rubber insert is to account for accidental misalignment and to reduce vibration. The more misaligned they are, the more rapidly the rubber insert will deteriorate due to friction.

Table 27-1 Shaft Alignment Tolerances

	Excellent		Acceptable	
RPM	Offset (inches)	Angularity (inches/coupling diameter)	Offset (inches)	Angularity (inches/coupling diameter)
600	.005	.001	.009	.0015
1800	.002	.0003	.003	.0005
3600	.001	.0002	.0015	.0003

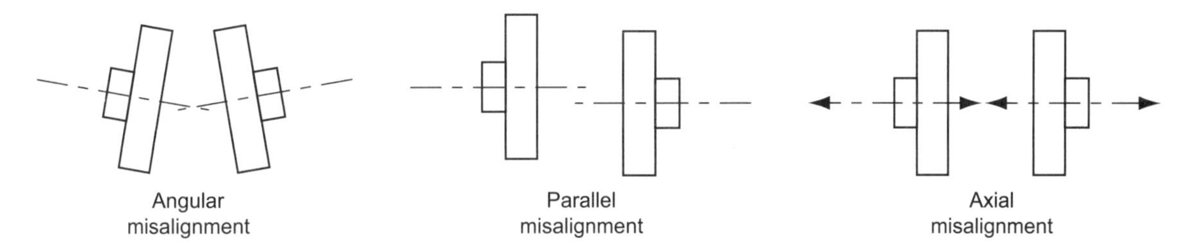

Figure 27–8
Three types of shaft misalignment.

As shown in Figure 27–8, there are three different types of misalignment: angular, parallel, and axial.

Angular misalignment in the vertical direction should first be corrected by placing shims under either the two front feet or the two back feet of the motor. Always perform this correction first, because it will either raise or lower the front of the motor. For best results tighten the motor mounting bolts before taking any measurements. Since the couplings may be mounted on the shafts at a slight angle, temporarily join the couplings with tape and take all measurements by turning both shafts. By taking all measurements from the same spot on the couplings, any coupling misalignment will be negated. The thickness of the shims can be calculated by the following formula:

$$\text{Shim thickness} = (\text{GD})\left(\frac{\text{CDMF}}{\text{CD}}\right)$$

GD = gap difference (top measurement–bottom measurement)
CDMF = center distance between motor feet
CD = coupling diameter

Example 4

Use Figure 27–9 and the formula above to calculate the shim thickness required to correct the vertical angular misalignment for the motor. Notice the gap is biggest at the top of the coupling, so the shims will go under both back feet.

$$\text{Shim thickness} = (0.020\text{–}0.010)\left(\frac{12''}{4''}\right)$$
$$\text{Shim thickness} = (0.010'')(3) = 0.030''$$

After placing 0.030″ shims under both back feet, the gap at the top of the coupling should now be the same or much closer to the gap at the bottom of the coupling. Now that the vertical angular misalignment is taken care of, the vertical parallel misalignment is next. This is shown in Figure 27–10, where the motor coupling is now sitting lower than the pump coupling. In this step, the height difference will equal the amount of shim thickness that is needed under all four feet.

Now that the vertical alignment has been taken care of, there is no need for additional shimming. The motor must now be moved side to side to eliminate any misalignment in the horizontal plane. Figure 27–11 shows how a simple straightedge can be used to check for horizontal alignment. Once the motor is in position, all four feet should be torqued down evenly.

The last misalignment, axial, can be adjusted by loosening the coupling on the motor and inserting the rubber component. The coupling manufacturer will recommend a separation distance.

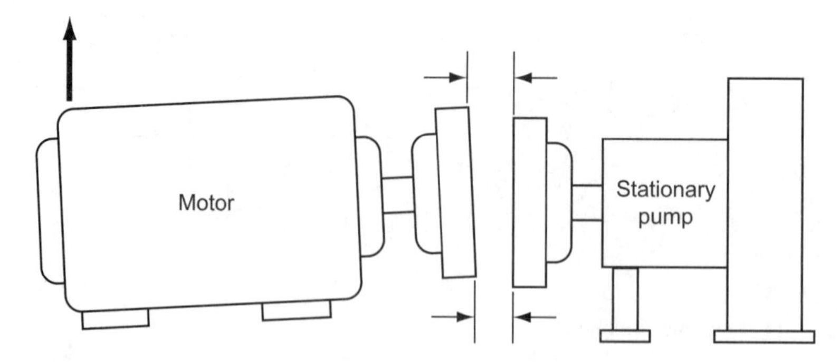

Figure 27–9
Notice that the gap is larger at the top; this means that shims need to be added to the rear of the motor. This will lower the motor coupling as well.

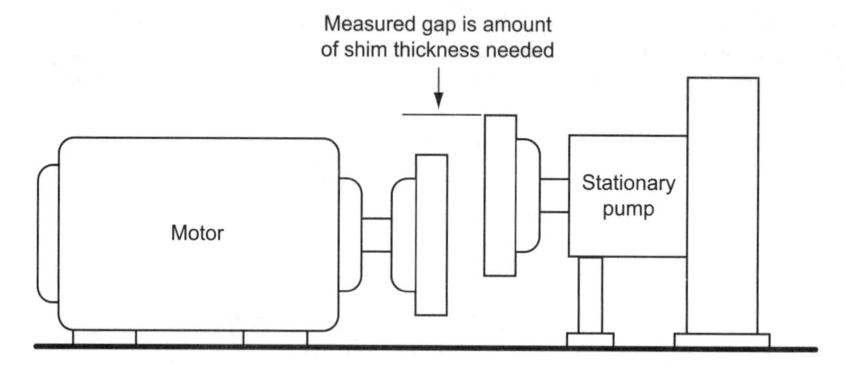

Figure 27–10
Vertical parallel misalignment is corrected by adding shims under all four motor feet. Do not remove the previous shims added to correct angular alignment.

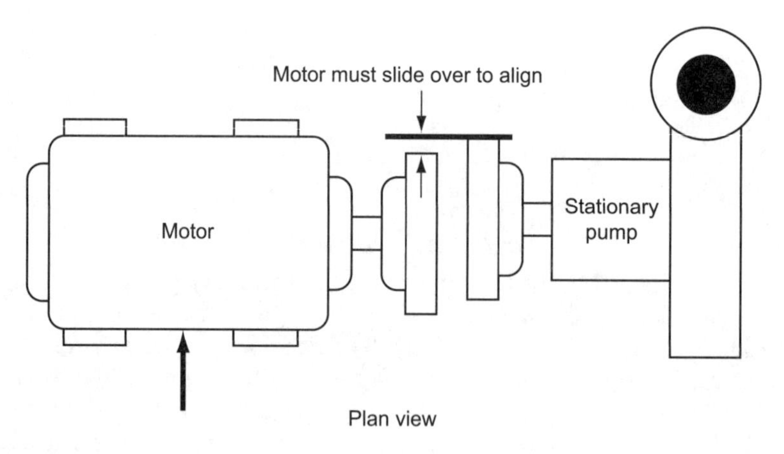

Figure 27–11
Horizontal alignment does not require any shims; simply slide the motor until it aligns with the pump coupling.

PIPING

The piping that connects to a base-mounted pump must line up with the pump connections. Do not draw up misaligned pipe by forcing the pipe to the pump. Doing so will introduce forces that will pull the pump out of alignment with the motor. In extreme cases, it could even result in breaking the pump flange or pump body. It could also put forces on the pump body and cause additional loads to be placed on the internal bearings.

Slope the piping to aid in natural venting through the pump. Large pumps may need a manual vent in the top of the pump volute to vent the air trapped there.

The final suction connection to a base-mounted centrifugal pump should be at least ten pipe diameters long so that the liquid entering the pump will fill the pipe and flow straight. Many pumps have connection sizes that are smaller than good practice would dictate for long runs of pipe at the flow capacity of the pump. The pipe size should be increased to the required line size on each side of the pump. In no case should the suction side line be smaller than the discharge line.

Accessories

Valves should be installed to permit servicing. If the system pump circulates water in a closed system, the valve on the discharge side should be capable of throttling. The valve on the suction side should be a gate, ball, or butterfly type for straight-through flow characteristics, and it should not be throttled. If the pump flow output must be reduced, throttle on the outlet only. If the inlet is restricted, the pump will not receive a full flow of liquid, and severe damage may result. Strainers may also be used to keep foreign objects out of the system.

Water-well supply pumps require foot valves at the bottom of the suction pipe to maintain the prime in the suction pipe at all times. The foot valve is a check valve and strainer assembly that holds the water in the suction pipe. The strainer screen must have adequate area to minimize the pressure drop across the strainer. The minimum acceptable area is equal to the cross-sectional area of the pipe size of the device.

Check valves are needed with pumps that lift water from a low level to a high level (water supply, sewage, sump, or storm water) and in any arrangement of parallel pumps. For pumps in lifting applications, spring-loaded check valves are preferred. These are referred to as non-slam or silent checks. The usual swing check valve does not close until the flow reverses. If the valve is 2 inches or larger, the reversal valve-close, sudden-stop, water hammer sequence makes considerable noise and produces large stresses in the check valve and adjacent piping. A spring-loaded check valve closes before the flow reverses, and thus there is no water hammer.

Most positive displacement pumps are equipped with a pressure regulating device on the output. These regulators are relief valves (with some means of adjusting them) that bypass the discharge fluid back to the inlet at the set point. Without such a regulator, operation of the pump could produce damaging pressures.

In the Field

If the pump output has to be throttled at all times, it is probably a sign that the pump is oversized. Talk with the pump manufacturer about either using a variable speed drive to reduce the pump rpm or reducing the impeller size. Either of these may be a more energy-efficient method to reduce flow.

Start-up

Piping should be flushed out and checked for leaks before starting the system. Lubricate the pump as recommended by the manufacturer. If it is accessible, turn the pump shaft manually before initial operation. Be sure the pump is full of liquid before turning on the power. Most pump designs should not be operated unless they are filled with the liquid to be pumped. Be sure all valves are open.

Check electrical circuits and check the motor nameplate for agreement. If proper rotation is required, turn power on and off quickly to check. Some pumps cannot be operated backward at all. In such a case, you will have to remove the coupling and check the motor. After you are assured the rotation is correct, you will have to reinstall the coupling.

After the above checks have been made, turn on the pump and check the system for leaks, proper circulation, undue noise, and other problems.

Once the pump is running under a steady flow of liquid, check the pump for pressure rise and check the motor for applied voltage and line current. Record all values and retain for future reference.

PUMP MAINTENANCE

Satisfactory service and long life for a pump can be ensured by a regularly scheduled maintenance program. By keeping accurate records of equipment conditions, early warning signs can help the technician predict when a piece of equipment is beginning to fail. This is known as predictive maintenance. Predictive maintenance allows companies to schedule equipment downtime so that repairs or adjustments can be made in a timely manner.

Lubrication

Lubricate pump bearings and motor bearings with the lubricant recommended by the manufacturer to obtain maximum service from the pump. Remember the oil used for lubrication is designed to keep the rotating parts separated so there is only minimal metal-to-metal contact. The lubricant must be thin enough to get between parts, but thick enough to keep them apart. Another important feature of lubricant is that it removes heat from the machine.

Contrary to popular opinion, it is possible to over-grease a bearing. When a bearing has the proper amount of grease, the grease will move around the rolling components and lubricate all parts. When too much grease is added, the extra grease can actually make the bearing run hotter than normal. A handheld grease gun can easily produce enough pressure to blow out a seal.

Some pumps are equipped with sealed bearings or water-lubricated bearings. Keep dirt and debris out of bearing areas. Any foreign material that enters the bearing will drastically reduce bearing life. Because these types of pumps use the pumping fluid as a lubricant, it is important to keep the pressure levels of the system well within the parameters of the pump. If a pump is operated at a higher than recommended pressure, it can cause premature seal failure.

For pumps with packing glands, the packing collar must be adjusted to permit a small leak. If the packing is tightened enough to stop the leak, the packing will overheat and score the shaft. This will result in permanent, excessive leakage and is the most common condition that requires major pump repair.

Periodically check pump/motor alignment and check the coupling for excessive wear. Since the temperature of the pumped fluid can have an effect on the pump itself, check while the pump and motor are at operating temperatures. Belt-driven equipment should be checked within 8 hours of installing a new set of belts. A new set of belts will stretch initially and then remain that length for some time. Overtightening belts during the initial installation will put additional stress on the bearing and shafts.

Measure pump hydraulic performance with gauges to be sure that the total pressure rise and flow are consistent with earlier pump history. Other tools such as infrared detectors, ultrasonic noise detection, and thermography can be used to detect problems before they get out of hand.

Maintenance Problems

Typical maintenance problems (and possible causes) frequently encountered include the following:

No Flow

- Pump may need priming
- Speed is low
- Pump rating is less than the system requires
- Suction line is obstructed
- Suction lift is too great
- Pump (centrifugal) is turning the wrong way
- Shaft key sheared off

Insufficient Flow or Inadequate Pressure

- Air leakage on suction side of the pump
- Speed is low
- Insufficient pump capacity for system
- Suction line is obstructed
- Suction lift too high
- Impeller obstructed
- Backward rotation (centrifugal pump)
- Excessive wear in impeller or pump housing (volute)

Loss of Suction

- Leak in suction pipe
- Defective casing seal
- Excessive suction lift required
- Defective or plugged foot valve

Excessive Power Consumption

- Insufficient capacity resulting in excessive operating time
- Pump is handling liquid with higher viscosity than originally planned
- Mechanical defects somewhere in the pump-motor combination

The manufacturer's instructions should be the final authority on the type and schedule of pump maintenance operations. Be especially careful with pumps that have any special characteristics or applications.

REVIEW QUESTIONS

1. Pump curves are drawn for different impeller diameters, for a given rotation speed, fluid, and temperature, and show flow versus ______.
2. Are there circumstances when you should operate a pump beyond the curve shown by the manufacturer?
3. Is it acceptable to throttle the valve on the suction connection to the pump?
4. The higher the ______ of the pump, the more critical the shaft alignment.
5. Doubling the pump rpm will ______ the pump pressure and could blow out pump seals.

CHAPTER 28

Blueprint Review and Shop Drawing

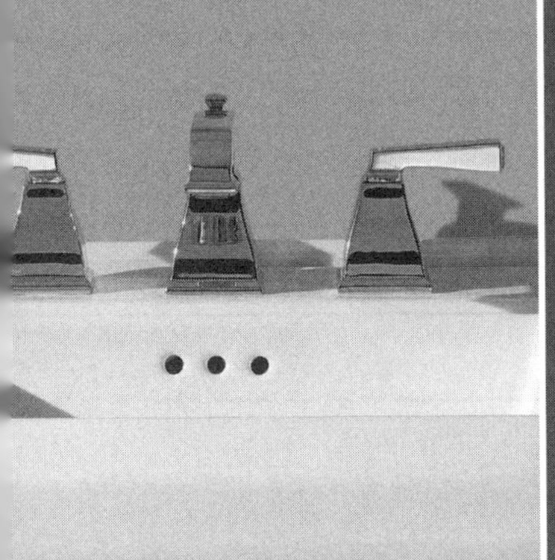

LEARNING OBJECTIVES

The student will:

- Describe the different line types used in blueprints.
- Interpret and draw isometric sketches.
- Create shop drawings from different orthographic building plans.

BLUEPRINTS

In this final section of drawing review chapters, we will review all of the previous blueprint information from this textbook series. The goal is to review the topics of:

- Blueprint reading
- Sketching
- Material estimating

The following is a breakdown of previous topics, by the book in which they appear.

Plumbing 101

- Lines
- Scale Rulers
- Sketching
- Symbols
- Detail Sketching
- Orthographic Projection
- Isometric Sketches
- Oblique Sketches

Plumbing 201

- Rough-In Sheets
- Single Line Drawings
- Detail Drawings
- Sections
- Exploded Views

Plumbing 301

- Specifications
- Floor Plans
- Drawing Types
- Site Plans
- Structural Plans and Elevations
- Plumbing Plans
- Electrical Plans
- HVAC Plans
- Detail Plans

Alphabet Of Lines (Figure 28-1)

Object Lines

Object lines are sometimes called visible lines because they are used to represent the edges of an object that are visible from our point of view. These are solid, dark lines that are wider than other lines, with the exception of cutting plane lines.

Center Lines

Center lines are used to indicate the center of circular (or cylindrical) objects and also to demonstrate the fact that the object is symmetrical. Center lines are thin dark lines

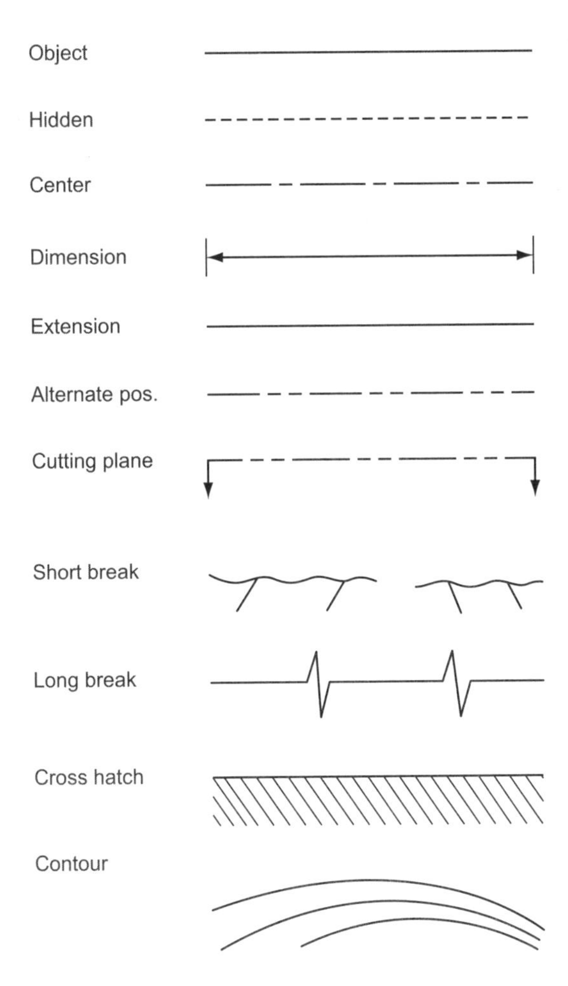

Figure 28–1
The alphabet of lines.

drawn as two long ($\frac{3}{4}$ inches to 1 $\frac{1}{2}$ inches) lines with a short dash ($\frac{1}{8}$ inch) between them. There should be approximately $\frac{1}{16}$ inch space between the long lines and the short dash.

Hidden Lines

Hidden lines indicate details about an object that we cannot see from the view represented. For example, on a side view of a water closet, hidden lines would be used to indicate the location and shape of the internal trap. Hidden lines are short dashes that are the same width and darkness as object lines.

Dimension Lines

Dimension lines show the extent of a length being displayed. They are thin, solid lines with an arrowhead at each end. The line is interrupted at a convenient midpoint with enough space to write in the dimension.

Extension Lines

Extension lines extend from close to the object line to just past the dimension line to make clear just what on the object is being dimensioned. They are the same thickness as dimension lines.

Leader Lines

Leader lines connect notes to an object to eliminate ambiguity. These lines are the same thickness as dimension lines.

Cutting Plane Lines

Cutting plane lines are used to show where a section is to be removed for examination. These lines are the thickest on the drawing. Some technicians use a series of equal dashes; others use a longer segment with two short dashes.

Phantom Lines

Phantom lines are used to show alternate positions of moveable parts. They are thin lines drawn using a long dash and two short dashes.

Broken or Break Lines

Broken or break lines are used to indicate that the drawing representing an object is incomplete. A portion of the object is removed, or the part continues but is not shown. For example, on a drawing of a long segment of pipe, a larger scale can be used to show detail of the ends by breaking out a section in the middle. Short break lines are drawn in irregular freehand to suggest a broken surface. Longer break lines are shown by thin line segments (from $\frac{3}{4}$ inches to $1\frac{1}{2}$ inches long) with a freehand "Z" connecting adjacent ends.

Contour Lines

Contour lines are used on site plans to indicate lines of equal elevation. Contour lines that indicate existing grades are freehand continuous fine lines with the elevations indicated in the line at some convenient place. Contour lines that indicate grades after the project is complete are freehand dashed fine lines with the elevation noted at a convenient place.

Cross-hatch Lines

Cross-hatch lines are used in various patterns to indicate a cut surface or to indicate any item(s) for emphasis or clarity.

Scale Rulers

Scale rulers are used to produce working drawings and perform take-offs of scaled drawings for a materials list. The most common scales are:

- $\frac{1}{4}'' = 1'0''$
- $\frac{1}{8}'' = 1'0''$

The three most common scale rulers include:

Architect's

Architect's scales equate different fractions ($\frac{1}{8}, \frac{1}{16}, \frac{1}{32} \ldots.$) of an inch to one foot. Architect's scales are commonly used for architectural drawings of buildings.

Engineer's

Engineer's scales equate a certain number of tenths of an inch to one foot. These scales are *not* in meters. The engineers scale is used for site plans and topographic uses.

Mechanical

Mechanical scales equate various fractions ($\frac{1}{50}, \frac{1}{100} \ldots$) of an inch to one inch. They are commonly used for machine tools and similar object drawings.

Isometric Sketches

Isometric drawings lay out the X, Y, and Z axes 120° apart, where the Z axis is always vertical. Remember, lines that are normally horizontal are now angled 30° above the horizontal on the drawing. See Figure 28–2.

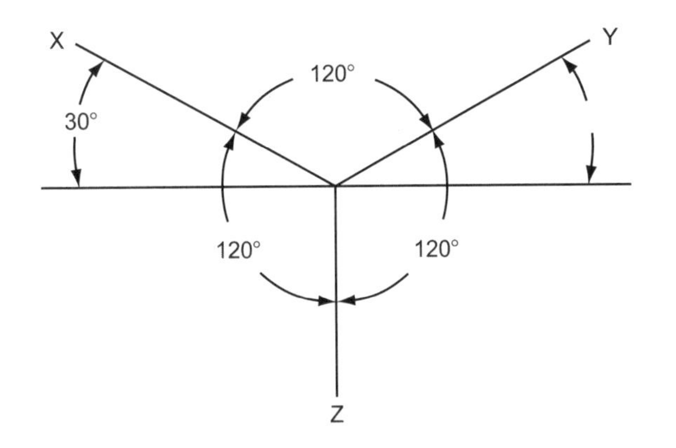

Figure 28–2
Isometric axes.

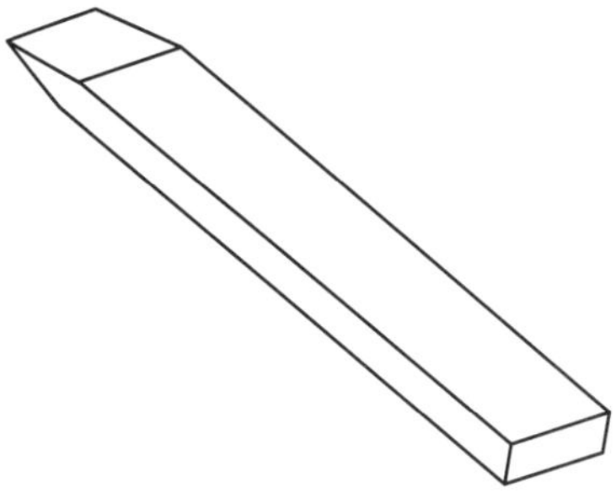

Figure 28–3
Isometric of brick chisel.

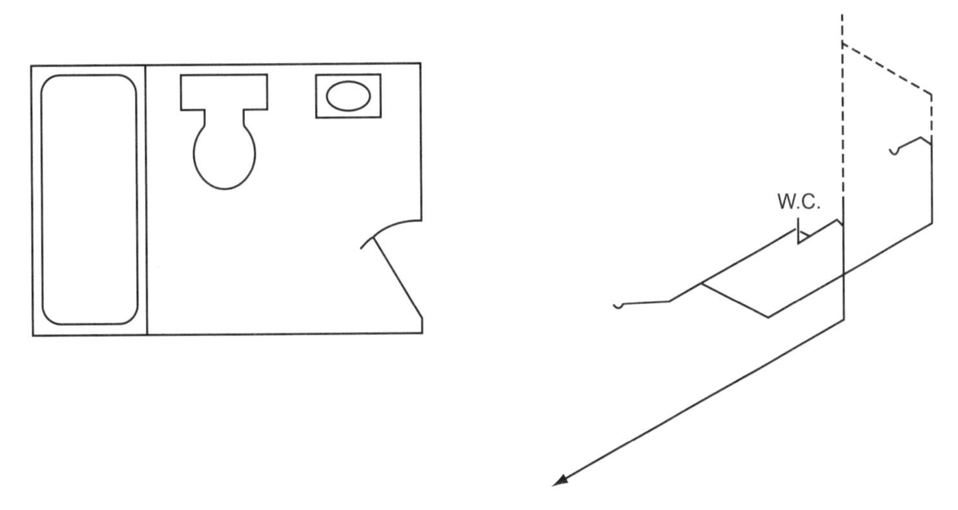

Figure 28–4
Isometric sketch of a typical bath.

Figure 28–3 shows an isometric drawing of a brick chisel. Figure 28–4 shows a plan view of a bathroom on the left side and an isometric view of the piping on the right.

Review any questions you may have regarding isometric sketches and blueprint reading in general. Tables 28–1 and 28–2 are included as a refresher on some common plumbing symbols used in blueprints.

SHOP DRAWING

A shop drawing is a detailed sketch of a plumbing or piping system or series of components that can be used for ordering materials, for laying out a job, or to show how the work fits into the general project. The shop drawing is developed from the blueprints and specifications.

Identify the section of the blueprint to be drawn based on the complexity of the plan. You may wish to indicate on the blueprint what sections of the plan pertain to which shop drawing. In order to simplify the procedure, you may wish to place a sheet of tracing paper over the plan view to be drawn and trace the section to be detailed so that you can then develop an isometric sketch. Another solution would be to scan the portion of the drawing with a flatbed scanner and save the image to the computer. If the plan is too large, a digital camera may work as well.

In the Field

Shop drawings are extremely handy for identifying conflicts in pre-construction meetings.

Table 28-1 Common Symbols for Mechanical Plans. Courtesy of Plumbing–Heating–Cooling–Contractors—National Association.

Symbol	Description	Symbol	Description
	Bath—recessed		Drain line
	Bathtub—free standing		Vent line
	Bathtub—corner		Cold supply
			Hot supply
	Shower stall		Hot recirculator
	Urinal—wall mounted	—— VAC —— VAC ——	Vacuum line
	Lavatory (vanity)	—— FG ——	Fuel gas
	Kitchen sink and drainboard	—— FP —— FP——	Fire protection
	Water heater	—— A —— A ——	Compressed air
	Water closet tank type		
	Laundry tubs		
	Floor drain		
	Roof drain		
	Meter		
	Hot water tank		
	Sillcock		

Just remember that in either case the new image will not be to scale. In any case, you will want to make an enlarged view of the section to be shown before attempting to prepare the isometric drawing.

Mark this new image with letters or numbers on the system's starting point, terminal point, and branches.

Prepare the isometric using these points as your guidelines.

Table 28-2 Table of Pipe Symbols. Courtesy of Plumbing-Heating-Cooling-Contractors—National Association.

Item	Soil pipe	Thread	Flange	Groove	Weld	Solder
Joint						
Ell-90%						
45%						
Reducing						
Tee						
Wye						
Reducer						
Gate valve						
Globe valve						
Check valve						
Gas cock						
Relief valve						
Safety valve						
Expansion joint						
Union						
Pump-centrif.		Indicate type of conn. with joint symbol				
Tank		Indicate type, application, and size				

- A change of direction of 90° will be shown by a deflection of 120° on the isometric representation.
- Vertical lines are drawn vertically in isometric drawings.
- After the basic diagram is prepared, note sizes and lengths for all piping, including branches. Also make special notes about valves, fittings, hangers, and special installation instructions.

Refer to Figures 28–5 through 28–17 for examples of various types of drawings. Note that pipe sizes, symbols, and similar notes are on the drawings. If need be, refer back to the appropriate chapters in *Plumbing 101*, *201*, and *301* to review.

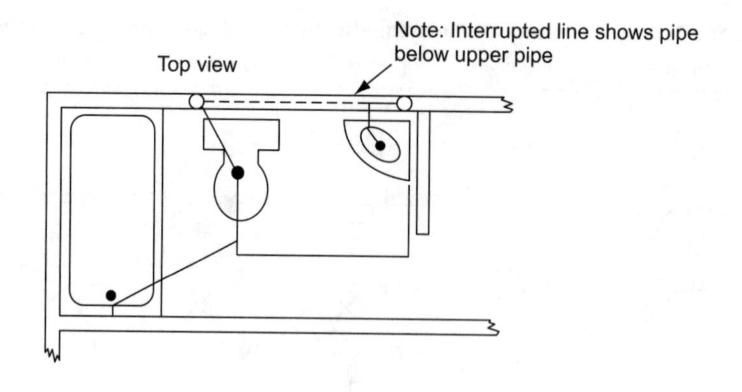

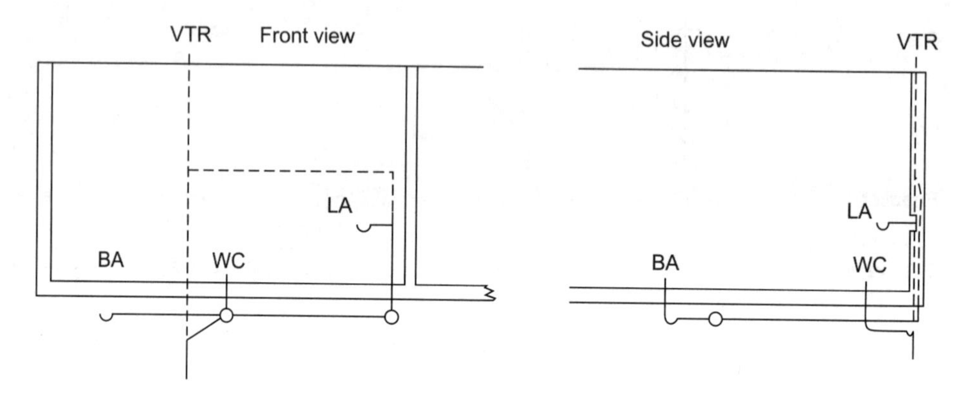

Figure 28–5
Orthographic projection of a typical bath showing top view, front view, and right side view.

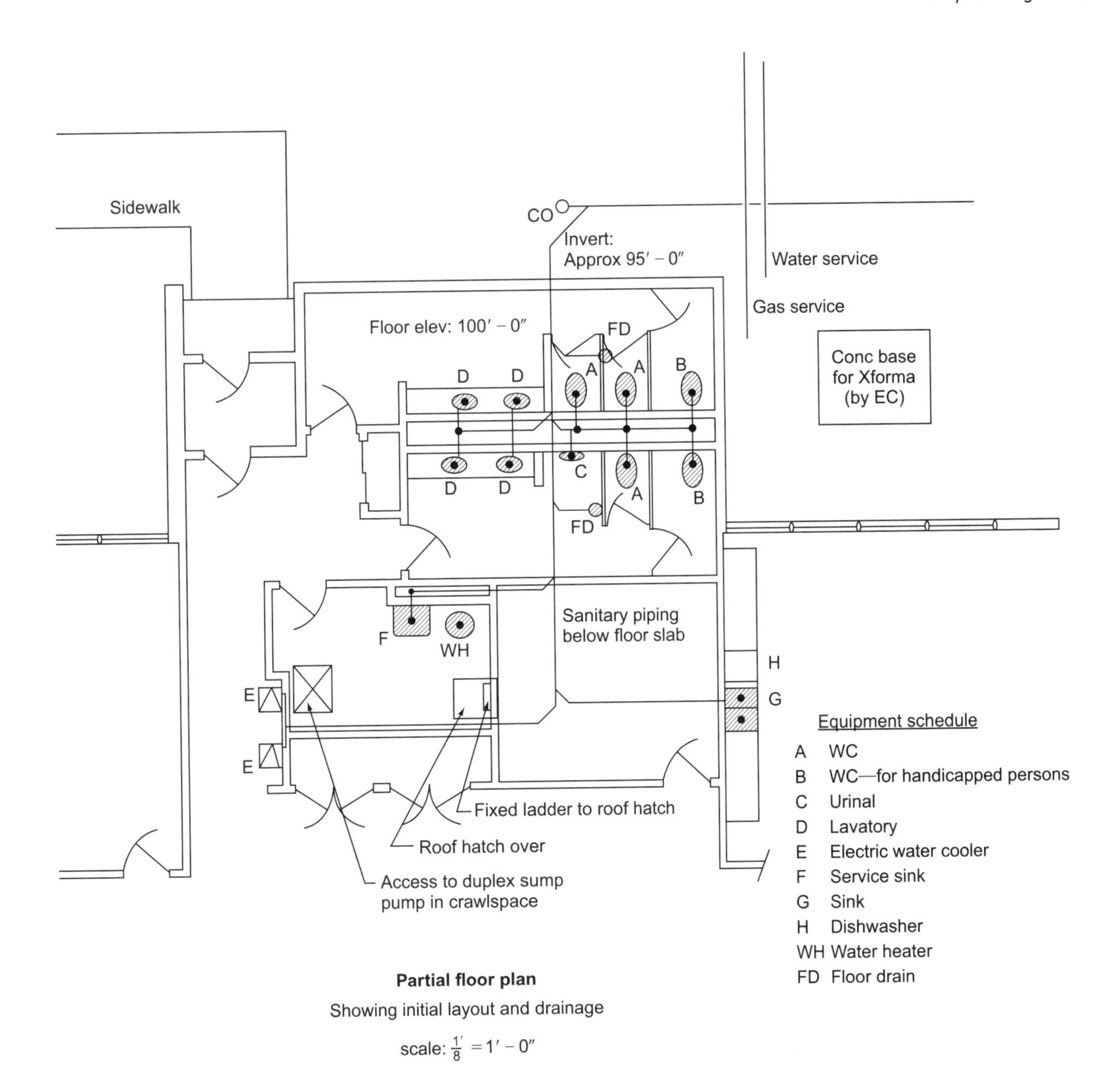

Figure 28–6
Partial drawing of a building floor plan.

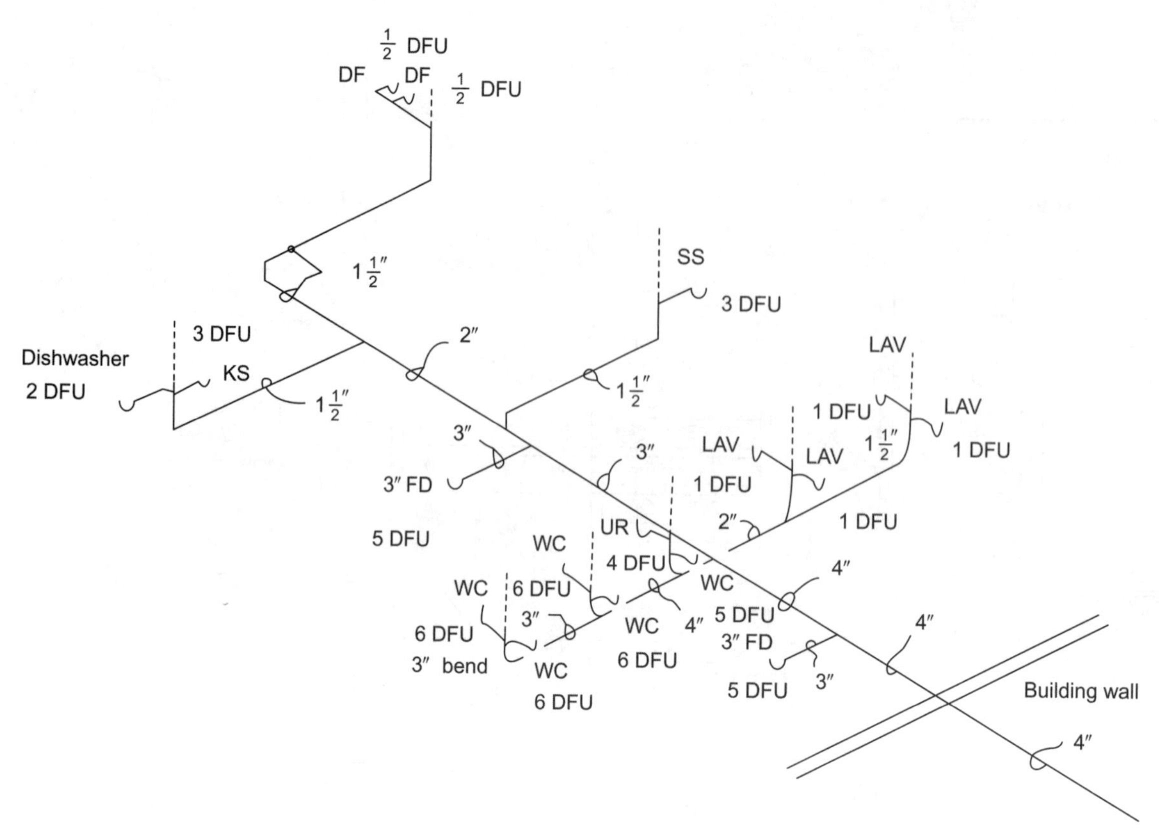

Figure 28–7
Isometric view of a plumbing system used for preliminary sizing.

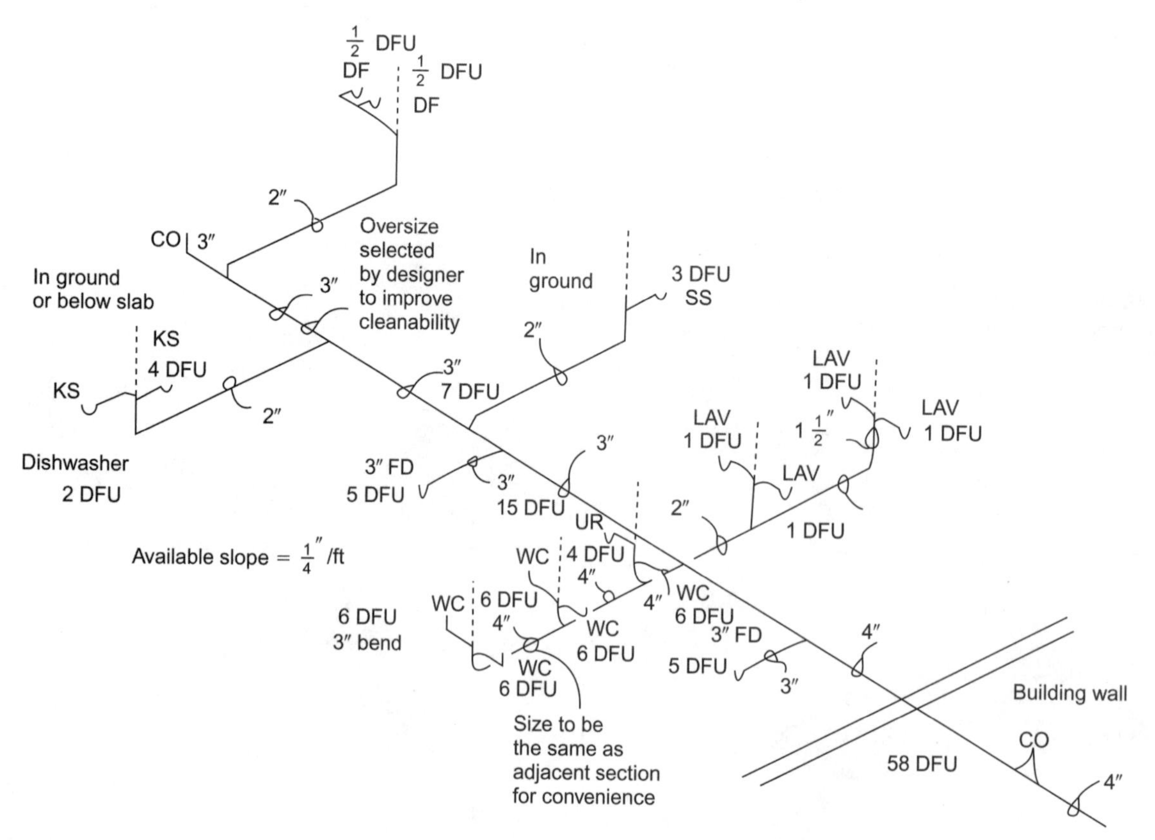

Figure 28–8
Isometric view of a plumbing system used for final sizing and cleanout locations.

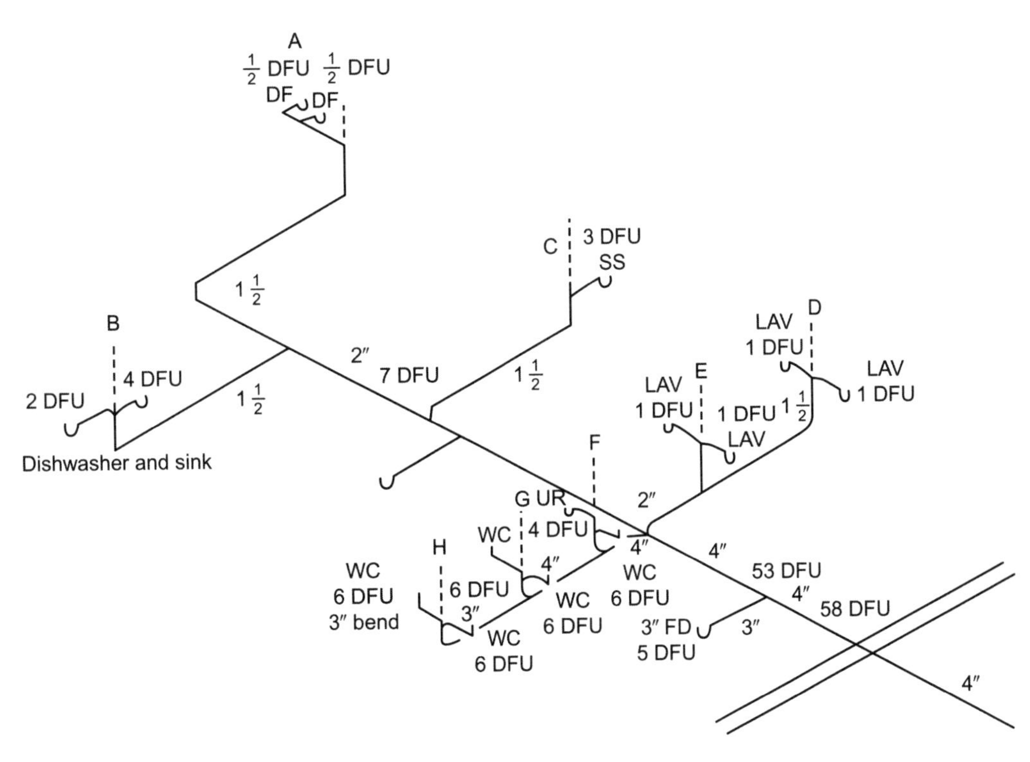

Figure 28–9
Isometric of a plumbing system, individual vents to outdoors, drainage based on initial sizing.

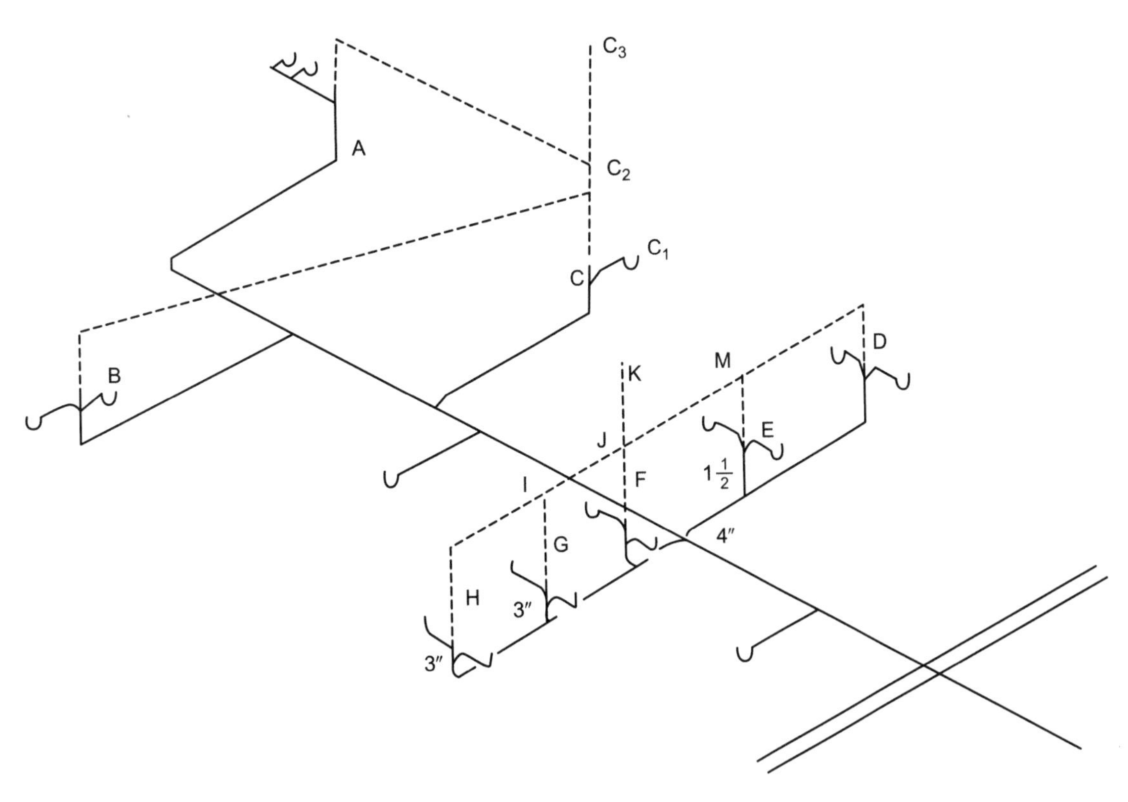

Figure 28–10
Isometric of a plumbing system including venting with headers.

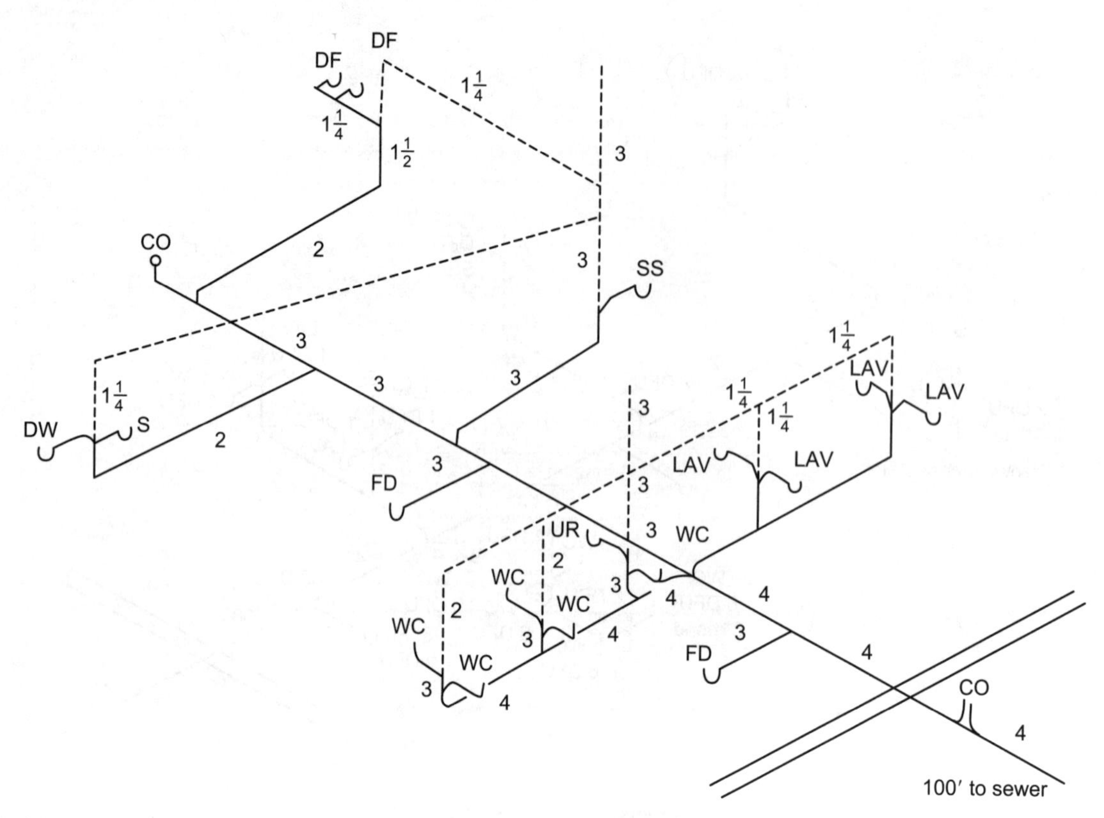

Figure 28–11
Isometric of a plumbing system, final drainage and vent design.

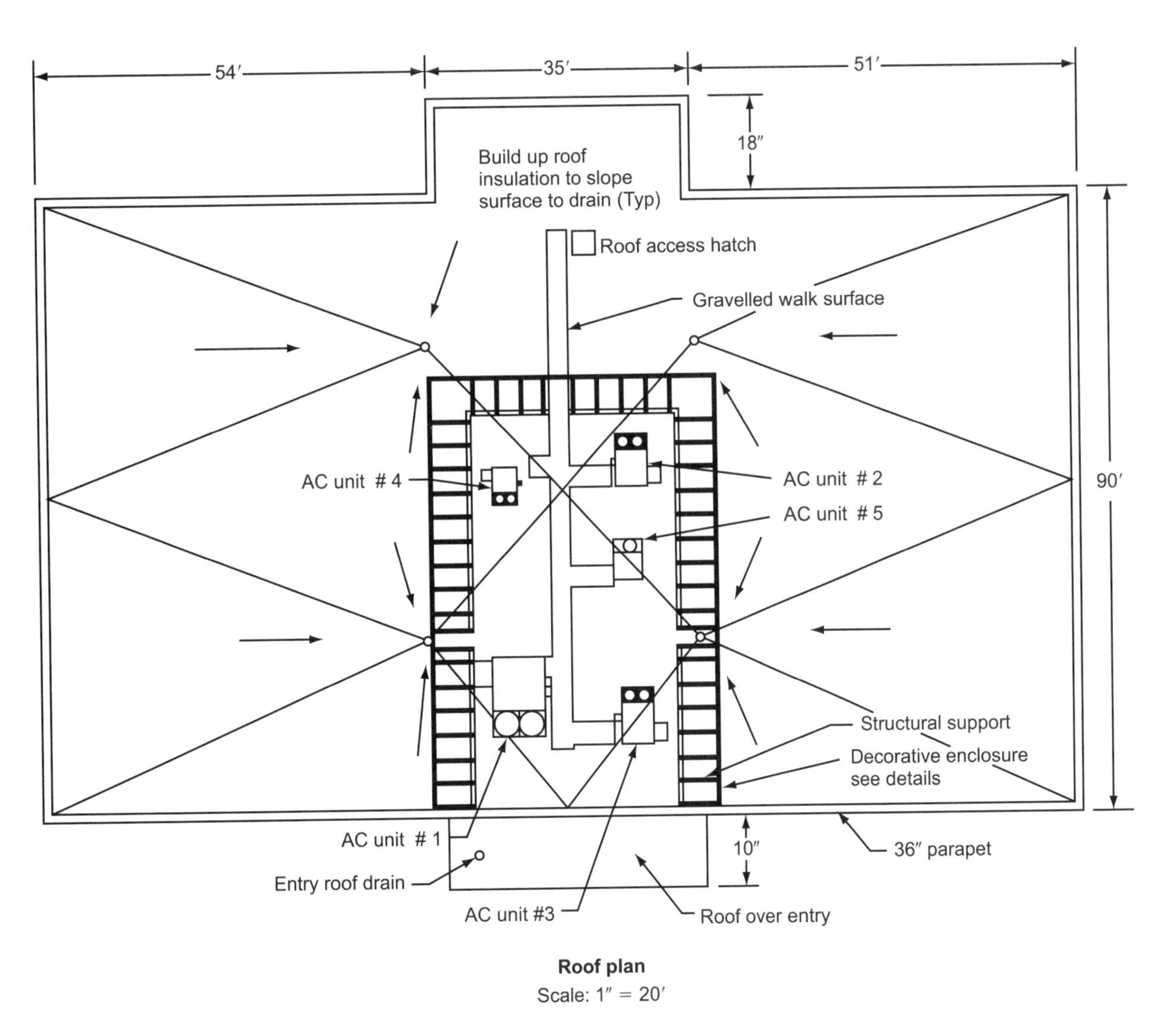

Figure 28–12
Roof plans give the plumber information about roof drainage and the location of peaks.

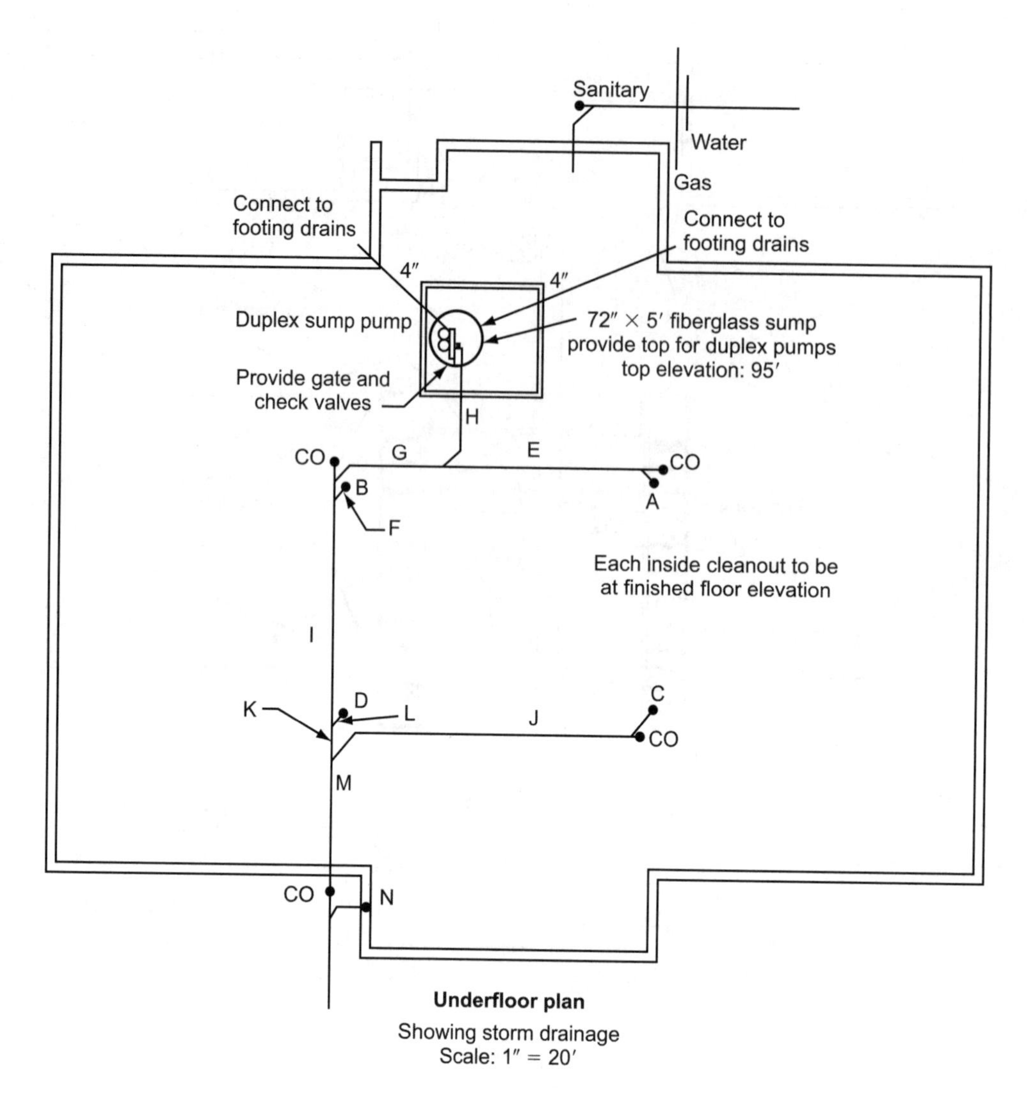

Figure 28–13
Underfloor piping and connections of roof drains.

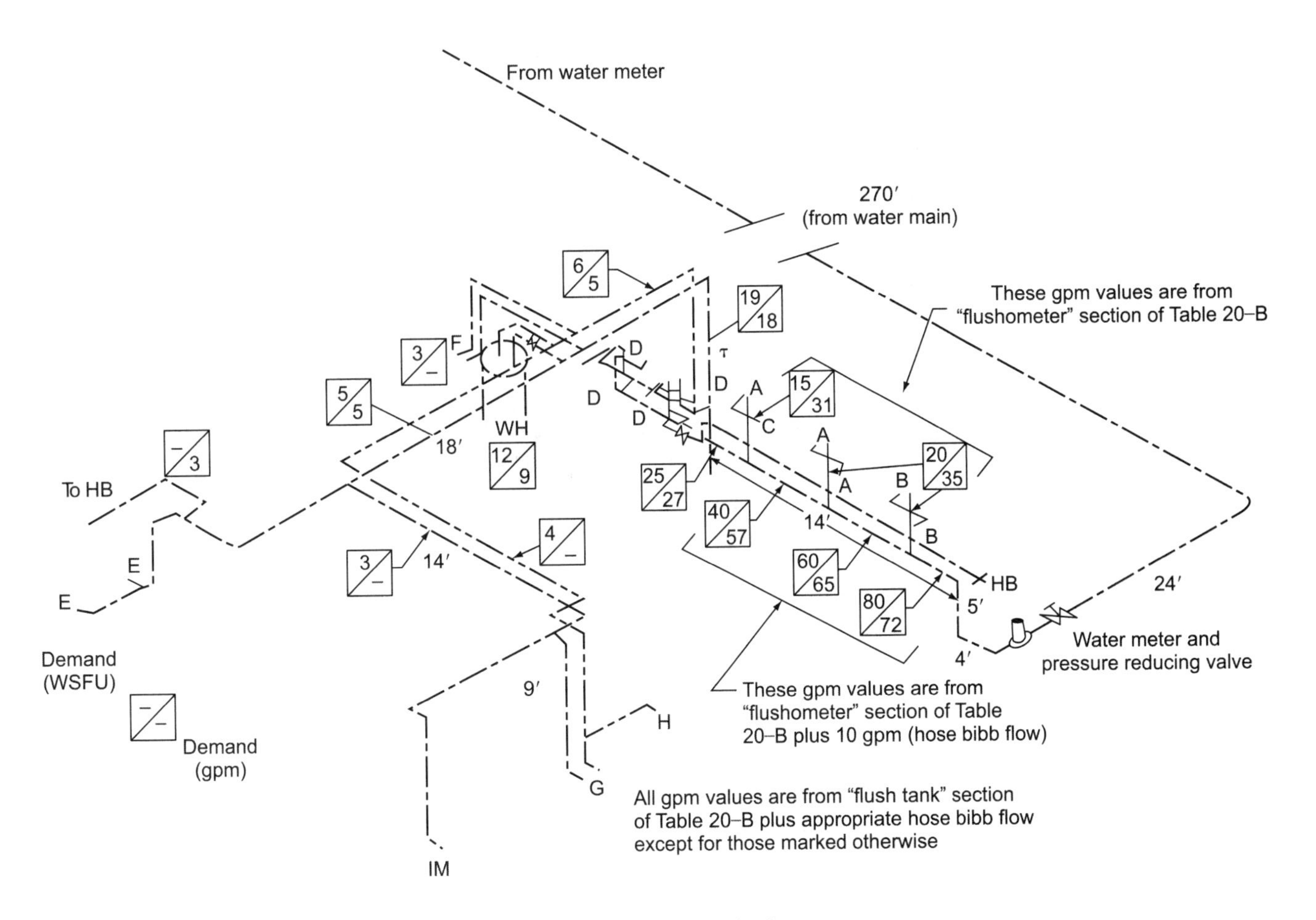

Figure 28–14
Isometric sketch of water supply and distribution system (not to scale).

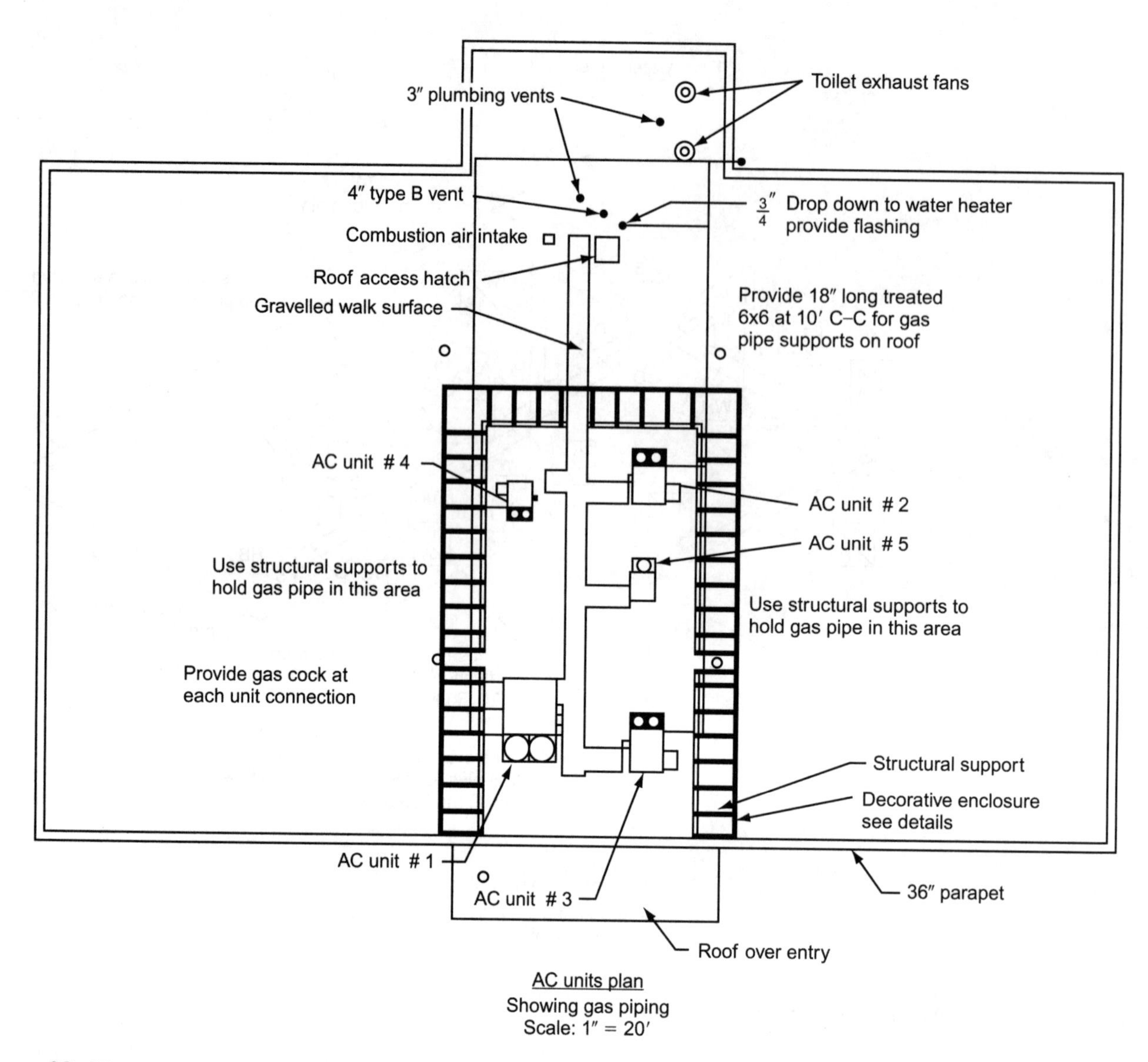

Figure 28–15
Roof plan showing the location of air-conditioning units and gas piping.

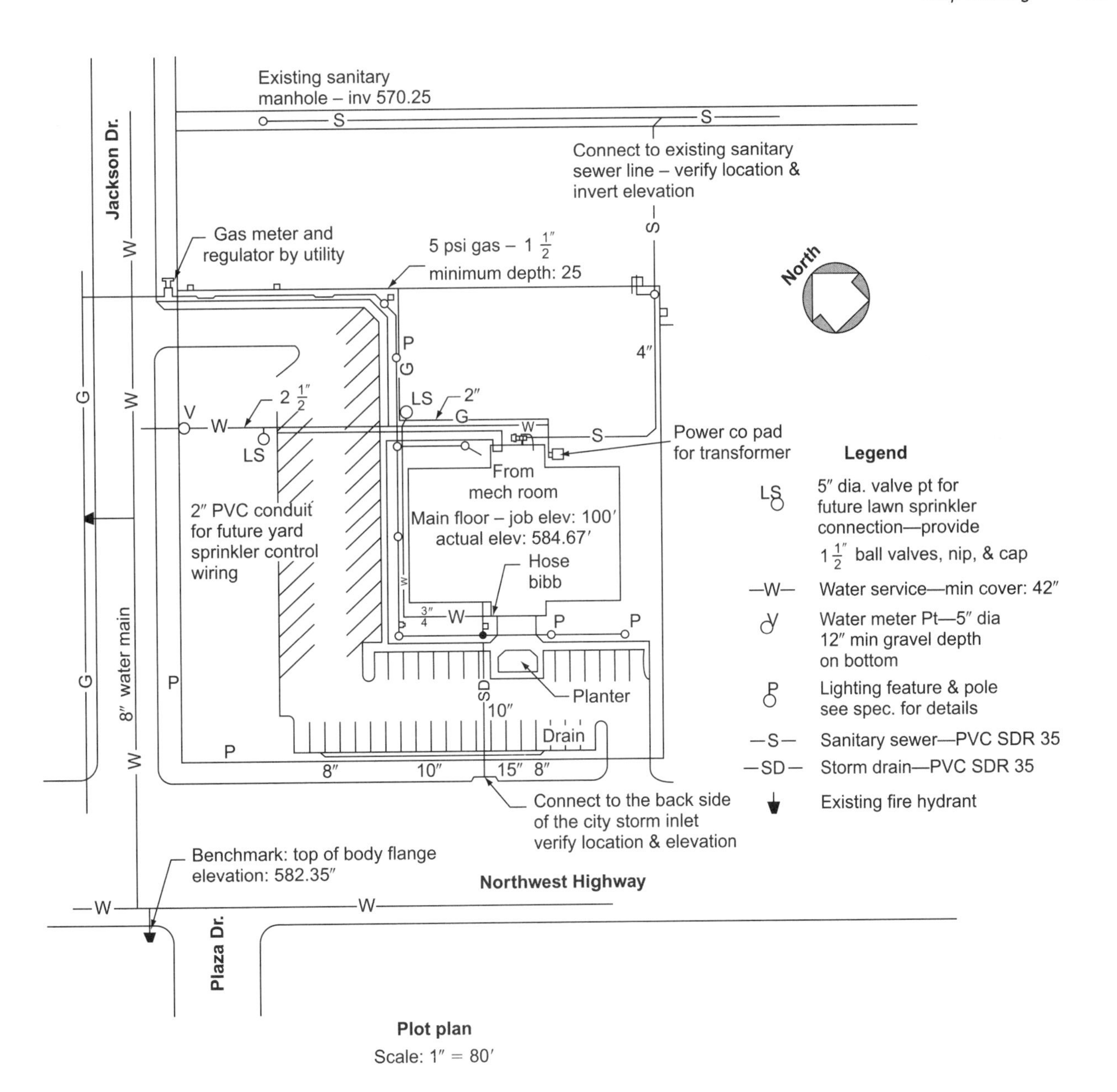

Figure 28–16
Plot plan.

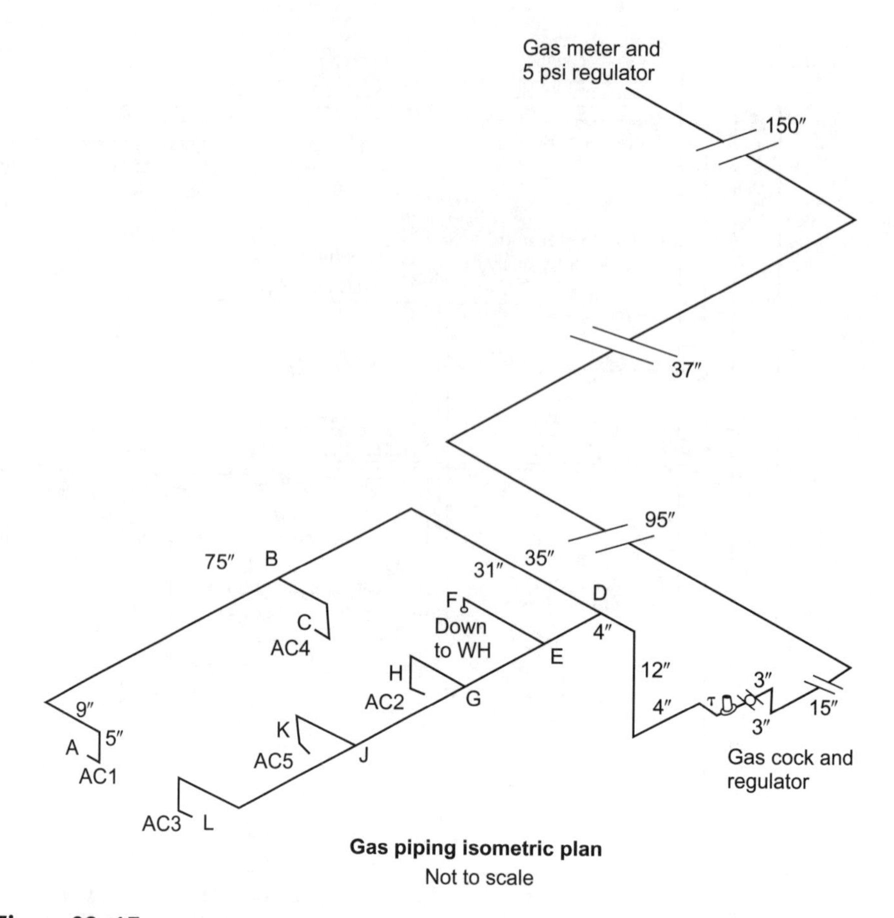

Figure 28–17
Isometric drawing of gas piping layout. Notice how break lines are used to allow the drawing to fit on a small page.

REVIEW QUESTIONS

1. Which line type is used on site plans to indicate lines of equal elevation?
2. Which line type is used to indicate the center of a circular (or cylindrical) object and also to demonstrate the fact that the object is symmetrical?
3. What is an advantage of using isometric sketches over orthographic ones?
4. Give one example of when to use a break line.
5. Why would a plumber need to look at the roof plan?

CHAPTER 29

Water Supply, DWV Isometric, and Storm Drainage Systems

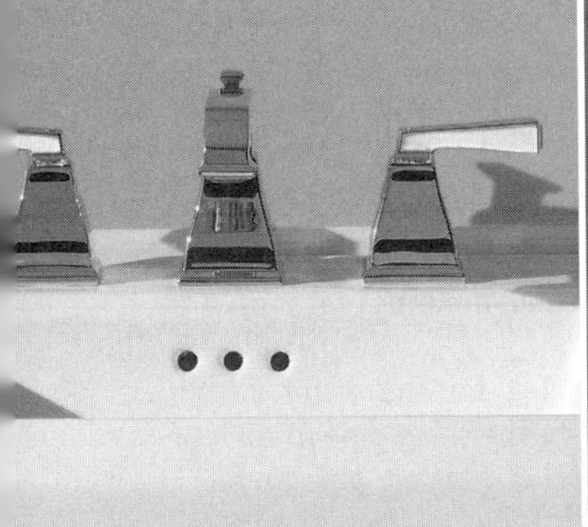

LEARNING OBJECTIVES

The student will:

- Prepare an isometric of a water supply system.
- Prepare an isometric of a DWV system.
- Prepare an isometric of a storm drainage system.

WATER SUPPLY ISOMETRIC

Remember from earlier discussions that pipe isometric drawings allow the user to see all three dimensions in one view. It is extremely handy for determining the necessary fittings and lengths of pipes.

When preparing an isometric drawing of the water supply system use the following steps:

1. Identify the section to be drawn.
2. Trace the section.
3. Mark the starting point, terminal point, and branches.
4. Prepare the isometric using these points.
5. Note changes of direction needed on the isometric.
6. Add pipe sizes and lengths, fittings, hangers, and notes pertaining to the drawing.

Pipe sizes will be determined from the plan riser diagram. Pipe lengths can be scaled from the plan. Meters, valves, and similar components should be shown on the isometric drawing. Be sure to provide the following data:

- Pipe sizes
- Pipe lengths
- Meter location
- Valve locations and types
- Type of pipe
- Joining methods
- Hanger requirements
- Insulation requirements
- Floor and wall penetrations
- Access door requirements
- Any other information necessary for the field technician to fabricate the system from the drawing.

Figure 29-1 shows a plan view of a simple residential building with only a kitchen and one bathroom.

Figure 29-2 shows an isometric take-off of Figure 29-1 for the water supply system. See if you can follow the piping layout and identify the same pipe in each of the two figures. As practice, redraw Figure 29-2 with another possible pipe route.

DWV ISOMETRIC

The same steps and methodology used to construct the water supply piping isometric can be applied to the generation of a DWV isometric drawing. Use Figure 29-3, a plan view of the DWV system, to prepare an isometric drawing of the DWV system for a residential building. Figure 29-4 shows the completed isometric drawing for Figure 29-3. Use the steps we have discussed before:

1. Identify the section to be drawn.
2. Trace the section or make a larger drawing.
3. Mark the tracing on its starting point, terminal point, and branches.
4. Prepare the isometric using these points.
5. Note the change of direction needed on the isometric.
6. Add pipe sizes and lengths, fittings, hangers, and notes pertaining to the drawing.

Be sure the following are noted on the sketches:

- Pipe sizes
- Pipe lengths
- Type of pipe

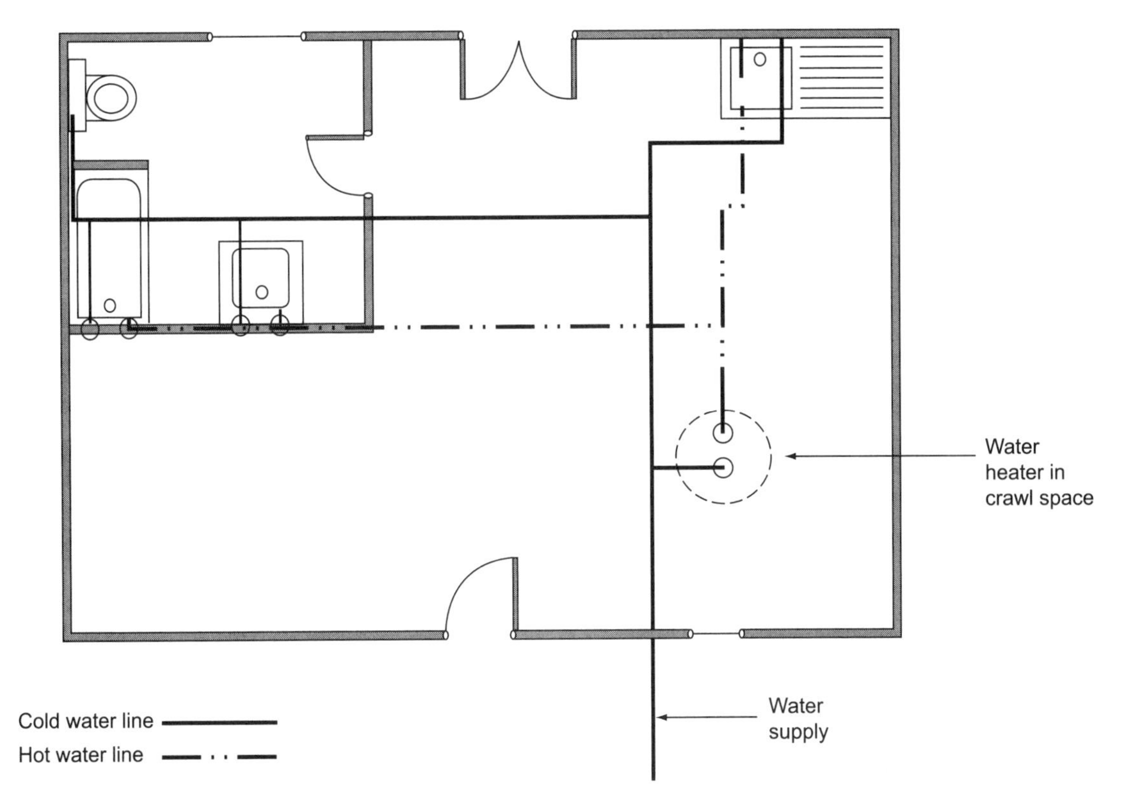

Figure 29–1
Plan view of the water supply system for a residential building.

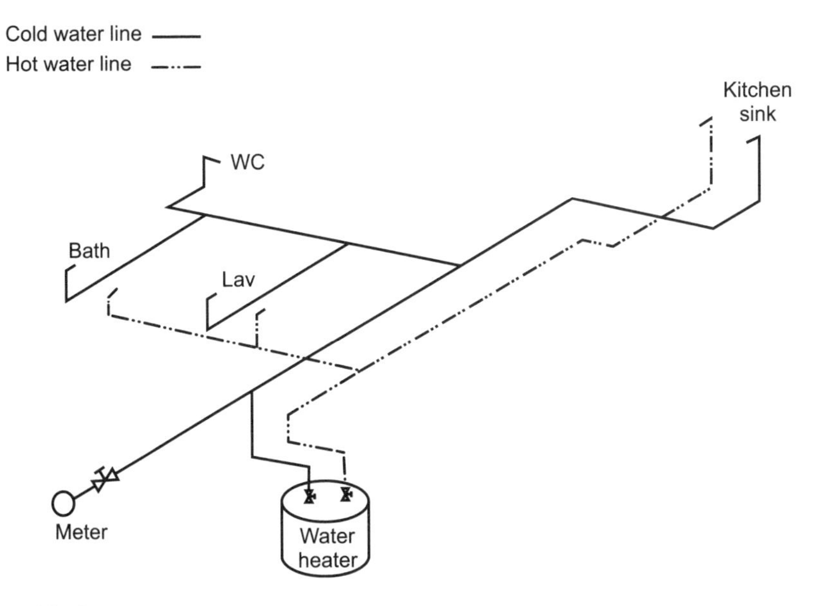

Figure 29–2
Isometric view of the water supply system for a residential building.

- Joining methods for the pipe
- Hanger requirements
- Trenching techniques and requirements
- Floor and wall penetrations
- Cleanout locations
- Pipe grade
- Any other information necessary for the field technician to fabricate the system from the drawing.

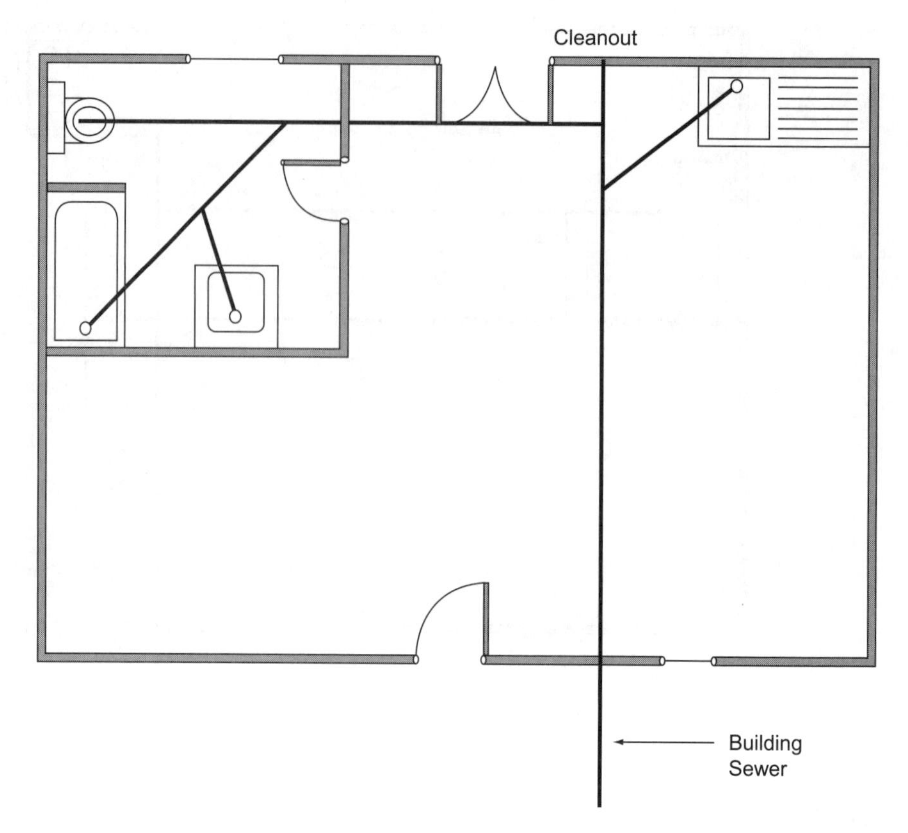

Figure 29–3

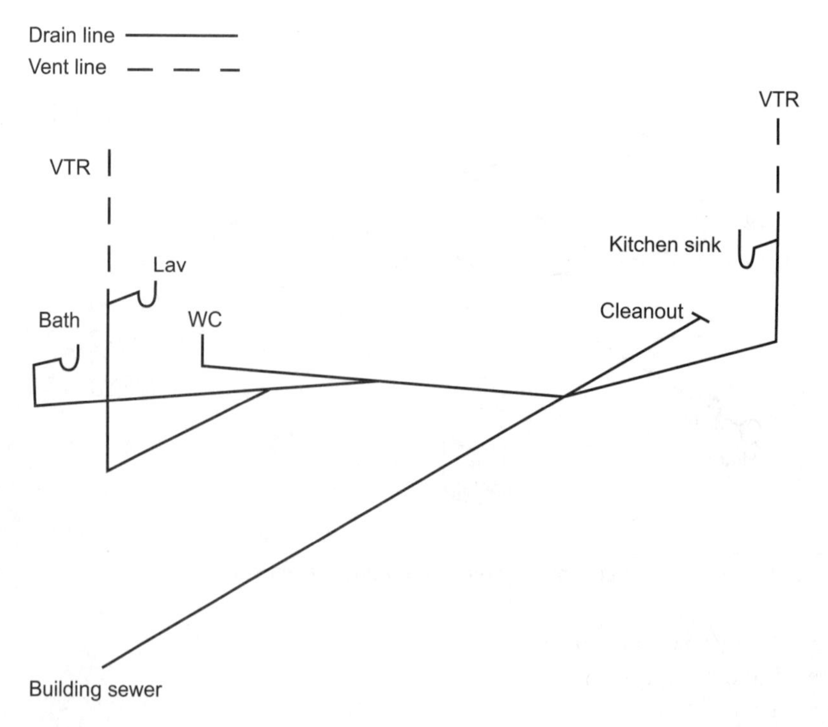

Figure 29–4
Isometric view of the DWV system for a residential building.

STORM DRAINAGE SYSTEM

Storm drainage for a flat roof is another good example of how an isometric drawing can be a powerful tool in generating the shop drawing and materials take-off. In the left side of Figure 29–5, drain locations are easy to identify, but there is no indication of the vertical component. The right side shows how the isometric view can give a better representation of the job.

Prepare an isometric drawing of the storm drainage system for the enclosed blueprints. Use the steps we have discussed before:

1. Identify the section to be drawn.
2. Trace the section or make an enlarged drawing of the section.
3. Mark the tracing on its starting point, terminal point, and branches.
4. Prepare the isometric using these points.
5. Note changes of direction needed on the isometric.
6. Add pipe sizes and lengths, fittings, hangers, and notes pertaining to the drawing.

Be sure the following are noted on the sketches:

- Pipe sizes
- Pipe lengths
- Conductor designs
- Type and number of roof drains needed
- Pipe grade
- Insulation for the conductors
- Minimum earth cover requirements
- Hanger requirements
- Types of joints used
- Cleanout requirements
- Any other information necessary for the field technician to fabricate the system from the drawing.

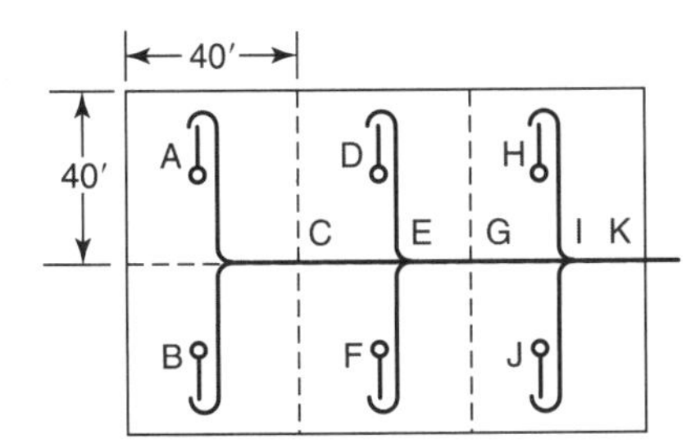

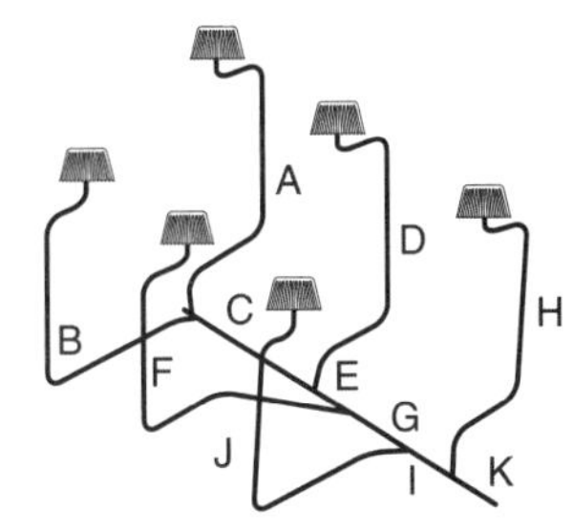

Figure 29–5
Left view is a plan view of a storm drain system. Right view is an isometric view of the same storm drain system.

REVIEW QUESTIONS

1. Must isometric drawings be made to scale?
2. Are plan views the best choice for measuring risers?
3. How can different piping purposes be shown?
4. Is it acceptable to add notes to drawings for additional information?
5. Which two pieces of information should be on the sketches regarding the person who prepared them?

CHAPTER 30

Gas Distribution Systems, Gas Appliance Venting, and Specialized Components

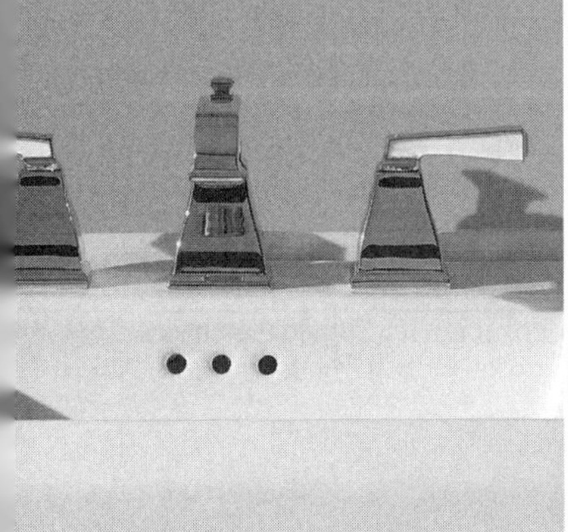

LEARNING OBJECTIVES

The student will:

- Prepare a shop drawing of a gas distribution system.
- Prepare a shop drawing for the gas appliance venting system.
- Prepare a shop drawing for specialized components.

GAS DISTRIBUTION SYSTEM

With the introduction of corrugated stainless steel tubing (CSST), the installation of gas distribution systems has become increasingly easier and quicker; this, however, does not mean that a plan does not need to be designed. No matter what material is being used, thought should be given to how the piping can be routed in the safest manner, using the least amount of pipe. Thought should be given to future expansion, if needed.

As a refresher, use the gas distribution plan shown in Figure 30–1 to prepare an isometric drawing. The main run is in a crawlspace with the furnace. Assume there is a 2 ft rise for the range and a 6 ft rise for the water heater. If you have questions concerning the sizing of gas pipe, refer back to *Plumbing 301*. Use the steps we have discussed before:

1. Identify the section to be drawn.
2. Trace the section or make an enlarged view.
3. Mark the tracing on its starting point, terminal point, and branches.
4. Prepare the isometric using these points.
5. Note the change of direction needed on the isometric.
6. Add pipe sizes and lengths, fittings, hangers, and notes pertaining to the drawing.

Be sure the following are noted on the sketches:

- Pipe sizes
- Pipe lengths
- Minimum earth cover requirements
- Pipe grade
- Hanger requirements
- Meter locations
- Valve locations
- Equipment locations
- Equipment ratings
- Any other information necessary for the field technician to fabricate the system from the drawing.

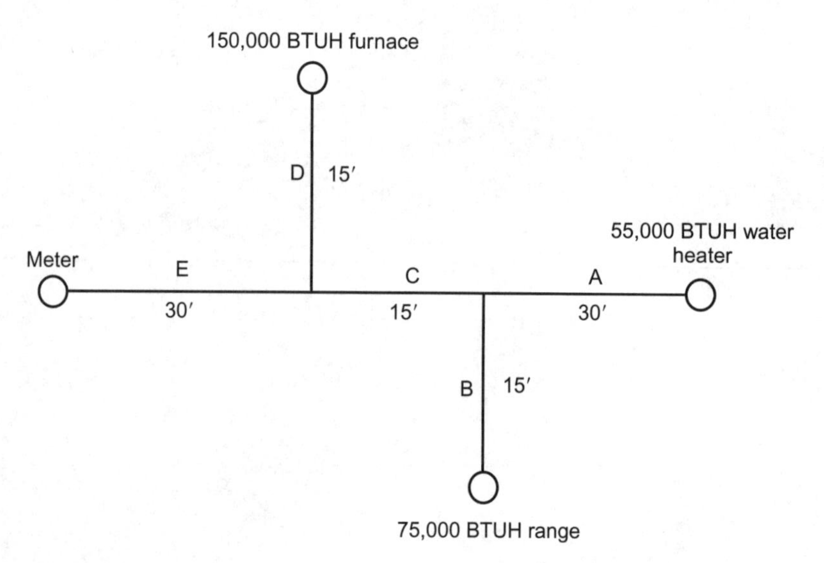

Figure 30–1
Plan view of a gas piping distribution system.

GAS APPLIANCE VENTING

Using Figure 30-1, prepare an isometric shop drawing of the vent system for the three appliances. Assume that for specific reasons the range must be vented. Always check to make sure that the appliances can have a combined vent system. In this case, vent each one separately. Use the steps we have discussed before:

1. Identify the section to be drawn.
2. Trace the section or make a larger drawing of the area.
3. Mark the tracing on its starting point, terminal point, and branches.
4. Prepare the isometric using these points.
5. Note the change of direction needed on the isometric.
6. Add pipe sizes and lengths, fittings, hangers, and notes pertaining to the drawing.

Be sure the following are noted on the sketches:

- Vent pipe sizes
- Pipe lengths
- Pipe grade
- Type of joint used
- Type of pipe used
- Wall and roof penetrations
- Vent terminal flashing requirements
- Hanger requirements
- Equipment locations
- Equipment ratings
- Any other information necessary for the field technician to fabricate the system from the drawing.

SPECIALIZED COMPONENTS

Specialized components can be any number of items that require any of the following: water supply, drain piping, fuel gas supply, or venting. The example shop drawing is for a water heater. See Figure 30-2. It should identify the following items:

- Tank
- Cold water supply
- Hot water outlet piping
- Gas piping
- Gas vent piping

Tank, valve, and piping specifications may also be listed on the figure.

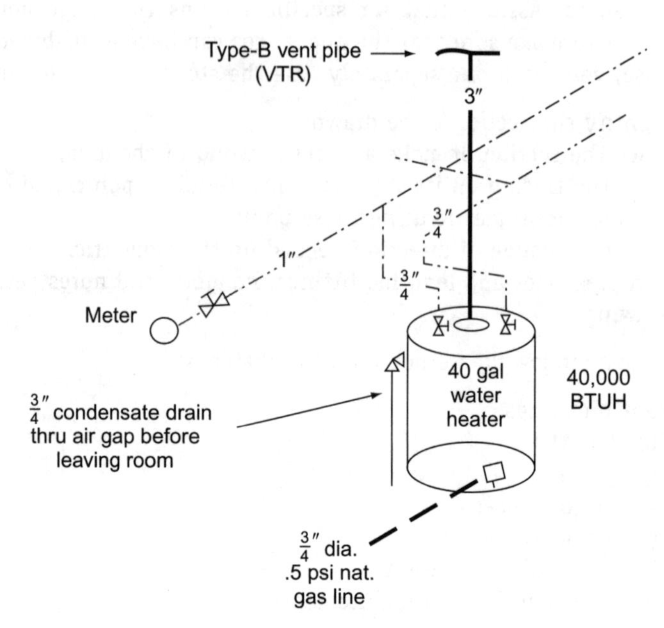

Figure 30–2
Example of an isometric shop drawing for specialized components.

REVIEW QUESTIONS

1. Using Figure 30–3, draw an isometric drawing for a two-story residential building with a gas-fired furnace in the crawlspace, a gas-fired water heater in the garage, and a gas furnace in the attic.

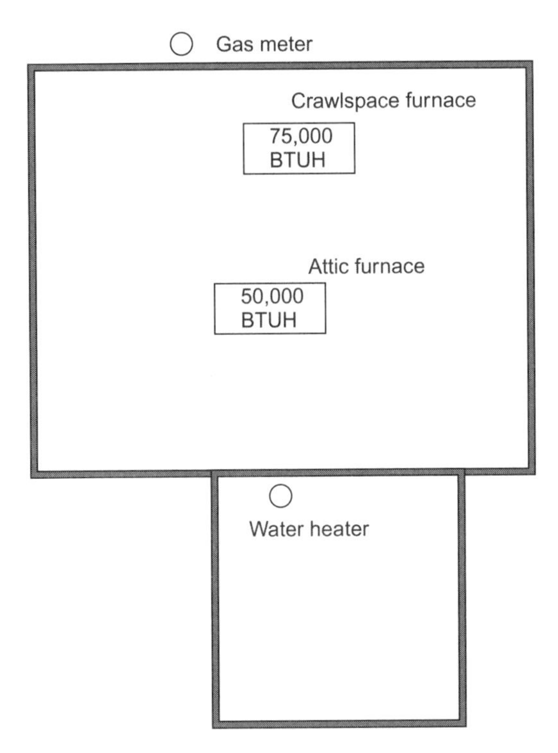

Figure 30–3
Two-story residential building.

2. In a remodel of an existing building, it is necessary to route the gas piping through the outside wall and across the attic before dropping down into the first story mechanical room; draw an isometric pipe layout to illustrate this scenario. Use Figure 30–4.

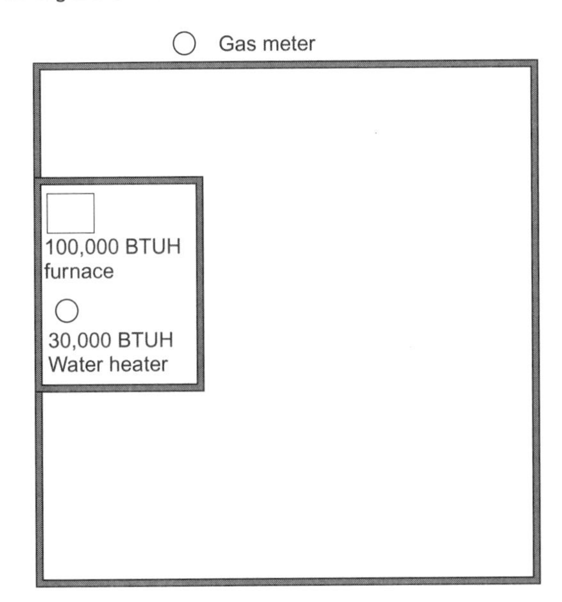

Figure 30–4
Single-story residential building on a concrete slab.

CHAPTER 31

Material Take-Off

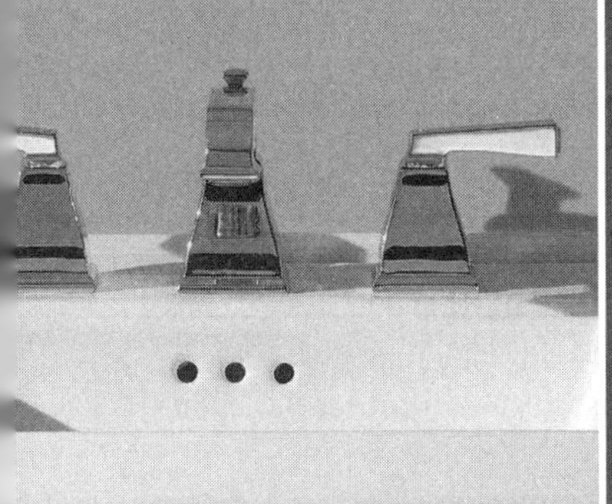

LEARNING OBJECTIVES

The student will:

- Prepare a basic material take-off.
- Explain how a take-off is related to labor units.

MATERIAL TAKE-OFF

Material take-off is the general term used to describe the process of determining the total material required based upon a set of plans and specifications. In this chapter, we will discuss proper procedures for performing a material take-off.

There are two approaches to a material take-off.

1. A material take-off for installation purposes involves the development of a complete listing of all materials required on a job, based upon the blueprints and specifications provided.
2. A material take-off for estimating purposes may be simplified, as the estimator may take shortcuts to accelerate the estimation task.

To develop your estimating skills, it is better to learn the material take-off process from the installation viewpoint. Since there is a possibility that the building plans are not complete, all components, whether shown on the plans or not, are logged on the material take-off sheet. Discrepancies can be discussed before the bid is submitted or during the preconstruction meetings. The procedure for conducting this analysis is as follows:

1. Review all specifications completely. Make notes of the materials to be used. Make notes on special installation standards that must be followed.
2. Study the architectural and structural plans. Note locations where the piping must be offset to avoid obstructions. Note special considerations such as beam locations, floors and wall obstructions, and special job requirements. The building materials illustrated on the plans will dictate the type of hangers and supports required for the pipe.
3. Study the plumbing plans. Make shop drawings and isometrics to ensure an accurate and complete material take-off.
4. Establish and use a system which will help you avoid omissions and/or duplications. Color coding, check marks, or similar demarcations can be used to identify groups of fixtures, piping, and so on.

Be certain to include all incidental requirements above and beyond those shown on the plan. In addition to pipe fittings needed to install a sanitary sewer, you must also include the following listings:

- Trenching.
- Shoring.
- Special backfill.
- Bedding materials.
- Any component necessary to complete the job.

Transfer all quantities listed on the take-off sheets to the appropriate requisition or estimate sheets. Use this recap process to double-check your material list.

Table 31–1 shows a labor and material take-off estimating sheet which may be used for plans and specification analysis. The following procedure may be used with this sheet:

1. Start in the middle with the description column. Move right, then left, and fill in the blanks.
2. The two columns to the right of the description column are for the unit price for the item and the extension total for the item based on quantity needed. To determine the extension value, multiply the quantity number by the material unit value.
3. The size column is determined from the plans.
4. The quantity column is also determined by the plan analysis.
5. The labor unit can be listed based on the time it takes a technician to install the particular component. (The PHCC Educational Foundation's Labor

Table 31-1 **Labor And Material Estimate Sheet**

Job Name:					System:			
Date:					Spec. Section:			
Labor Estimation		Labor Units	Quantity	Size	Description	Material Unit		Extension
						Page ______ of ______		

Calculator is an excellent reference tool for noting the labor hour units for different component assemblies.)

6. The labor extension is determined by multiplying the labor unit by the quantity amount.

The following estimating procedure should aid you in the development of an accurate take-off. The emphasis in this section is the overall job characteristics. Use this list as a checklist for all estimating.

Since it is customary to have multiple jobs occurring at the same time or long delays between the bidding of a job and its occurrence, use a master estimating or bid sheet, commonly known as a top sheet, to keep tract jobs separate. The following items should be included:

- Name of job, owner, and location.
- Architect and engineer, including address, phone number, and fax number.
- Number and nature of special requirements.
- Section of specifications.
- Plan sheet numbers.

Mark the specifications and plans with a colored pencil indicating the number and page of any addenda involved.

Study the architectural drawings and list the following items:

- Number of floors and type of construction.
- General layout of the building—check the vertical distance between floors.
- Square feet of floor space.
- Location of mechanical rooms and toilet facilities.
- On the plot plan, study the relation of the building to the property site and to the streets.

- Check location of utilities, water, gas, sewers, etc., noting if rock is indicated.
- Carefully read all notes on the plans.

Study the mechanical plans. Note the following:

- Number and size of sewer connections.
- Size of water meter and fire connections.
- Location of gas meter.
- Underline with colored pencil all special notes.
- Check legends for symbols and abbreviations.

Read the specifications, noting on the summary sheets the type and kind of pipe for each system.

List on each equipment sheet the manufacturer, type of equipment, size, and other identifying data.

When performing an actual estimate for a bid, the estimator notifies potential vendors of the need to obtain price quotes for special products or materials during this stage.

Note any subcontractors required and indicate what they will furnish and/or install.

- Insulation.
- Sheet metal.
- Controls.
- Electrical.
- Utility work.
- Fire protection.
- Sterilization of system.
- Utility company charges.
- Determine the license, bonding, and permit fees.
- List equipment rental requirements and rates.

The fixtures and equipment material take-off is performed using a multicolumn sheet. Make sure to take off the number of each type of fixture or equipment for each floor or section of the building, noting particularly the following data:

- Type of fixture by symbol number.
- Size and make of water heaters.
- Size and type of circulating pumps.
- Type of house pumps.
- Sewage ejectors.
- Storm pumps.
- Specialties such as water hammer arresters, backflow preventer, etc.
- Total the number of units per floor.
- Total the number of similar units for the job.
- Add the totals of the individual floors. (This should be the same as the total for the job.)
- Summarize the number of fixtures, drains, etc. on the plumbing estimate sheet.

As the complexity of a job increases, each system should be taken off separately, such as the following:

- Cold water.
- Hot water.
- Gas.
- Waste and vent.
- Storm drains.

Color-coding each system will simplify checking your take-off. Also make a note whether the underground materials are different from those used in above-ground applications; if so, separate them.

The following is an example of a color-coding scheme.

- Line out waste and vent in red
- Line out cold water in green
- Line out hot water in orange, etc.

An added advantage to writing up systems independently is ease in keeping track of the shipping of materials. In writing up hot water separately, you have an automatic insulation take-off.

Take-off for Pipe and Fittings

In this section, we will discuss the development of an accurate material take-off for pipe and fittings. A systematic procedure has been prepared for this purpose. After discussing the procedure, continue to practice the material take-off until you feel comfortable with it.

The pipe and fitting take-off requires that you prepare a checklist before determining quantities. You should:

- Note scale on pipe and fitting tabulation sheet.
- Note the source of the line and quickly trace it out.
- Note the number of typical units.
- Beware of photographically reduced size drawings as scale noted is incorrect.

The next step is to determine the quantities needed:

1. Always try to start at a common starting point. The connection to the water main or the sewer main is as good a point as any. Follow along the main, checking off and noting the number of valves and tees for that size, as well as the size and number of risers.
2. Mark the point where the size of the line changes.
3. Return down the same line, checking and taking off the number of fittings. (Note: 45s, couplings, adapters, reducers, etc. are counted as ells when estimating for a bid, but are listed separately in an installation take-off.)
4. Start again from the beginning. Scale the footage of pipe with a scale wheel (map measurer), and round the footage off to the nearest ten feet (10 ft). After noting the number of feet, mark the line with a colored pencil to indicate you have completed the take-off.

You may prefer to use a scale ruler with your colored pencil to take off and color at the same time.

Follow the same procedure for each pipe size, continuing to the end of the main or riser. (If you missed a fitting or specialty, you should note it.)

Work from the end of the main, return down the line, and take off the branches. Mark valves and tees first, then ells, and finally, the pipe to the last outlet.

Recheck the length of the risers for distance between floors from the architectural plans.

Carefully check the drawing for unmarked pipelines to make sure that nothing has been missed.

CHAPTER 32

Plumbing Code Administration and Licensing

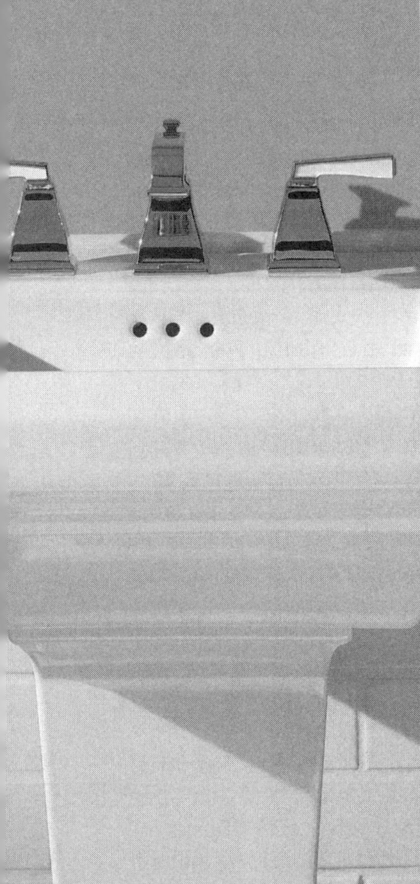

LEARNING OBJECTIVES

The student will:

- Discuss how the plumbing codes and licensing programs are dependent on each other.
- Discuss how the basic plumbing principles serve in guiding the installation of plumbing.

AN OVERVIEW OF CODES AND LICENSING

The last unit of *Plumbing 401* examines plumbing codes, which are closely related to installation and licensing functions. All plumbing codes begin with administrative sections and then define terms related to installation. Details are then provided for materials and connections that address water distribution; fixtures with related waste discharge, including vents; and many other matters, such as storm systems. It is important to understand that the code identifies the *minimum* requirements to ensure the health and safety of the occupants. Codes are published to provide legally enforceable documents identifying safety levels with material and installation methodology details.

Public health and safety matters are also ensured by periodic assessment of every installer's competencies and skills. A major goal in understanding the code that affects you as the technician is the ability to move up in licensing status—for example, from apprentice to journey status. These improved licenses come through the examination process. Plumbing codes are the principal source of assessments in the plumbing licensing examination process, with occasional practical skills exams.

A third aspect in the importance of code conformance is the issue of finances. Technicians who have inadequate knowledge of the code are prone to make installation errors, which can be very costly to correct. In addition to affecting the employer's profit, these errors jeopardize the employer's reputation. A company's reputation leads to additional work, which helps to ensure continued employment.

Codes

Plumbing codes have a great history but, more importantly, are always changing to address new technology. New products and methods of construction are the primary reasons why codes change. Piping product changes from the plastic industry have accounted for the majority of code changes, with plumbing fixture improvements being a close second. These new products require new code sections to identify how the proper connections should be made and to list the conformance standards. Product conformance standards are created by the plumbing industry to ensure that the product meets minimum performance criteria, giving everyone a level playing field to start from.

The basic plumbing codes in this country are the *National Standard Plumbing Code (NSPC)*, the *International Plumbing Code (IPC)*, and the *Uniform Plumbing Code (UPC)*. All are very similar but each has a slightly different style and may provide greater detail on specific code matters. For example, the NSPC provides excellent insight into what driving forces must be considered by the code in its 22 Basic Principles. The plumbing codes vary from one jurisdiction to another because of technical amendments based upon local conditions, differing interpretations, or concerns of enforcement bodies such as local health departments. There are several states that have an entire family of statewide legally enforced codes. Areas with this uniformity of code enforcement contribute to construction cost savings for structure owners, designers, and technicians.

The three major codes are updated every three years by a process of strict consensus. The consensus processes are set in place to ensure fair and equitable code development. All parties—whether industry manufacturers, code officials, or technicians—are given a chance to provide input. In the update process, plumbing codes should be considered as part of a larger family of codes, which includes the parent building code, mechanical, electrical, and several other common codes, as well as uncommon areas such as existing property maintenance codes. The building code adopted for the enforcing agency is considered the parent because it will address the needs of the entire structure, such as fire safety, which includes fire stopping penetrations and means of egress (exiting). The means of egress leads to building occupancy numbers, which are then interpreted by the plumbing code to dictate how many plumbing fixtures must be provided for the individuals in the building. Everything ties together, and a change in one area will often cause costly, wide-ranging, and time-consuming changes throughout the structure. It is critical to get things right on the first try.

The three year code change update is generally divided into two cycles. The first-cycle changes are commonly called supplements and are published. They are generally not enforced until the conclusion of the second cycle, because of the high cost to enforcement agencies of adopting new codes. Each of the cycles is composed of code change submittals, which are considered and voted on by code change committee members after they hear public testimony. These results are then published, which allows concerned parties to challenge the committee suggestions in an open meeting. Public meetings composed of voting members are then able to hear testimony and vote as a body to establish the final outcome for the next code cycle.

Licensing

Plumbing licenses in this country vary in detail from one enforcing agency to another. The enforcing agency could be a statewide licensing program to a large metropolitan area with a city license. Licenses are issued following successful examinations, which are heavily composed of code content and some practical testing. These examinations usually allow open book testing.

License laws usually recognize levels from apprentices all the way to business owners through a continuing development of skills and responsibilities. Registered apprentices are individuals learning the trade in programs four to five years long. They must complete the program prior to being allowed to test for a journeyman license. They must have at least one licensed person on the job site to ensure proper supervision. The path from apprentice to licensee varies in accordance with different licensing laws. The first licensed step is the journey status. Individuals who have passed this examination are recognized as competent, skilled technicians who may work on their own.

Generally, most licensing programs call the next step a master plumber who has served at least two years as a journeyman and successfully passed the master's examination. This individual provides supervision, commonly referred to as "the ability to plan, direct, and control the installation or service/repair of plumbing." Often they are authorized to obtain plumbing permits and may even be classified as the business owner.

Some licensing laws have a status by examination for a plumbing contractor. This individual has been designated as the "person financially responsible for the business" and is generally a sole proprietor or listed as one of the principals in a corporation. The licensing agency may take administrative action against any individuals, at any stage of certification, for gross negligence or incompetence.

In the Field

Individuals involved in the plumbing profession recognize that increased skills and responsibilities based upon license measurements lead to greater financial rewards. This fact alone is the greatest motivation to study the codes.

Installation Errors

Installation errors could jeopardize the health and safety of the building occupants. However, plumbing is unique, in that mistakes could also affect individuals outside the structure who share the same water supply. Installation errors drastically increase costs and, as stated earlier, filter down to the installing technician. These costs could vary from an hour of lost time correcting a small code item, such as a pipe hanger misplacement, to litigation in a burn incident caused by incorrect temperature control selection and installation.

To be more positive, a major cost consideration is technicians' seeking to meet new code conformance specifications. New products accepted in the code may provide your employer with an excellent cost advantage if they are ready to use them and competitors are not. Another example would be a competing contractor who vents every fixture unnecessarily, while your company chose a code-compliant group-venting system.

In the Field

Codes are organized by chapter number, section header, a possible subsection header, and title. For example, from the NSPC, 1.4.1 Addition or Repair, which is from the Administration chapter pertaining to Applicability, 1.4.

ADMINISTRATION

The administrative details of code enforcement management is discussed in the first chapter of the code. It is established in accordance with the identified jurisdictional authority. Plumbing laws are a different matter, addressed elsewhere. General enforcement items are best presented and clarified by a list of the applicable sections (the three major codes vary somewhat here) with summaries as follows:

1. Title and scope—This section identifies the name of the code or jurisdiction and states the degree of involvement, which more than likely includes new and remodel work. It may then go on to explain whether the code applies to commercial or residential occupancies and whether gas piping, venting, and other areas are included.
2. Applicability—The type and extent of work affected by the code is specifically explained, such as new or maintenance alteration. A unique clarification in this section states that all structures that are moved must be brought up to the current code.
3. Approvals—The acceptance criteria for materials and methods of installations are set forth here. Some codes address alternative engineered designs in this area, with the conditions of acceptance.
4. Organization and Enforcement—The authority having jurisdiction is named, along with their powers. Code officials are considered here. Code officials and plumbing inspectors are frequently former technicians.
5. Violations and Penalties—Violations are clarified and possible responses are discussed, such as placing a stop-work order on the construction process. Most jurisdictions will address penalties by other means, such as local enforcement ordinances.
6. Permits—This section will clarify when permits are necessary and is often modified by local agencies in accordance with their licensing law. Fees could be included here, as well as the process for obtaining permits, such as by plan review.
7. Inspections—The details of inspections, such as frequency and procedures, are clarified here.
8. Appeals—Governmental agencies generally have methods to deal with practical, acceptable difficulties. The process for these actions will be spelled out. These are commonly referred to as appeals or variances.

The meetings where code changes are discussed are usually subject to the "Open Meetings Act" and "Freedom of Information Act," to ensure fairness in the process. Your local governmental agencies can clarify these regulations for you.

BASIC PRINCIPLES

Almost every individual code requirement is directly traceable to basic plumbing code principles that are the foundation of the code. Technicians who understand those fundamentals will grasp the code and its requirements far better than those who try to memorize the code. Some codes have not included their core principles because of what they refer to as unenforceable language.

The *National Standard Plumbing Code* "22 Basic Principles of Plumbing" are listed here. The code describes the principles as defining the code intent, not addressing jurisdictional issues. The last phrase is significant since some jurisdictions may license well drillers under another code or requirement.

1. All occupied premises shall have potable water—All premises intended for human habitation, occupancy, or use shall be provided with a supply of potable water. Such water supply shall not be connected with unsafe water sources, nor shall it be subject to the hazards of backflow.

2. Adequate water required—Plumbing fixtures, devices, and appurtenances shall be supplied with water in sufficient volume and at pressures adequate to enable them to function properly and without undue noise under normal conditions.
3. Hot water required—Hot water shall be supplied to all plumbing fixtures which normally need or require hot water for their proper use and function.
4. Water conservation—Plumbing shall be designed and adjusted to use the minimum quantity of water consistent with proper performance and cleaning.
5. Safety devices—Devices for heating and storing water shall be so designed and installed as to guard against dangers from explosion or overheating.
6. Use public sewer where available—Every building with installed plumbing fixtures and intended for human habitation, occupancy, or use, and located on premises where a public sewer is on or passes said premises within a reasonable distance, shall be connected to the sewer.
7. Required plumbing fixtures—Every family dwelling unit shall have at least one water closet, one lavatory, one kitchen-type sink, and one bathtub or shower to meet the basic requirements of sanitation and personal hygiene. All other structures for human habitation shall be equipped with sufficient sanitary facilities. Plumbing fixtures shall be made of durable, smooth, non-absorbent and corrosion-resistant material and shall be free from concealed fouling surfaces.
8. Drainage system—The drainage system shall be designed, constructed, and maintained to guard against fouling, deposit of solids and clogging, and with adequate cleanouts so arranged that the pipes may be readily cleaned.
9. Durable materials and good workmanship—The piping of the plumbing system shall be of durable material, free from defective workmanship and so designed and constructed as to give satisfactory service for its reasonable expected life.
10. Fixture traps—Each fixture directly connected to the drainage system shall be equipped with a liquid seal trap.
11. Trap seals shall be protected—The drainage system shall be designed to provide an adequate circulation of air in all pipes with no danger of siphonage, aspiration, or forcing of trap seals under conditions of ordinary use.
12. Exhaust foul air to outside—Each vent terminal shall extend to the outer air and be so installed as to minimize the possibilities of clogging and the return of foul air to the building.
13. Test the plumbing system—The plumbing system shall be subjected to such tests as will effectively disclose all leaks and defects in the work or the material.
14. Exclude certain substances from the plumbing system—No substance which will clog or accentuate clogging of pipes, produce explosive mixtures, destroy the pipes or their joints, or interfere unduly with the sewage-disposal process shall be allowed to enter the building drainage system.
15. Prevent contamination—Proper protection shall be provided to prevent contamination of food, water, sterile goods, and similar materials by backflow of sewage. When necessary, the fixture, device, or appliance shall be connected indirectly with the building drainage system.
16. Light and ventilation—No water closets or other fixture shall be located in a room or compartment which is not properly lighted and ventilated.
17. Individual sewage disposal systems—If water closets or other plumbing fixtures are installed in buildings where there is no sewer within a reasonable distance, suitable provision shall be made for disposing of the sewage by some accepted method of sewage treatment and disposal.
18. Prevent sewer flooding—Where a plumbing drainage system is subject to backflow of sewage from the public sewer or private disposal system, suitable provision shall be made to prevent its overflow in the building.

19. Proper maintenance—Plumbing systems shall be maintained in a safe and serviceable condition from the standpoint of both mechanics and health.
20. Fixtures shall be accessible—All plumbing fixtures shall be so installed with regard to spacing as to be accessible for their intended use and for cleaning.
21. Structural safety—Plumbing shall be installed with due regard to preservation of the strength of structural members and prevention of damage to walls and other surfaces through fixture usage.
22. Protect ground and surface water—Sewage or other waste shall not be discharged into surface or subsurface water unless it has first been subjected to some acceptable form of treatment.

Definitions

All plumbing codes have a chapter that clarifies plumbing code definitions to pinpoint the specific terms pertinent to the code. Some codes have what is referred to as a common plumbing format; these will have the definitions in Chapter 2 similar to building codes. The location is unimportant; the concern is understanding the definitions in order to use the code successfully.

A very small portion of the many listed code definitions are discussed below:

- Approved—Acceptable to the authority having jurisdiction.
- Branch—Any part of the drainage system other than a riser, main, or stack.
- Branch interval—A vertical distance of 8 feet or more from the top down on a stack between the connections of horizontal branches on a drainage stack. The reference is critical when using the drainage fixture unit tables in order not to overload stacks. The code DFU tables generally address one, between one and three, and then greater than three branch intervals.
- Building drain—The lowest part of the building drainage system that receives the discharge from stacks and horizontal fixture drains and conveys sewage, waste, or other drainage to the building sewer. The code will define building sanitary drain and building storm drain.
- Building sewer—That part of the building drainage system that extends from the end of the building drain and conveys its discharge to a public sewer, private sewer, individual sewage-disposal system, or other point of disposal. The code will define building sanitary sewer and building storm sewer.
- Building subdrain—That portion of a drainage system that does not drain by gravity into the building drain. Automatic pumping is required here.
- Circuit vent—A vent that connects to a horizontal drainage branch and vents from two to eight traps or fixtures connected in battery.

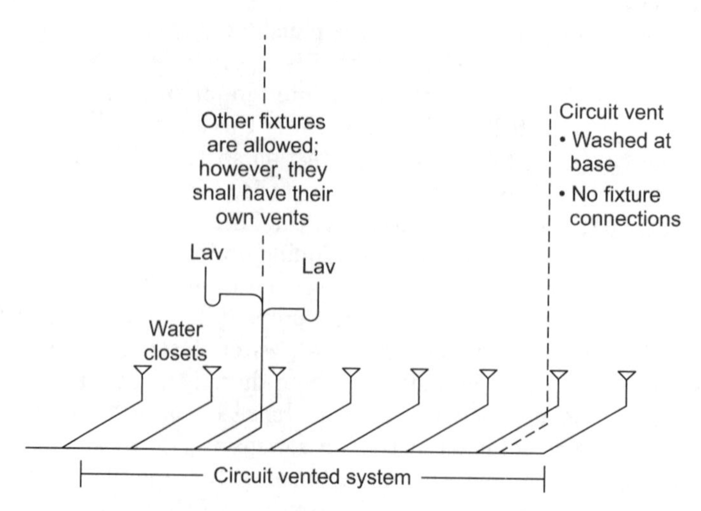

Figure 32–1
A basic circuit vented system serves up to eight battery vented fixtures at a time.

- Combination waste and vent system—A designed system of waste piping using the horizontal wet venting of one or more sinks or floor drains by means of a common waste and vent pipe adequately sized to provide free movement of air above the flow line of the drains.
- Common vent—A vent connected at the common connection of two fixture drains and serving as a vent for both fixture drains.
- Contamination—The impairment of the quality of the potable water that creates an actual hazard to public health through poisoning or the threat of diseases by sewage, industrial fluids, or waste.
- Cross connection—Any connection or arrangement between two otherwise separate systems, one that contains potable water and the other that contains water of questionable safety, steam, gas, or chemicals, whereby there may be a flow from one system to the other, the direction of flow depending on the pressure differential between the two systems.
- Drainage fixture unit (DFU)—The measurement of probable discharge for drainage fixtures based on volume and discharge time.
- Dry vent—A vent that does not receive the discharge of any fixture.
- DWV—An acronym for "drainage, waste, and vent."
- Fixture branch—A drain serving two or more fixtures that discharges to another drain or to a stack.
- Fixture drain—The drain from the trap of a fixture to the junction of that drain with any other drain pipe.
- Flood level rim—The edge of the receptor or fixture over which water flows if the fixture is flooded (a plugged drain).
- Flow pressure—The pressure in the water supply pipe near the faucet or water outlet while the faucet or water outlet is fully open and flowing.

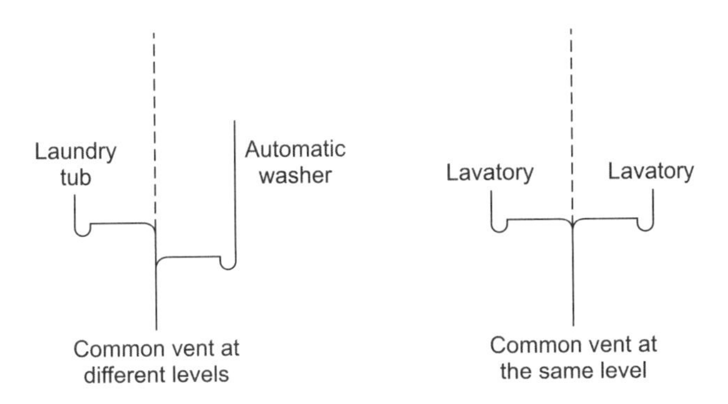

Figure 32–2
A common vent serves two fixtures and allows the lower fixture to receive venting air while discharging the upper fixture by ensuring the proper size according to code.

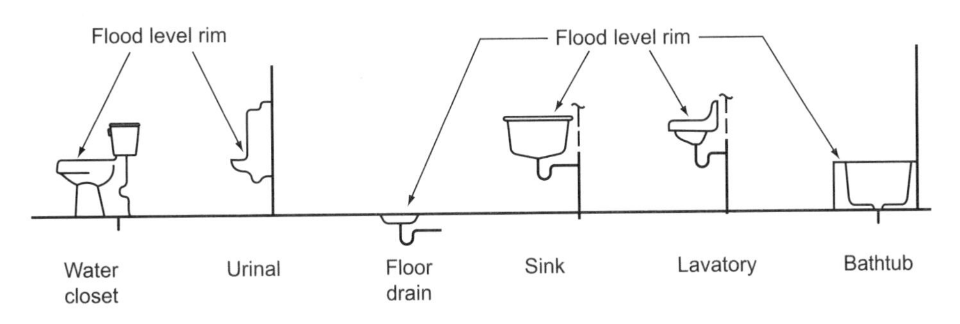

Figure 32–3
The flood level rim of fixtures is the reference point understood to be the termination point of contamination potential. Courtesy of Plumbing-Heating-Cooling-Contractors—National Association.

- Horizontal branch drain—A drain pipe extending laterally from a soil stack, waste stack, or building drain with or without vertical sections or branches that receives the waste discharged from one or more fixture drains and conducts the waste to a spoil stack, waste stack, or building drain.
- Hot water—Potable water that is heated to a temperature of 110°F or above.
- Indirect waste pipe—A waste pipe that does not connect directly with the drainage system, but which discharges into the drainage system through an air break or an air gap into a trap, fixture, receptor, or interceptor.
- Individual vent—A pipe installed to vent a single fixture drain.
- Invert—The lowest portion of the inside of a horizontal pipe. This is often used as an installation reference point when discussing grade or pitch during installation layouts.
- Main—The principal pipe artery to which branches may be connected.
- May—The word "may" is a permissive term when dealing with code items.
- Medical gas systems—The complete system used to convey medical gases for direct application from central supply systems through piping networks with pressure and operating controls, alarm warning systems, and other typical features, and extending to station outlet valves at use points.
- Plumbing—The practice, materials, and fixtures within or adjacent to any building structure or conveyance, used in installation, maintenance, extension, alteration, and removal of any piping, plumbing fixtures, plumbing appliances, and plumbing appurtenances in connection with sanitary or storm drainage, venting systems, and private or public water supply.
- Plumbing appliance—Any one of a special class of plumbing fixtures intended to perform a special function. Examples include clothes washers, water heaters, dishwashers, water softeners, and the like.
- Plumbing appurtenances—A manufactured device, prefabricated assembly, or on-the-job-site assembly of component parts that is an adjunct to the basic piping system and plumbing fixtures. It demands no additional water and does not add any discharge load to the drainage system. Examples include water filters, backwater valves, dilution tanks, interceptors, backflow preventers, and so on.
- Plumbing fixture—A receptacle or device connected to the water distribution system of the premises and that demands a supply of water therefrom; or discharges used water, liquid-borne waste materials, or sewage either directly or indirectly to the drainage system of the premises;

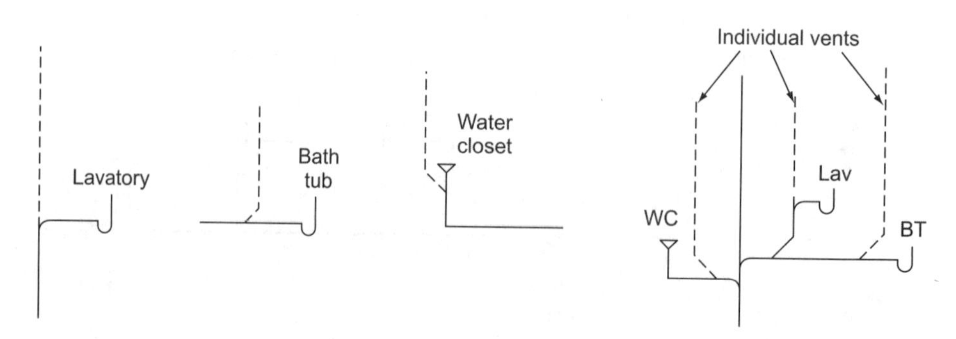

Figure 32–4
An individual vent is the basic method of ensuring proper drainage and trap seal protection. Courtesy of Plumbing-Heating-Cooling-Contractors—National Association.

or which requires both the water supply and a discharge to the drainage system of the premises.

- Pollution—An impairment of the quality of potable water to a degree that does not create a hazard to the public health but does affect the aesthetic qualities of the water.
- Potable water—Water free from impurities present in amounts sufficient to cause disease or harmful physiological effects and conforming in its bacteriological and chemical quality to the requirements of the public health standards having authority.
- Relief vent—An auxiliary vent that permits additional circulation of air in or between drainage and vent systems. They are commonly used in circuit vent systems when upper floors are discharging on stacks having circuit-vented restrooms.
- Sewage—Liquid containing human waste and/or animal, vegetable, or chemical waste matter in suspension or solution.
- Shall—Is a mandatory term when used in the code.
- Slope—The fall (pitch/grade) of a piping in reference to a horizontal plane.
- Stack—A term for any vertical line of waste or vent piping.
- Stack vent—The extension of a soil or waste stack above the highest horizontal drain connected to the stack.
- Subsoil drain—A drain that connects below the surface of seepage water and conveys it to a place of disposal. These lines are generally around the structure perimeter at footing level.
- Tempered water—Water having a temperature between 85°F and 110°F.
- Trap—A fitting or device that provides a liquid seal to prevent the emission of sewer gases without materially affecting the flow of sewage or waste water through it. Traps isolate the building sewer system odors and health hazards from habitable areas.
- Trap seal—The maximum vertical depth of liquid that a trap will retain, measured between the crown weir and the top of the dip of the trap.
- Vacuum—Any pressure less than that exerted by atmospheric pressure, which is 14.7 psi at sea level.
- Vent stack—The vertical vent pipe installed for the purpose of providing air circulation to any part of the drainage system.
- Waste—The discharge from any fixtures that does not contain fecal matter.
- Yoke vent—A pipe connecting upward from soil or waste stacks used to equalize pressure in the stack.

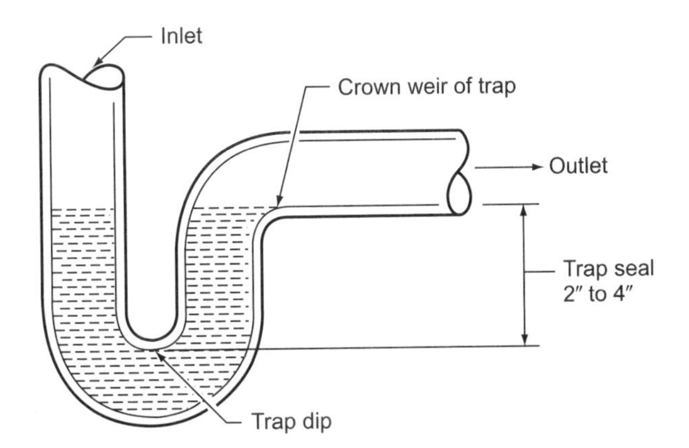

Figure 32–5
The trap seal depth of 2 inches minimum is established by code and has double the venting design protection demanded by proper venting procedures which is 1 inch of water column pressure differential. Courtesy of Plumbing-Heating-Cooling-Contractors—National Association.

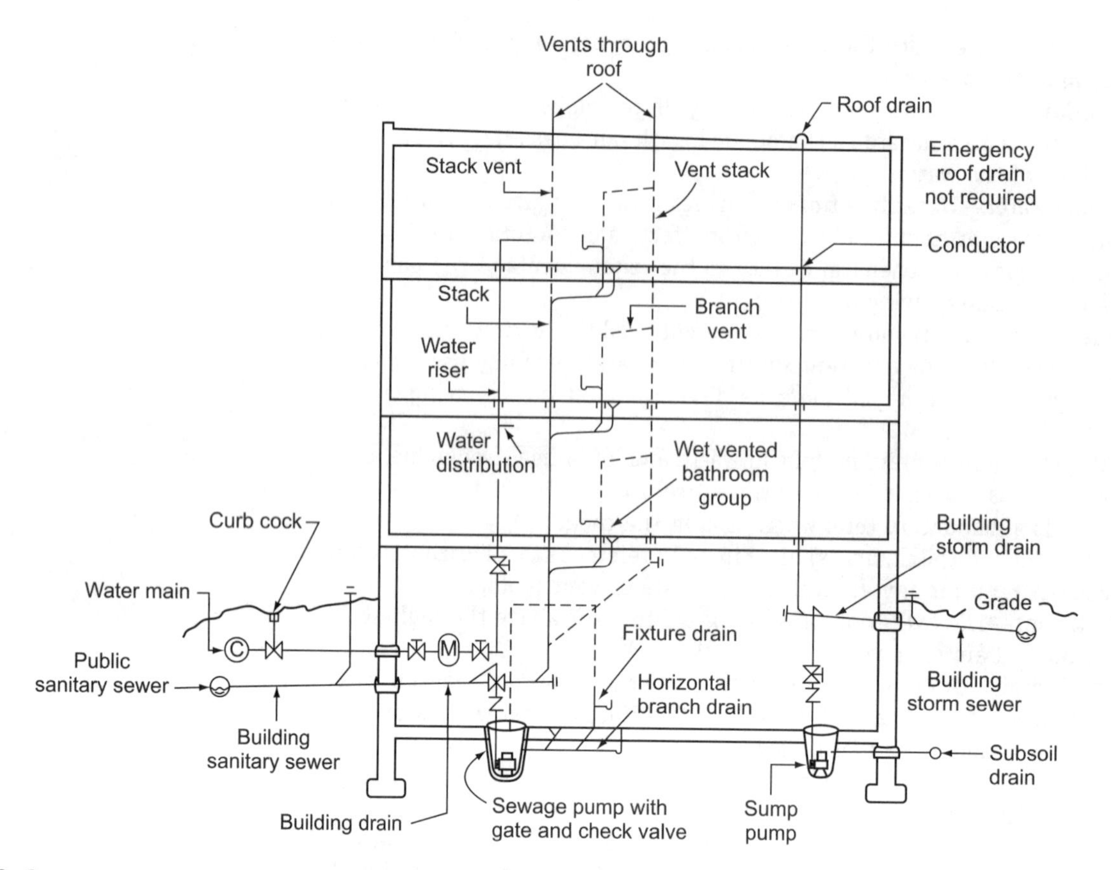

Figure 32–6
This plumbing system components illustration provides a basic overview of a DWV installation.

> **In the Field**
>
> When you have difficulties in finding rarely referenced code-mandated items, first check the general regulation sections.

GENERAL REGULATIONS

The general regulations of a plumbing code list a series of miscellaneous requirements for installation of plumbing systems to ensure the installations conform to the intent of the code. Information in the remainder of this chapter addressing general requirements will cover those areas of the code that are not normally covered in the water supply, sanitary drainage, venting, or storm areas of the codes.

General regulations of greater value in common use are listed below in alphabetical order with summaries to aid in proper code enforcement. Although the summaries are brief, the reader must understand that each of the items—and several others not referred to here—has many subsections that address the details.

- Changes in direction of piping—Fittings shall be installed to guide waste in the direction of flow and minimize obstructions. Plumbing codes have tables to clarify what fittings may be used in directional changes from horizontal to vertical, vertical to horizontal, and horizontal to horizontal. The *National Standard Plumbing Code* has a very detailed table that addresses the specific types of materials used in the fitting manufacturing.
- Detrimental sewage materials—The codes prohibit materials and chemicals that will be detrimental to the systems. Local treatment jurisdictions may have local technical amendments in this area of the code also.
- Excavations, trenching, and bedding—Detailed information on the proper methods of installing underground piping will be addressed in these sections. Trenching safety is not covered in this general area of the code.
- Fitting changes—Heel and side inlet fittings connection methods are clarified in these sections. Sanitary crosses are addressed in the sanitary drainage chapter.

- Materials—Material identification is addressed briefly in these sections. The vast majority of the standards referenced in the codes already detail these requirements. Product information is extremely important as it aids in obtaining repair parts, manufacturer recalls or warranties, and the guaranteeing of defective products. Some codes also have information in this general section to discuss the acceptance of products by the authority having jurisdiction, listing programs, or third party testing and certification. All codes are different in this area, with the major decision being based upon the authority having jurisdiction. For that reason, test reports and good record-keeping skills are a must.
- Protection of pipes and system components—Several areas are addressed here, such as temperature protection (freezing and overheating protection), breakage, corrosion, sleeves, water proofing, and many other common sense items.
- Structural safety—Drilling, cutting, and notching to install plumbing systems has always been a closely regulated item. Recent improvements in sophisticated building materials have increased technicians' awareness of meeting building code requirements in these matters. The parent building codes with the other various codes insist on meeting the proper fire stopping requirements.
- Tests—Clear information on how and when to perform required tests on plumbing drainage and water systems is identified in these sections. Some of the codes will explain the frequency and necessary methods for testing water supply backflow preventers.
- Toilet room requirements—Most codes will reference the ANSI Z4.3 for construction workers in the code section area. Other codes will discuss added information here, such as privacy partitions for water closets and urinals. Fixture clearances will be discussed in other areas of the code, generally under the fixture chapters.

REVIEW QUESTIONS

1. What is a code appeal?
2. Provide two examples of a plumbing appliance.
3. Name the licensing designations found in the plumbing profession.
4. What are the miscellaneous requirements of the code called?
5. What is the correct keyword for a mandatory term used in the code?

CHAPTER 33

Code Materials and Referenced Standards

LEARNING OBJECTIVES

The student will:

- Recognize the different standard writing agencies and which standards may be applied to a particular product.
- Locate standards for a particular product.
- Describe the approval process for new standards.
- Identify products that may be installed in different locations based on local code tables.

MATERIALS AND STANDARDS

Material product standards are of extremely great importance to the plumbing code and the successful installation of plumbing. The standards define the material composition, test methods, and rating and set the performance exceptions of the various products. These products range from plumbing fixtures to pipes and numerous other system components.

The plumbing industry is unique regarding products and installations because plumbing codes utilize standards to a great degree. Other codes, such as the mechanical code, rely heavily on code officials' approval and manufacturers' installation recommendations. Using standards in the plumbing code creates a level playing field, so to speak, and promotes uniformity in different geographical areas.

The plumbing codes numerous standards appear to address most of the products in the installation of plumbing. Of course, the plumbing industry is always utilizing new products ahead of their code inclusion. Other times products with very small markets, such as specialized healthcare bathing tubs for burn patients, do not require standards enacted for their acceptance. In the absence of a standard, the code official (the authority having jurisdiction) will be required to make an administrative approval decision. Code officials often view this acceptance as a very serious matter.

When codes are adopted, the specific language becomes the jurisdiction requirement. Another consideration by code officials is their required commitment to the standards published in their adopted code. The standards always have edition dates, which require conformance to any specific standard's printing editions. Standards often change due to new products and methods recognition. Referencing earlier editions of standards may diminish an inspector's ability to allow the installing contractor to use a new method. In these cases, the inspector will act on an appeal or variance procedure. An example would be the change in the *ICC Plumbing Code* from the 2003 edition, which recognized trap seal primers conforming to ASSE 1044-86 (1986), to the 2006 edition, which recognized ASSE 1044-01 (2001). That small recognition change was of great benefit to contractors because it meant they could use a water-supplied, solenoid-operated, trap seal primer based on code acceptance.

Conformance to code consensus standards is far more important then you might assume. There have been very rare cases where unscrupulous manufacturers or their representatives distributed products that later failed, resulting in litigation. The majority of failures have had causes ranging from unforeseen problems resulting from poor installation procedures to excessive chemical contents within pipes.

THE STRUCTURE OF STANDARDS

Standards are written by national consensus bodies comprised of experts in the given area and various competitors. Involving competitors ensures that single interest groups will not control the standards-writing process. To have a consensus process, it must be open to all interested parties, have a formal review timeframe, provide ample notification to all interested parties, and have a formal appeal process. This promotes a fair and level playing field. Code bodies adopting the standards in their documents will have a formal process with established criteria. Those criteria are generally that the standard shall be clear, enforceable, and written in mandatory language. An example of mandatory language would be the use of *shall* rather than *may*. The standards are performance orientated.

Standards are generally organized in the following manner:

1. Title

The specific product name or methodology will be stated with its numerical identification and edition year. The numerical designation is extremely important and often used in place of the standard name.

2. Introduction

A short history will be provided on the compelling need for the standard's development, and a timeline usually will be incorporated into the discussion. The committee participants might be identified with the name of the organization they represent. The introduction aids the reader by providing greater background information.

3. Scope

A brief clarifier is provided to narrow down what the standard addresses. For example, the clarifier may name the application, material requirements, workmanship dimensions, strength, tolerance, test method, and so on. The specifics will follow; this is simply an outline. Some standards will also have a section here named *Purpose*, which defines the standard's purpose.

4. Referenced Documents

This area of the standard lists all the applicable standards applying to the new standard. For example, a new standard may reference the Test Methods for Determining External Loading standard.

5. Terminology

This section defines the appropriate terms found in the standard, for example, the relation between standard dimension ratio, stress, and internal pressure.

6. Materials

This section defines what the basic materials are composed of and, at times, identifies if excess materials can be reused.

7. Requirements

This section addresses the workmanship, dimensions, and tolerances.

8. Test Methods

This area of the standard addresses how to test the product.

9. Product Markings

It is important to note here that the method and manner of markings are dependent upon the standard, not how technicians and inspectors think it should be marked.

10. Quality Assurance

Retesting, labeling, and certification issues are addressed in this area.

11. Appendix

This last area of the standard is composed of nonmandatory language, such as storage of the product and installation advice.

STANDARDS ORGANIZATIONS

When you have read the code and various trade magazines and taken the journeyperson examination, knowledge of the standards acronyms is essential. Acronyms are the abbreviations for standards-writing organizations. They are listed for your benefit.

AHRI: Air-Conditioning, Heating & Refrigeration Institute (drinking fountains)
ANSI: American National Standards Institute (plastic fixtures)
ASME: American Society of Mechanical Engineers (fixtures and fittings)
ASSE: American Society of Sanitary Engineering (backflow preventers)
ASTM: American Society for Testing Materials (pipe and fittings)
AWS: American Welding Society
AWWA: American Water Works Association (water piping)
CISPI: Cast Iron Soil Pipe Institute (cast iron products)
CSA: Canadian Standards Association
FS: Federal Specification
IAPMO: International Association of Plumbing and Mechanical Officials
ICC: International Code Council (codes)
ISEA: Industrial Safety Equipment Association
NFPA: National Fire Protection Association (fire safety and medical gas piping)
NSF: National Sanitation Foundation (public health)
PDI: Plumbing and Drainage Institute (grease interceptors)
UL: Underwriters Laboratories, Inc. (electronic controls)

STANDARDS DIVISIONS, I.–XI. (NSPC)

The *National Standard Plumbing Code (NSPC)* in Section 3.1.3, Standards Applicable to Plumbing Materials, sums up the majority of ways codes deal with standards when it states "material shall be considered approved if it meets one or more of the standards cited in table 3.1." The NSPC lists its standards in Chapter 3 and the *International Plumbing Code (IPC)* list its standards in Chapter 13.

The NSPC table organizes its numerous standards from the many organizations previously listed in a logical manner, which follows.

1. Ferrous pipe and fittings: 11 items
2. Non-ferrous metallic pipe and fittings: 16 items
3. Nonmetallic pipe fittings: 63 items
4. Pipe joints, joining materials, couplings, and gaskets: 28 items
5. Plumbing fixtures: 27 items
6. Plumbing fixture trim: 23 items
7. Plumbing appliances: 15 items
8. Valves and appurtenances: 23 items
9. Backflow prevention devices: 21 items
10. Miscellaneous: 15 items
11. Recommended practices and standards for qualifications, installation, and testing: 19 items

Other code agencies that list all their standards together in the back of their books specify the applicable standard reference by document number in the code text. For example, the *International Plumbing Code* in Section 608.13.8, "Spillproof vacuum breakers. Spillproof vacuum breakers (SVB) shall conform to ASSE 1056. These devices . . ." The IPC Chapter 13 will then list the complete ASSE standard and further specify all sections of the code where that same standard will be included.

STANDARDS TABLES BY APPLICATION

Another unique function of the *National Standard Plumbing Code* is the organized manner in which it applies standards to potable water materials, sanitary waste and drain, vent piping, storm drainage, and foundations' drains materials. The various products are addressed separately in Chapter 3 and then product requirements are provided. For example, Section 3.4, Water Piping states that water service pipe shall be pressure rated to no less than 160 psi at 73°F and plastic water piping materials shall conform to NSF 14. Sections for the products listed above then reference the specific material tables with code application standards. This is best understood by reviewing the tables (see Table 33-1 through Table 33-5). Please note only the first item is illustrated. The actual NSPC should be consulted for the facts.

Standards may seem like an insignificant and overly detailed topic to the average installer, but nothing could be further from the truth. These details must be understood because installing the wrong product will needlessly increase costs to all parties involved.

Table 33-1 Materials for Potable Water (seven notes are referenced in the original*). Courtesy of Plumbing-Heating-Cooling-Contractors—National Association.

Item	Standard	Water Service	Cold Distribution	Hot Distribution
1	ABS pipe ASTM D2282	A	X	X
Continues to Item 28	Ends with PVC ASTM D2241	(Depended upon the item)	(Depended upon the item)	(Depended upon the item)

(A indicates Approved)
(X indicates Not Approved)
*Notes provide greater detail for the entire table

Table 33-2 Materials for Sanitary Waste and Drain (six notes are referenced in the original*). Courtesy of Plumbing-Heating-Cooling-Contractors—National Association.

Item	Standard	Sewers Outside	Underground Inside	Aboveground Inside
1	ABS pipe ASTM D2661	A	A	A
Continues to Item 26	Ends with Vitrified Clay Pipe ASTM C700	(Depended upon the item)	(Depended upon the item)	(Depended upon the item)

(A indicates Approved)
*Notes provide greater detail for the entire table

Table 33-3 Materials for Vent Piping (two notes are referenced in the original*). Courtesy of Plumbing-Heating-Cooling-Contractors—National Association.

Item	Standard	Underground	Aboveground
1	ABS pipe ASTM D2661	A	A
Continues to Item 16	Vitrified Clay Pipe ASTM C700	(Depended upon the item)	(Depended upon the item)

(A indicates Approved)
*Notes provide greater detail for the entire table

Table 33-4 Materials for Storm Drainage (six notes are referenced in the original*). Courtesy of Plumbing-Heating-Cooling-Contractors—National Association.

Item	Standard	Sewers Outside	Underground Inside	Aboveground Inside
1	ABS pipe ASTM D2661	A	A	A
Continues to Item 25	Vitrified Clay Pipe ASTM C700	(Depended upon the item)	(Depended upon the item)	(Depended upon the item)

(A indicates Approved)
*Notes provide greater detail for the entire table

Table 33-5 Materials for Foundation Drains and Subsoil Drainage. Courtesy of Plumbing-Heating-Cooling-Contractors—National Association.

Item	Standard	Subsoil
1	Clay Drain Tile ASTM C498	A
Continues to Item 8	Vitrified Clay Pipe ASTM C700	All listed items are acceptable

(A indicates Approved)

REVIEW QUESTIONS

1. What does the acronym ASTM stand for?
2. What does the scope of a standard address?
3. What is one advantage in understanding the use of code standards?
4. Explain why the standard edition (year of publication) is important to code administrators and technicians.
5. What standards agency is primarily responsible for the development of backflow prevention standards in this country?

CHAPTER

34 Joints, Connections, Fittings, and Appurtenances

LEARNING OBJECTIVES

The student will:

- Discuss the importance of the consensus standards when making plumbing system connections.
- Compare the differences between water and drainage waste and venting connections for similar and dissimilar connections.
- Discuss the importance of traps and cleanouts in the plumbing system.
- Identify the required code distances for interceptors.

INTRODUCTION

Plumbing codes have long been dedicated to providing accurate, mandated information to ensure proper joining methods are used in the installation and repair of connections. The importance is heightened by the fact that failures in connections are readily identified by leaks; in short, their very visibility identifies installation errors or the choice of inappropriate materials. Plumbing professionals understand the consequences of joining failures. First, there are the underlying health and safety factors. Second, damage caused by unsatisfactory joining can be extremely expensive. This chapter will analyze code requirements addressing piping, fixtures, and appurtenances.

Plumbing professionals have understood for many years the health and safety matters associated with leakages. Leakages could be in waste lines, exposing sewage to food and water sources or directly to individuals. System failures could also result in airborne bacteria without the presence of a liquid indicator. Recently, public concern has focused on structural mold problems caused by water leaks.

Destruction of property caused by unsatisfactory joining can be extremely expensive. Inexperienced individuals attempting to install plumbing or make repairs have not only created amusing stories, they have illustrated the correction cost. The average technician considers water damage to homes and overlooks the less common, but more costly, effects to commercial spaces. All of us who consider computer equipment and operations shudder to think of water damage. Computer and electrical components are now common to all structures. Consider the manufacturing industry, where the old standard of pneumatic operations has now been replaced with electronic robotic equipment.

JOINTS BETWEEN SIMILAR PIPING MATERIALS

The term *joint* in the plumbing code is understood as the process of making a piping connection. Dictionaries and thesauruses have numerous entries for *joint*, one of which is *to mingle*. To mingle connotes the differences that may occur in piping connections. The differences may be minor, but they can increase the opportunity of connection failures.

Understanding the standards presented in the previous chapter ensures that fittings have greater standardization, thereby decreasing failures. Many pipe standards include information about fittings and address materials and pattern lay-lengths. Further, the previous chapter stated that the *National Standard Plumbing Code*, in Section 3.1.3, Standards Applicable to Plumbing Materials table, organizes the numerous standards from many organizations in a logical manner that addresses fittings and connections. The headers are listed here as a reminder.

1. Ferrous pipe and fittings: 11 items
2. Non-ferrous metallic pipe and fittings: 16 items
3. Nonmetallic pipe and fittings: 63 items
4. Pipe joints, joining materials, couplings, and gaskets: 28 items

In the Field

The basic premise of all plumbing codes is that connections shall be gas and watertight.

The plumbing code describes and lists many different types of approved connections that have proven to be very successful under typical conditions. These different materials are listed below, with brief comments where applicable.

DWV, Similar Piping Materials

The codes will list the types of DWV materials and then describe the types of connections and what standards apply.

1. ABS plastic: Connections shall be made with referenced standards by mechanical joints, solvent cementing, or threaded joint. See Figure 34-1.
2. Asbestos cement: Connections shall be made with a sleeve coupling of the same composition as the pipe, sealed with an elastomeric ring.

3. Brass: Connections shall be made with referenced standards by brazed joints, mechanical joints, or welded joints.
4. Cast iron: Connections shall be made with referenced standards by caulked joints, compression gasket joints, or mechanical joint couplings.
5. Concrete joints: Connections shall be made with referenced standards with an elastomeric seal.
6. Copper pipe: Connections shall be made with referenced standards by brazed joints, mechanical joints, soldered joints, threaded joints, or welded joints.
7. Copper tubing: Connections shall be made with referenced standards by brazed joints, mechanical joints, or soldered joints.
8. Lead: Connections shall be made with a burned or wiped process in accordance with code thickness.
9. Polethylene plastic pipe: Connections shall be made with referenced standards by heat-fusion joints or mechanical joints.
10. Polyolefin plastic: Connections shall be made with referenced standards by heat-fusion joints or mechanical or compression sleeve joints.
11. PVC plastic: Connections shall be made with referenced standards by mechanical joints, solvent cementing, or threaded joint.
12. Steel: Connections shall be made with referenced standards by threaded joints or mechanical joints.
13. Vitrified clay: Connections shall be made with referenced standards with an elastomeric seal.

Water, Similar Piping Materials

The codes list the types of water-supply materials and then describe the type of connections and what standards will apply. Another important conformance requirement is that water-supply materials, by code, shall conform to NSF 61.

1. ABS plastic: Connections shall be made with referenced standards by mechanical joints, solvent cementing, or threaded joints.
2. Asbestos cement: Connections shall be made with a sleeve coupling of the same composition as the pipe, sealed with an elastomeric ring.
3. Brass: Connections shall be made with referenced standards by brazed joints, threaded joints, or welded joints.
4. Copper pipe: Connections shall be made with referenced standards by brazed joints, mechanical joints, soldered joints, threaded joints, or welded joints.

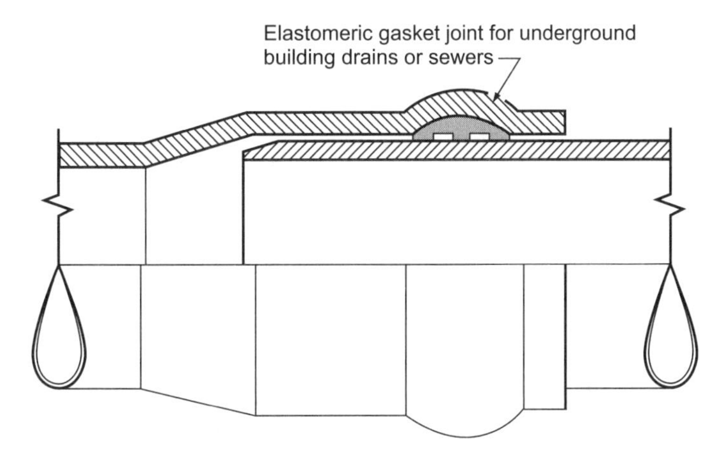

Figure 34–1
This mechanical elastomeric gasket joint for underground plastic-to-plastic piping illustrates a connection for similar piping. The presence of gaskets rather than bolts or threads still classifies this connection as mechanical. Courtesy of Plumbing–Heating–Cooling–Contractors—National Association.

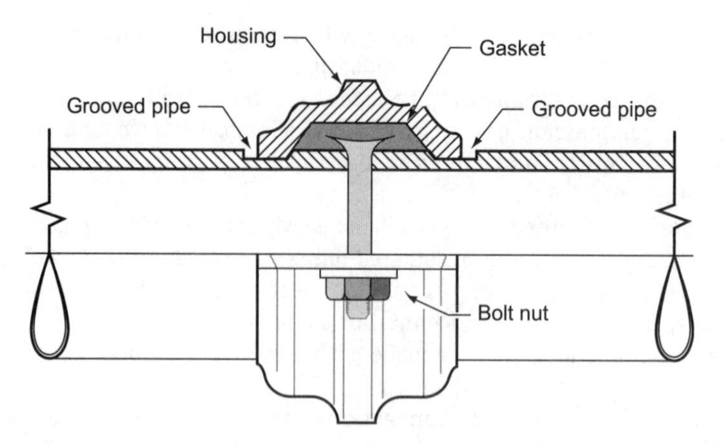

Figure 34–2
A mechanical joint for similar galvanized steel lines found in commercial water distribution systems. Courtesy of Plumbing-Heating-Cooling-Contractors—National Association.

5. Copper tubing: Connections shall be made with referenced standards by brazed joints, flared joints, mechanical joints, or soldered joints.
6. CPVC plastic: Connections shall be made with referenced standards by mechanical joints, solvent cementing, or threaded joints.
7. Cross-linked polyethylene plastic: Connections shall be made with referenced standards by flared joints or mechanical joints.
8. Gray iron and ductile iron joints: Connections shall be made with referenced standards in accordance with the manufacturer's instructions.
9. Polybutylene plastic: Connections shall be with referenced standards by flared joints, heat-fusion joints, or mechanical joints.
10. Polyethylene plastic pipe: Connections shall be made with referenced standards by heat-fusion joints or mechanical joints.
11. Polyproplene plastic: Connections shall be made with referenced standards by heat-fusion joints or mechanical or compression sleeve joints.
12. PVC plastic: Connections shall be made with referenced standards by mechanical joints, solvent cementing, or threaded joints.
13. Stainless steel: Connections shall be made with referenced standards by mechanical joints or welded joints.
14. Steel (galvanized): Connections shall be made with referenced standards by threaded joints or mechanical joints. See Figure 34–2.

JOINTS BETWEEN DISSIMILAR PIPING MATERIALS

Connections between different types of piping materials require greater attention to detail than similar piping materials. The difference in sizing is an apparent consideration; however, methods such as soldering on one side of the connection and compression on the other side are also of great consideration. The methods vary even more than the materials. The final goal is a joint that will not fail and meets the code provisions. The best defense against failed connections, resulting in damage and possible legal action, is adherence to code requirements.

There are many new, different types of fittings on the market today commonly called *push-connect types*. They are manufactured for convenience and have many benefits. Among the benefits is reduced installation time, especially when water is present in lines. Water in lines is a condition in which the pipe normally would have to be completely drained and heated, for example, when soldering. Very few of the push-connect products have recognized consensus standards at this time. The dissimilar connections are discussed below; remember, nothing is better than reviewing your current code book for dissimilar connections.

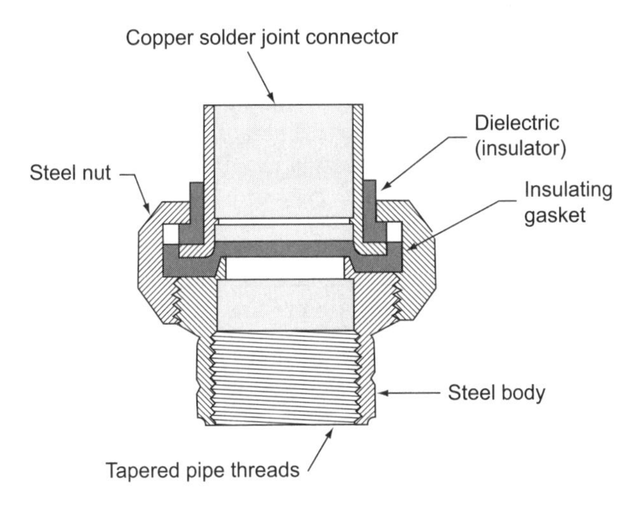

Figure 34–3
A dielectric union fitting used as an isolator from copper to galvanized steel lines in water distribution systems. Courtesy of Plumbing-Heating-Cooling-Contractors—National Association.

DWV, Dissimilar Piping Materials

Codes will reference the manufacturer's installation instructions and referenced standards for the compression seal. A partial list of the referenced connections is provided to aid in your search.

1. Cast iron pipe to galvanized steel or brass pipe.
2. Copper tubing to cast iron hub pipe.
3. Copper tubing to galvanized steel pipe.
4. Lead pipe to other piping material.
5. Plastic pipe to other material.
6. Stainless steel to other material.

Water, Dissimilar Piping Materials

Generally, dissimilar joint connections in water systems follow the DWV language. Again, codes will reference the manufacturer's installation instructions and referenced standards for the compression seal. A list of the referenced connections is provided to aid in your search; note the choice is smaller than the DWV application.

1. Copper tubing to galvanized steel pipe (see Figure 34–3).
2. Plastic pipe to other material.
3. Stainless steel to other material.

PROHIBITED JOINTS AND CONNECTIONS

Some plumbing codes will list prohibited connections, which is helpful. The following information was gathered from the *International Plumbing Code (IPC)*. Exercise caution when considering the prohibitions because the International Code Council has an International Residential Code for one- and two-family dwellings. Their residential code has greater flexibility than the IPC by not prohibiting saddle valves on a water line. This is important because many small saddle valves on water lines serve icemakers and humidifiers.

DWV

The following joints and connections are prohibited in DWV applications.

1. Cement joints, which were used previously with cement pipe.
2. Hot poured bituminous joints, which were used previously with clay tile.

3. Rolling O-rings, which were commonly used in wet sewer connections.
4. Solvent-cement joints between different plastics, a very controversial issue because suppliers distribute these cements.
5. Saddle-type fittings, not controversial because saddles are used on sewer mains in the public right-of-way, out of the jurisdiction of the plumbing code.

Water

The following joints and connections are prohibited in water applications.

1. Cement joints, which were used previously with gray iron water mains.
2. Hot poured bituminous joints, which were used previously with clay tile.
3. Solvent-cement joints between different plastics, a very controversial issue because suppliers distribute these cements.
4. Saddle-type fittings, a controversial issue because saddles are used often on icemakers.

FITTINGS

The fittings referenced here are not for connecting pipes and drains. These fittings perform very specific functions in the plumbing system for isolation (traps) and maintenance access (cleanouts). Little has changed in the design and application of the two; however, their importance and performance have improved.

Traps

Drainage-system traps separate noxious sewer fumes from habitable spaces. Hydraulics and system operations for the plumbing industry have created uniformity of code requirements. All fixtures and drainage openings shall have traps. The openings referenced here refer to floor drains, floor sinks, or similar uses. Some plumbing fixtures have built-in traps called integral traps, such as water closets.

Traps perform their isolation function by containing liquid from the required 2- to 4-inch level. Some codes allow deep seal traps to slow down the seal loss from evaporation. Other codes mandate seal replenishment from a water supply or drain supply called trap seal primers. For years, plumbing exams have asked about the basic reasons traps lose seals.

The basics are back pressure, capillary action, evaporation, leakage, momentum, aspiration, oscillation, cracked pipe, and siphonage. The 2- to 4-inch requirement is noteworthy because the codes, in their venting section, state that proper venting shall be provided to prevent the drainage system from being subjected to more pressure than 1 inch of water column, which equates to .0375 psi. A better perspective comes from considering that the drainage system is tested with a 10-foot head, about 5 psi for discussion purposes. The obvious consideration here is that a loss of 1 inch might be tolerated, but the code's built-in safety factor of a minimum of 2 inches provides double the normal operating protection.

The distance from trap to fixture is closely regulated by the code. Vertical distances of 24 inches and horizontal distances of 30 inches are the standard. Some codes allow running traps a vertical distance of 30 inches when a trap must be moved down the horizontal drain line away from the fixture because of obstructions. For example, when a bathtub is installed on an overhang there is not sufficient height available to install a trap directly under the bathtub. Traps are required to be as close as possible to the fixture to allow a minimum of untraped pipe, thereby reducing odors.

All codes prohibit certain traps, most of which are antiquated. Previously traps with interior partitions were prohibited. These were commonly called *bottle traps*, and they could corrode and lose the ability to maintain a trap seal. There is a need

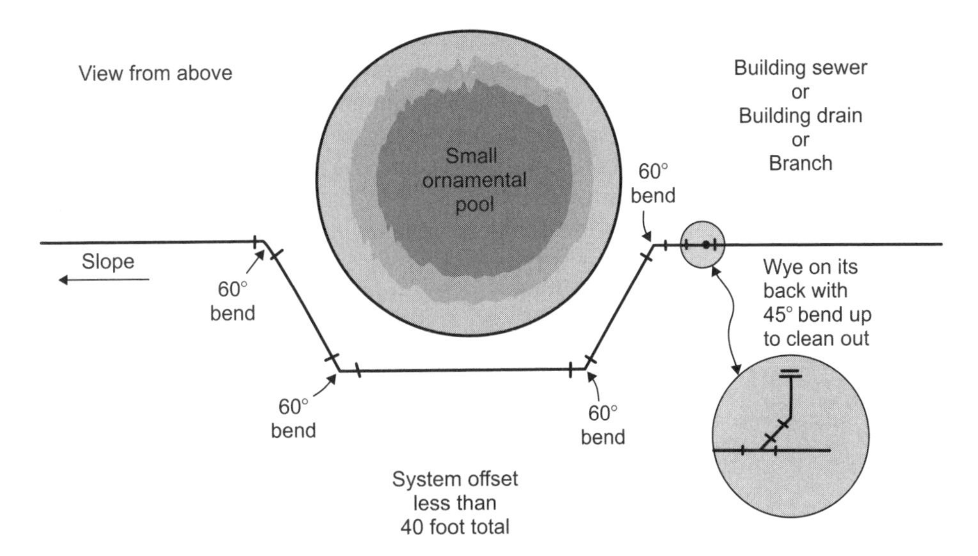

Figure 34–4
This illustration represents the necessary cleanouts in a drainage system when changes in direction greater than 45° occur within 40 feet of each other. Courtesy of the International Code Council.

for these narrow traps under pedestal lavatories to reduce the trap's visibility. Manufacturers and code officials have addressed this need by manufacturing and accepting traps with noncorroding partitions (plastic bottle traps), which are coated for appearance's sake. Other changes are being considered, such as canisters of fluid less dense than urine, which allows the urine to drain through without a water supply, and a type of duckbill check for waterless urinals.

The most common discussion for some time related to traps is the prohibition of S-traps. S-traps are subject to momentum problems and prevent air from the venting system from reaching the trap seal. This aspect will be discussed later when our venting discussion addresses hydraulic gradients and why sanitary tees are used when feeding sinks from a vertical riser, rather than a wye and $\frac{1}{8}$ bend.

Cleanouts

Building sanitary and storm drainage systems continues to require drainage cleanouts (see Figure 34–4). Some previous distance requirements have been relaxed due to new, improved drain-cleaning equipment. Drain-cleaning contractors have become very specialized in their field. The majority of licensing programs do not include this trade, although its technicians often remove traps, drains, and fixtures such as water closets to gain access to the drainage system. Recent changes in the plumbing code allow removable traps and water closets to be recognized as system cleanouts. Other improvements have been the recognition of a two-way cleanout at a building entrance. That two-way cleanout with today's new equipment reduces cleanup costs for the inside of structures during drain-cleaning processes. Consult your code book for other details when taking your examination.

APPURTENANCES/INTERCEPTORS

Interceptors have been discussed in other portions of this book related to installation and service work. Chapter 32 included a discussion of code definitions and addressed a plumbing appurtenance as a manufactured device, prefabricated assembly, or an on-the-jobsite assembly of component parts that is an adjunct to the basic piping system and plumbing fixtures. It demands no additional water and does not add any discharge load to the drainage system. Examples included water filters, backwater valves, dilution tanks, interceptors, and backflow preventers. This chapter's discussion will address interceptor code requirements for grease, oil, and sand interceptors.

The major function of an interceptor is to stop and accumulate unwanted materials in the drainage system. The unwanted materials can harm and eventually stop the drainage system. Further, the materials can eventually disrupt and hinder the function of sewage treatment for public or private systems.

All interceptors are sized according to the maximum volume and rate of discharge of the plumbing fixtures, equipment, or area flowing into the device. The interceptor's basic venting purposes are to keep the materials from becoming air bound and to prevent the loss of trap seals.

Grease Interceptors

All interceptors designed for specific functions in the plumbing system will have a descriptive title attached to the interceptor designation, for example, grease, oil, and sand interceptors. Kitchen equipment is the most common area for grease interceptors to be installed. Frequently designers have indiscriminately required all equipment to flow through interceptors. The interceptors were commonly placed outside the structure because of their large sizes, which were based on the large volume of discharge.

Grease interceptors have become far more sophisticated from their original design. Many years ago, devices referred to as *grease traps* had a cold-water supply connected to a water-cooled jacket area around a great trap. The purpose was to cool the grease-bearing liquid, which improved the ability to capture the grease by allowing it to congeal. It was a dangerous practice because the water supply did not commonly have backflow protection. Today's devices operate efficiently and have modern-day advantages, such as electronic warning devices and skimmers for removing and recovering the grease.

The location of grease interceptors is of code importance, but of greater importance related to their successful operation. Individual manufacturers will provide detailed instruction regarding how far the interceptor can be located from the fixture discharges. One concern is that the interceptor may be placed below the same floor elevation as the fixture. Common sense indicates that as the interceptor becomes lower, the discharge head-volume velocity will be increased, which reduces the interceptor's operational design.

The interceptor's distance from the fixture also becomes a factor when considering the trap. The code's concern for proper sanitation requires a trap at the fixture when interceptors are farther than 30 inches (IPC) or 48 inches (NSPC) horizontally and 30 inches vertically from the interceptor. Considering the additional trap, double trapping would cause the fixture to become air bound and hinder proper drainage.

Double trapping does not take place, as addressed above, due to two different venting provisions. First, the vast majority of interceptors have a flow-control device designed to be mounted on the outside horizontal inlet of the interceptor. These flow-control devices slow down the incoming flow to allow grease separation in the interceptor. The flow controls shall be vented, which allows additional air to enter the flow (a process called *air entrainment*), which benefits interceptor separation. Second, other provisions of the code require a vent on the fixture branch between the fixture trap and the flow control. Where and how these two different vents are connected are of great importance to separate noxious sewer fumes from the habitable space. This is best described by two direct quotes from Chapter 6 of the 2006 *National Standard Plumbing Code, Illustrated*, as seen in the sidebars.

The vent system connection discussed in Figure 6.2.2-A of the 2006 *National Standard Plumbing Code, Illustrated* (see In the Field sidebar), which serves the flow control and fixture branch along with the fixture trap, will keep sewer fumes from employees' work area.

In the Field

"Figure 6.2.2-B, A Grease Interceptor Serving as a Fixture Trap, Notes: 1. When a fixture that is connected to a grease interceptor is within 4 feet horizontally and 30 inches vertically from the inlet to the interceptor, a fixture trap is not required. *If the fixture is not trapped the air intake for the flow control device must not be connected to the vent piping system.* It must be terminated at a 180 degree return bend with its inlet at least 6 inches above the flood level of the fixture." (Emphasis added by the author.)

In the Field

"Figure 6.2.2-A, A Grease Interceptor Serving Trapped and Vented Fixtures, Notes: 2. When the fixtures connected to the grease interceptor have traps, the air intake on the flow control device must be connected to the fixture vent piping system."

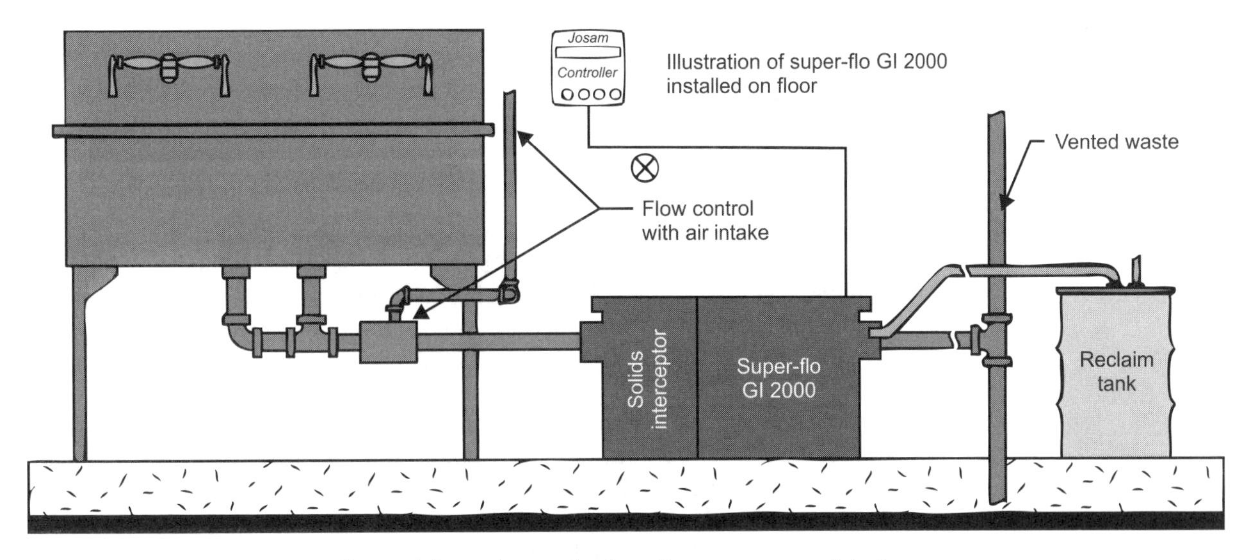

Grease interceptors and automatic grease removal devices

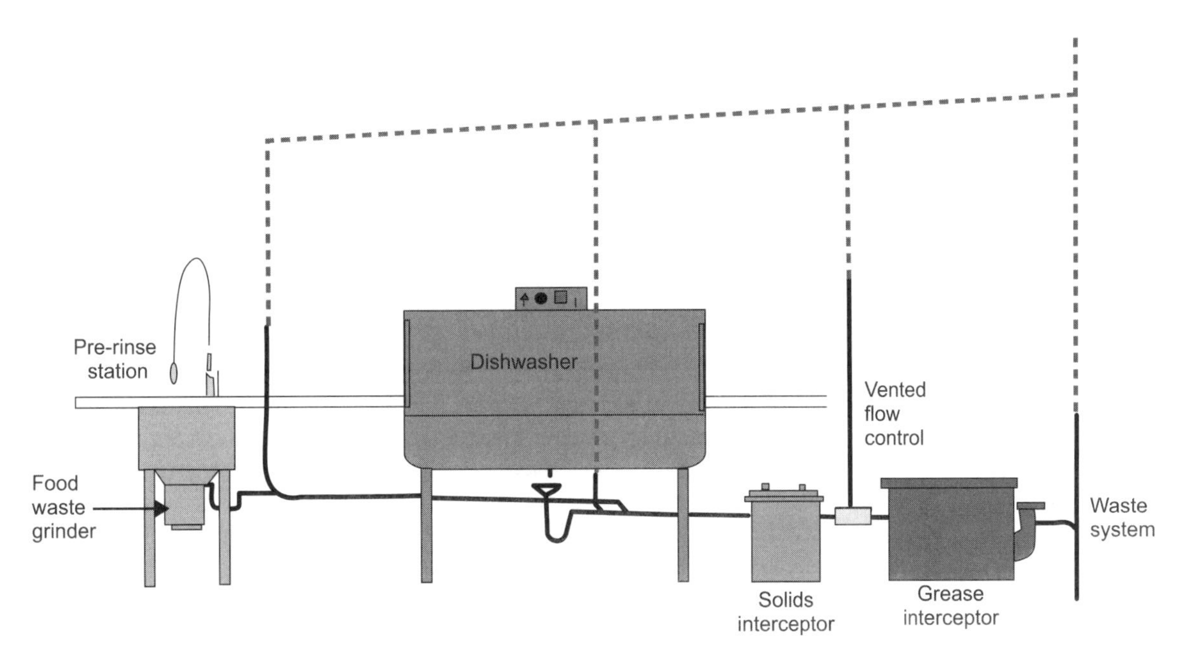

Food waste grinder with solids interceptor

Figure 34–5
A grease interceptor that serves a trapped fixture with a vented flow control connected to the sanitary venting system. These connections ensure that sewer fumes will not be present in the employees' work area. Courtesy of the International Code Council.

The codes also discuss food waste grinders and dishwasher discharges to interceptors. If a food waste grinder discharges through a grease interceptor, a solids interceptor shall collect the food waste products to keep them out of the grease interceptor. Commercial dishwashers may discharge through interceptors, but the interceptors' designs must be evaluated to ensure correct retention operations. See Figure 34–5.

Oil Separators

Oil separators are devices designed to stop and displace oil and grease in order to remove flammable products from the sewer system and to ensure proper operations of the sewer system. These appurtenances are referred to as separators rather than interceptors because the goal is not to only separate and retain materials in the interceptor, such as with a grease interceptor, but to separate and constantly siphon off the harmful product to another container. The other container is also considered part of the process and must be closely monitored.

Oil separators are required when a commercial process could produce harmful substances such as petroleum products, which are then directed to the sanitary drainage system. Examples of these processes would include repair and service garages, car washes, and manufacturing processes. Facilities that collect and contain their harmful liquid waste by means other than drainage systems, such as special safety vacuum systems or oil-dry absorbent products, are not required to have separators. Separators are also not required in parking areas or garages, and these areas are not required to be washed down for cleaning. It is common to find parking areas discharging to storm systems.

Oil separators are sized based on the area flowing into the separator and gallons per minute of rated flow. Consideration of flow area is wise because it includes not only nearby water discharging through the unit, but also the materials or products that are available to discharge with the water supply. For example, washing an engine would discharge water and petroleum products. For that reason, codes will consider the floor area as the sizing factor. Codes require 6 cubic feet for the first 100 square feet of drained area, plus another square foot for each 100 square feet of drained area. This could translate to huge devices for commercial garages with large floor areas. Common sense practice has addressed this issue by only requiring sizing inclusion for areas where service actually takes place, by methods such as pitching the floor in the service bay and not including (by pitching) the parking or traffic-lane areas.

Oil separators are either manufactured commercially or field fabricated. The code's sizing information requires at least 24 inches of depth below the discharge pipes inversion and a trap water-seal depth of 18 inches. The draw-off or skimmer piping, including the storage device, is commonly governed by the authority having jurisdiction. The *National Standard Plumbing Code* carefully addresses the vent piping and identification. The separator's venting shall always be independent of the sanitary system to avoid flammability issues. That consideration should also remind technicians that areas draining into the separators shall be trapped to avoid combustible vapors moving from the separator to the work area. Work area floor drains and sand interceptors shall be trapped. These traps are often a point of discussion with individuals stating that solids such as sand are difficult to remove from common traps. Sand interceptors with a trapping ability are the answer, and they avoid the flammability issues.

Sand Interceptors

Sand interceptors have been a part of plumbing systems for many years and have different construction methods. Some plumbing codes see the sand interceptor with its accumulator area as requiring a trap with a trap seal depth of no less than 6 inches. These smaller units generally have an open grate, which receives the water and sand debris above the circular containment area. Cleaning these units is a fairly simple process accomplished with periodic maintenance by standing on the floor above the interceptor and reaching down with a shovel or scooping device. Other codes appear to view the sand interceptor as a much larger containment device and detail its size and flow requirements. All sand interceptors shall be watertight.

In the Field

Servicing all tanks and containers is dangerous. Always follow confined-space entry procedures to ensure safety.

REVIEW QUESTIONS

1. What is the difference between interceptors and separators?
2. What are the two major reasons for requiring proper joining methods?
3. Why have drainage piping cleanout locations changed in the code from earlier codes?
4. Name the two major areas of consideration for water and DWV connections.

CHAPTER

35 Plumbing Fixtures and Minimum Fixture Requirements

LEARNING OBJECTIVES

The student will:

- Discuss the importance of providing accessible and usable fixtures in a structure.
- Identify the fixture discharge amounts required by code to address water conservation concerns.
- Summarize the overall code requirements for the various major fixture categories.
- Compare the various methods used to obtain occupancy numbers in the calculation of required fixtures.

PLUMBING FIXTURES

Plumbing fixtures in sufficient numbers, operating correctly, are one of the most important aspects of code enforcement. When classifying areas of overall importance, the code follows the safe supply of water and adequate drainage closely. Having adequate numbers of fixtures would be of no significance if they did not perform in a safe manner, which is why the code is greatly detailed, in addition to addressing the required standards.

This chapter examines code requirements in alphabetical order for ease of understanding. All the information is presented in universal code matters to provide a broad base of understanding for the code examinations. The majority of plumbing codes in this country that are updated on a regular basis have greater uniformity than may be expected because the same manufacturers and designers are involved in writing the codes. The final goal is to have sanitary and hygienic cleaning, washing, and food preparation for building occupants.

Accessible and Usable Buildings and Facilities

Accessible plumbing facilities and fixtures are of great importance to all individuals, because they may need to access the facilities themselves. Statutes and other regulations also require this consideration. *Accessible* describes a facility as being usable by persons having physical disabilities. One of the more commonly recognized standards in this matter is ICC/ANSI A117.1, Accessible and Usable Buildings and Facilities. Plumbing codes generally do not list dimensional details to meet the accessibility requirements. Code officials typically agree that the requirements of the parent building code, in conjunction with the authority having jurisdiction, in the fewest publications best serve the public. Later discussions in this chapter will not accurately apply to accessibility matters such as dimensional issues.

Installation Locations

The goal of codes in addressing fixture locations and materials was outlined by the twenty-two principles (see Chapter 32). Fixtures shall be installed with their minimal clearances to ensure proper use (this includes privacy) and cleaning. The following list identifies the required minimum clearance for fixtures.

1. Bidets, lavatories, urinals, or water closets shall not be set closer than 15 inches from their center to a sidewall or partition.
2. Bidets, lavatories, urinals, or water closets shall not be set closer than 30 inches center to center between adjacent fixtures.
3. Bidets, lavatories, urinals, or water closets shall have at least 21 inches of clearance in front of the fixture lip.
4. Water closet compartments shall be no fewer than 30 inches wide and 60 inches in length.

Some of the codes address residential occupancies with different details based on greater flexibility for the residents, who are familiar with their surroundings.

Additionally the codes clarify installation details related to fixture supports and sealing around the fixture-to-floor or fixture-to-wall surfaces. Access to slip-joint connections is another area that provides options when considering the type of drainage connections used.

Fixture Water Conservation Requirements

Several governmental agencies, local ordinances, and codes have established requirements in addition to what the national consensus standards require for maximum flow rates and consumption of plumbing fixtures and fixture fittings. All individuals are now familiar with the mandated water closet discharges of

Table 35-1 Maximum Flow Rates and Consumption of Plumbing Fixtures

Plumbing Fixture or Fitting	Maximum Flow Rate or Quantity
Private lavatory	2.2 gpm at 60 psi
Public lavatory, metering	0.25 gallon per cycle
Public lavatory, non metering	0.5 gpm at 60 psi
Shower head	2.5 gpm at 80 psi
Sink faucet	2.2 gpm at 60psi
Urinal	1 gallon per flush
Water closet	1.6 gallon per flush

Note: 1 gallon per minute = 3.785 liters per minute

1.6 gallons per flush. Recently, one state has mandated a smaller flush requirement of 1 gallon. Code officials have also become familiar with waterless urinals, and some have viewed demonstrations of water closets flushing with .7 gallons per flush. The discharges of lower-consumption fixtures have forced the codes to adjust by recognizing lower drainage fixture unit (DFU) values in sanitary drainage sizing table requirements. Fixture conservation requirements are often placed in different areas of the codes, making them more difficult to understand. Table 35-1 is provided for your convenience.

FIXTURE REQUIREMENTS

The code requirements associated with the following fixtures are summarized in a list below for your convenience. Fixture standards may vary from code to code and will not be listed except when other areas may be affected, such as water temperature requirements. Figure 35-1 illustrates the maximum water temperatures to fixtures as listed in standards.

Automatic Clothes Washers

Automatic clothes washers are fixture appliances and have water supply protection from an air gap built into the unit and do not require vacuum breakers on the hose-connected water supply. The drain for these washers discharges into a standpipe served by a trap through an air break. Standpipes must be at least 18 inches and no more than 42 inches in height. The height is regulated, which results in the standpipe being at least as high as the unit's top to prevent the basket from siphoning itself dry. Commercial washers dump their contents by gravity in a trough at the back of unit. The most recent concern about residential applications is that new, high-volume discharges may affect drain line performance. Check your local code to see if an increase in line size is required for branch headers located downstream of the automatic washer discharge branch. Another area of the code for consideration is that the supply lines shall be served by a water hammer arrestor to compensate for the solenoid-valve operation, which would shock the water distribution system.

> **In the Field**
>
> A word of caution: healthcare bathing tubs, with their associated sophisticated equipment, vary a great deal. For that reason, code officials may require additional backflow protection on the water supply, such as reduced pressure zone backflow preventers.

Bathtubs

Bathtubs are constructed of plastic, enameled cast iron, and enameled steel, which conform to various consensus standards for performance. Healthcare bathing tubs may also have pressure-sealed side doors.

All bathtubs shall have at least $1\frac{1}{2}$ inch waste and overflow. The slip joint connections on the drains shall be accessible unless they are permanently fused together.

Whirlpool tubs may also be considered with bathtubs and have their own appropriate standard. A significant code change related to whirlpool tubs is that some codes now identify specific access openings in order to service the pump. If the pump is within 2 feet of the access panel, the opening may be 12″ × 12″. If the pump is more than 2 feet from the access panel, the access opening shall be at least 18″ × 18″ in size.

The temperature of the water supply from tub valves is a major safety concern and is addressed by codes. Some codes have new language that requires bathtubs and whirlpool tubs to have temperature protection even when a shower outlet is not present. This new requirement is an ASSE 1070 device set to a maximum temperature of 120°F. The temperature requirements are mandated to protect individuals from burns, which is of considerable importance because only 7 minutes of exposure to 120°F water will cause burns.

Bidets

Bidets are widely accepted fixtures with two major safety concerns: the water supply with a submerged inlet and the temperature supply control. These fixtures have gained wider acceptance recently and have become imitated by water closet seats with supply jets directed upward to clean the perineal area of the body. The submerged inlets are isolated from the water distribution system by the appropriated backflow preventers. Water temperature concerns have recently been addressed by the codes mandating an ASSE 1070 device set to a maximum temperature of 110°F.

Dishwashers

Dishwasher consideration in the code is based on residential units and commercial units. The water supply to both the units shall have proper backflow protection through either an air gap or an approved backflow preventer. Commercial units often have pipe-applied backflow preventers, and residential units will have an integral air gap. Dishwasher drains for commercial units have an air gap or air break drain connection. Residential dishwashers may drain continuous waste into a kitchen sink after looping under the countertop. When a high loop, which prevents drainage siphoning off the unit, is not practical, an air gap that fits on the drain shall be provided.

In the Field

It is advisable for installers to review the requirements of the authority having jurisdiction when dealing with kitchen equipment connections. Local health departments commonly have different or more restrictive requirements than national plumbing codes.

Drinking Fountains

Drinking fountains have specific conformance standards and do not always have to be refrigerated devices. Protecting the bubbler head from contamination has always been a concern, especially in schools. Past practices of school designers have included installing fountain bubblers on hand sinks in individual classrooms. Recently, many local health boards have prohibited that practice. A leading manufacturer of these devices has recently marketed a unique bubbler that has a cap that pivots over the bubbler when the valve is closed and opens when the hand valve is turned on.

Some codes may allow water coolers or bottled-water dispensers as substitutes for a small percentage of the required number of drinking fountains. This is very controversial because officials understand that building owners can decide to stop their supplier's contract, leaving the building without the required drinking fountain.

Emergency Equipment

Emergency showers and eyewash stations shall comply with ISEA/ANSI Z358.1. The standard requires temped water supply, not tempered as defined in the plumbing codes. Tempered water is water with a temperature adjusted to an optimum temperature based on the chemical encountered in order to ensure that the chemical is washed away properly and safely. The code does not require drains to the sanitary system for these fixtures because of the infrequency of use.

Faucets and Flushing Devices

Several conformance standards ensure proper health and safety conformance for building occupants and other public-water-supply users when addressing faucets and flushing devices. Faucets and fittings supplying water for human ingestion shall meet the requirements of NSF 61. This National Sanitation Foundation code measures products for acceptable criteria related to health matters commonly described as toxicological concerns. It is not intended to affect backflow preventers and boiler-drain type faucets.

Flushing device standards apply to flushometer valves, flushometer tanks (pressure-assisted), and flush-fill valves.

Floor Drains

Floor and trench drains have performance standards referenced in the code and require accessible covers and strainers. Older codes allowed deep seal traps and trap seal primers to protect against trap seal loss. A deeper trap seal only delays the trap seal process. For that reason, several codes now only accept trap seal primers that conform to ASSE 1018 or ASSE 1044. The code has a minimum size mandated as 2 inches.

Traps are an important part of the plumbing system and are not installed indiscriminately. They serve a purpose and are placed by designers to perform specific functions. Occasionally conservationists or treatment operators will seek to develop ordinances that terminate presently installed floor drains. Their motives are to stop the discharge of unacceptable chemicals down commercial floor drains. Those well-meaning actions are not acceptable and will have a negative affect of the structure's performance as designed. These suggested programs would instruct enforcers to fill

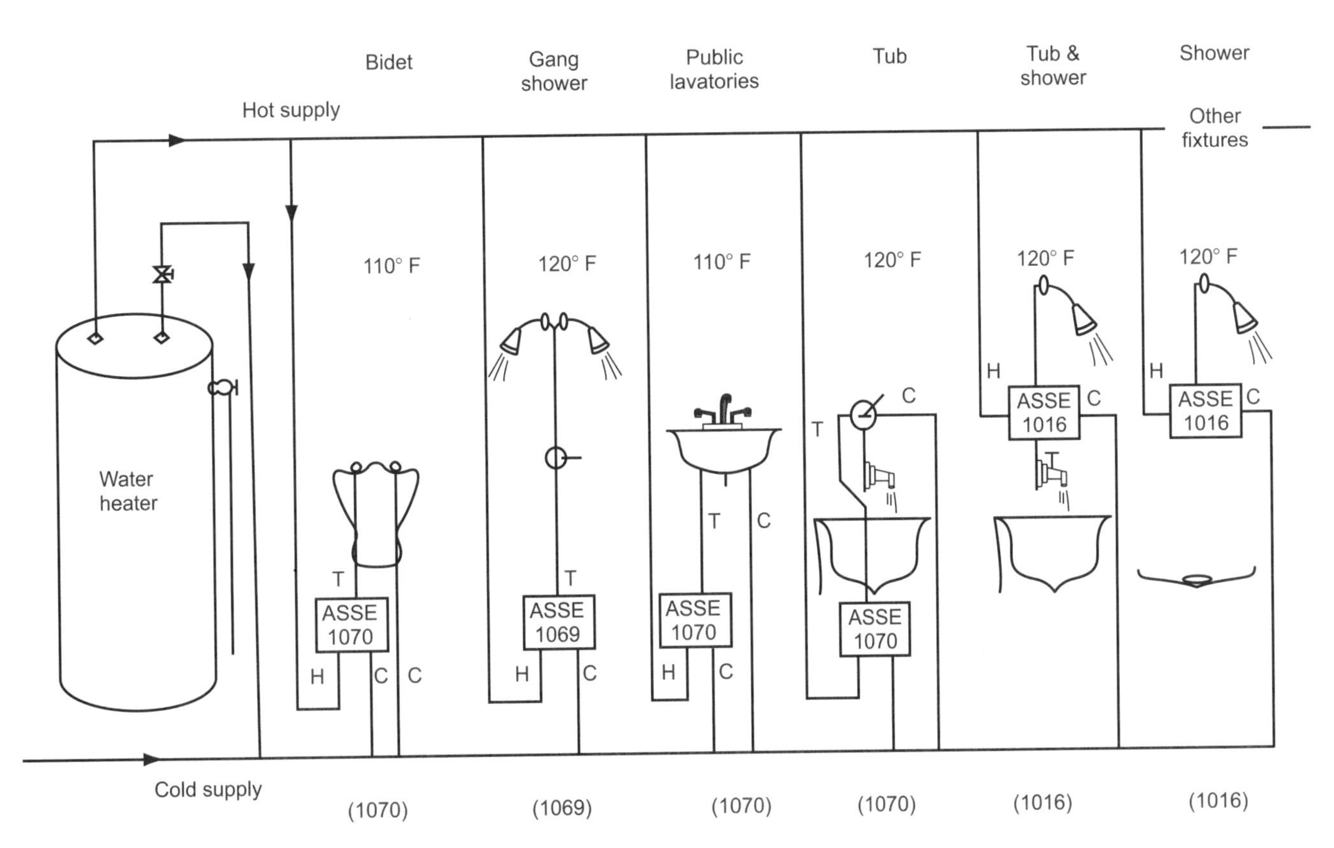

Figure 35–1
Fixture temperature limitations by standards designation.

the drains with concrete to stop their use, which further aggravates the problem. Concrete is not an acceptable plumbing connection. Its shrinkage will allow sewer gas to enter the building after the trap seal evaporates.

Food Waste Grinders

Food waste grinders, commonly referred to as garbage disposals, have code conformance standards. Although they are classified as fixtures, they are fixture appurtenances and do not have assigned DFU loads for line sizing. The minimum size drain for domestic units is $1\frac{1}{2}$ inches, and 2 inches for commercial units. Some commercial units have a water supply, which must have acceptable backflow protection. The cold water supply generally connects to the top basket area of the unit to aid in washing down food-scrap contents. A pipe-applied, atmospheric vacuum breaker commonly protects this water supply. A solenoid valve wired with the disposal switch supplies the cold water. Remember the solenoid valve cannot be downstream to the vacuum breaker. That would be a violation of the code standard by inhibiting the movement of the vacuum breaker to a closed position when not in use for extended periods of time. That lack of movement allows the vacuum breaker to "setup" (become inoperable).

Healthcare Fixtures

Healthcare fixtures include a broad spectrum with many different uses and requirements. Technicians must apply the other fixture requirements first and then seek out additional water supply and drainage code principles. Examples would include clinical sinks with bedpan washers and sterilizers, which have water, drain, and steam connections. All are best dealt with on a case-by-case basis.

Lavatories

Lavatories are manufactured from several different materials and have referenced conformance standards listed in the code. The minimum size waste outlet shall be no less than $1\frac{1}{4}$ inches, and overflows are not always required. Healthcare facilities do not approve of these overflows because of their concern for potential unwanted bacterial growth. In addition, healthcare facilities may prohibit the use of aerators at the faucet discharge. Group wash fountains in all use-group areas recognize that one occupant is assigned 18 inches of space.

The temperature of the water supply in lavatories in public facilities is a major safety concern and has been addressed by codes for some time. Some codes require protection by an ASSE 1016 device adjusted to a maximum temperature of 110°F. Newer language requires by an ASSE 1070 device set to a maximum temperature of 110°F. The temperature requirements are mandated to protect individuals from burns and are often referred to as tempered water. These discussions will focus on public hand-washing facilities. Several codes that have localized technical amendments to the national code further expand the public hand-washing designation to include facilities for individuals who do not have the ability to discern excessive temperatures. Those facilities may include accessible fixtures, elementary schools, and adult congregate homes for the aged.

Showers

Recent years have seen showers increase in popularity because of their convenience and reduced water consumption for bathing. Recent years have also provided code changes. One of these changes is the temperature of the water supply from shower and tub valves. This is a major safety concern, and mandates protection by an ASSE 1016 device adjusted to a maximum temperature of 120°F. The ASSE 1070 device required for tubs and lavatories protects individuals from burns. The ASSE 1016 device protects individuals from burns and thermostatic shock. Thermostatic shock results from pressure differentials in the water supply system that could take place when someone flushes the adjoining water closet fixture. The shock may cause an individual to quickly move back and possibly slip in the shower, resulting in serious injury. ASSE 1016 valves provide the necessary safety to protect standing individuals. There are several of these devices, which are adjusted by the installing technician and are listed by their manufacturers to meet both ASSE 1070 and ASSE 1016.

The size of showers is regulated by the code to be a minimum of 900 square inches, which allows the user practical movement space for cleaning. Another code change recently clarified that one of the dimensions could be reduced below 30 inches (30″ × 30″ = 900 sq. in.). This change clarified that a side width of 28 inches would be permitted. The change targeted homes where individuals had their 28″ × 60″ bathtub-shower combination removed in order to install a 28″ × 60″ shower base. The change allowed less mobile individuals to have shower access without stepping over the tub threshold, thereby reducing the possibility of slipping.

Other changes specified that shower doors shall not be less than 21 inches wide to ensure acceptable access. The measurement was taken from comparable emergency-exit numbers based on the average person's shoulder width.

Showers, which are constructed on site with shower liners, have been carefully addressed by the codes. Two factors are considered here: that early liners were constructed of lead, for which plumbers were installation experts, and that fixture leaks caused significant damage. New liner products and methods of installation, along with testing language, will be addressed in future code cycles.

Sinks

Sinks have numerous conformance standards addressing their construction materials and the items connected to them, such as faucets and drains. All sinks have a minimum drain size requirement of $1\frac{1}{2}$ inches and require a strainer or cross piece to catch or retain waste products.

Some codes include discussions of service/mop sinks in the sink area, which have drains that vary in size from $1\frac{1}{2}$ inches to 3 inches. The code generally requires at least one service sink per structure to address cleanup issues in a commercial building.

Urinals

Vitreous china and plastic urinals are addressed by code conformance standards. Some of the codes do not reference the new standard for waterless china urinals, which requires code officials to make acceptance decisions. There are several different types of urinals, from free standing to trough urinals, including urinals for men and women. Trough urinals are not included in conformance standards and do not have the ability to wash the entire unit by flushing. For that reason, they are generally not acceptable.

The codes vary concerning the number of required fixtures by considering how many water closets can be replaced by urinals. Replacement consideration varies from 50 percent to 67 percent based on the building's use-group designation. Consideration of the required number of urinal fixtures involves other practical matters. Many men visiting a restroom for urinating prefer the privacy of a water closet compartment and use the water closet. This has a negative effect on the design numbers of water closets. For that reason, some codes require small privacy partitions between urinals. Generally, partitions extend from the wall 18 inches or from the lip of the urinal 6 inches; the height is also to be detailed. The wall and floor surfaces surrounding fixtures are also addressed by requirements of dimensional water proofing in plumbing codes.

Water Closets

There are several different types of water closets and flushing mechanisms recognized by codes with accompanying standards. Backflow protection for the water supply is always critical. Codes also address the issues of privacy and seats. Privacy partition dimensions are 30″ × 60″. Ironically, the codes do not mandate the door swing arrangement or locks on the privacy compartment or restroom doors. Locks on restroom doors for multiple occupants generate the following discussions.

1. Are the restrooms closed to unauthorized individuals without keys to prevent an intruder from hiding in the restroom?
2. Are the code-required restrooms locked to avoid letting customers use them, in an attempt to address store security and theft issues rather than customer needs?
3. Can restrooms be locked from the outside allowing building staff to overlook an occupant and lock down the unit? This is devastating related to fire safety matters.

The codes also address water closet seats by mandating open front seats with elongated bowls for public restrooms. This a proven requirement to address proper sanitation concerns, primarily for men. Public restrooms do not require seat lids and hotel/motel room water closets are not considered public.

The *National Standard Plumbing Code* has an excellent design chart for water closet heights. Fixture heights (including sinks and lavatories) are not always addressed by codes and vary. Consult the job specifications and fixture manufacturers' installation recommendations for greater detail.

MINIMUM FACILITIES

Several factors must be considered when researching the required number of plumbing fixtures in a structure. Factors include the number of occupants expected in the building, the percentage of each sex in the structure, the distances to the restroom, and if the occupants are employees or guests.

There are several methods to obtain the number of occupants in a building. They are listed numerically below to improve clarification.

1. Actual-occupant method: The easiest method is for a structure designed as an assembly use group A-1, such as a theater with fixed seating. You would consider the number of seats and factor in the structure's employees. Every other building requires far more consideration.
2. Engineering estimate: This method is based on designers' past experience and documentation. It establishes the number of occupants based on square feet per person. Code officials will verify these facts in their plan review prior to inspections.
3. Legal limit: Some codes use the occupancy numbers generated by the parent building code for means of egress (exiting) established for fire safety. These numbers consider the use-group designation, which has an established number of square feet for each person. That square footage number would then be divided into the total square footage available in the building, resulting in the maximum number of occupants.

   Exercise caution depending on which plumbing code you use. The *National Standard Plumbing Code* states, "the number of occupants for plumbing purposes shall be permitted to be reduce to two-thirds of that required for fire or life safety purposes" (two-thirds of the fire code occupants). The *International Plumbing Code*, in its code commentary, states, "The means of egress occupant loads of a building do not always reflect typical day-to-day occupancy loads; however, the table takes this into account by modifying the values for determining the number of fixtures." The table they reference is their Minimum Number of Required Plumbing Fixtures table.

4. Owner/Agent Affidavit: This method is available in some cases and must be reviewed carefully because it is to the owners' benefit to be very conservative in their projected number. This would benefit them in reduced costs by enabling them to reduce the number of plumbing fixtures provided.

Generally speaking, the codes obtain the number of users by gender through common sense division of the total occupancy load by two. Some codes allow consideration of higher numbers of a specific sex when backed up by documentation, for example a dress shop. This may seem very basic, but consideration must be given to the fact that a building is being constricted per code and the next owner may question why the code and actual fixture numbers are different.

Plumbing codes, especially those prone to having different technical amendments, differ in single-occupant, separate-sex restroom criteria. Some codes state that when the occupant load is fifteen or fewer, one unisex restroom is acceptable. Other codes have embraced the concept that when the occupant load is fifty or fewer, one unisex restroom is acceptable. The majority of the time it is dependent on the use group and code. All users, whether male or female, appreciate the privacy of a single-occupant restroom dedicated to their sex. For example, some codes will not allow a urinal in a single-sex restroom because females have expressed displeasure with what they perceive as an unnecessary and dirty fixture.

The majority of plumbing codes use the same allowances for travel distance. Restrooms for each sex are allowed to be at every other floor level based on code wording that, "accessibility to the fixtures shall not exceed one vertical story" for multiple-floor buildings. The travel distance to centrally located restrooms in large structures is limited to 500 feet. The NSPC has very clear requirements at the end of their fixture tables addressing several specifics, and ICC has specific language that prohibits customers from walking through storage or kitchen areas in traveling to the customer restroom. In nearly every case, customer and employee restrooms may be combined.

MINIMUM FIXTURE NUMBERS

Tables 35-2 and 35-3 have been created from ICC information as very simple illustrations for discussion purposes.

Example

What are the required fixtures for a small fixed-seating theater with an occupancy load of 640 patrons and ten employees?

The total occupancy is 650.
650 ÷ 2 = 325 men and 325 women
Water closets for men: 325 ÷ 125 = 2.6 (rounded up) =3 water closets for men
Water closets for women: 325 ÷ 65 = 5 water closets for women

Note: the different divisors for men and women produce a ratio of 40:60 in order to address "potty parity" issues recognized for many years. This creates a more equitable distribution. Studies have shown that the lines at female restrooms were a result of more time required per restroom visit and more frequent visits to the restroom for women than men.

Lavatories: 650 ÷ 200 = 3.25 (rounded up) = 4; 2 for men and 2 for women

Drinking fountains: 650 ÷ 500 = 1.3 (rounded up) = 2

Service sinks: 1 is required

Table 35-2 Minimum Number of Required Plumbing Fixtures

Classification	Occupancy	Description	Water Closets		Lavatories		Bathtubs	Drinking Fountains	Other
			M	F	M	F			
Assembly	A-1	Theater	1 per 125	1 per 65	1 per 200		None	1 per 500	1 Service Sink

Table 35-3 Minimum Number of Required Plumbing Fixtures for 650 Occupants

Classification	Occupancy	Description	Water Closets		Lavatories		Bathtubs	Drinking Fountains	Other
			M	F	M	F			
Assembly	A-1	Theater	3	5	2 per each restroom		None	2	1 Service Sink

REVIEW QUESTIONS

1. What is the definition of *accessible facilities*?
2. What is the minimum required clearance in front of a water closet?
3. What is the current maximum discharge acceptable by code for a water closet?
4. What is the maximum allowable temperature for a tub and shower valve, and what device is allowed to control that temperature?
5. What is the maximum allowable temperature for a public hand-washing sink, and what device is allowed to control that temperature?

CHAPTER 36

Hangers and Supports, Indirect Waste Piping, and Special Waste

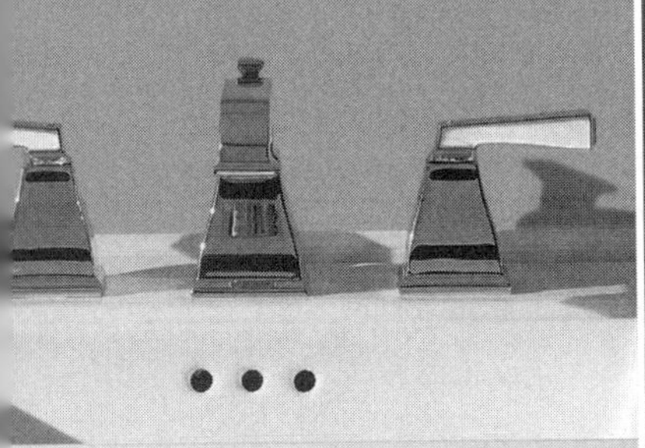

LEARNING OBJECTIVES

The student will:

- Discuss the importance of pipe hangers and their required distances.
- Describe why piping sleeves are necessary for piping protection of plumbing piping systems.
- Explain how indirect piping connections ensure proper sanitation in drains.
- List the acceptable waste receptors with their applications.

In the Field

Occasionally technicians will discuss the importance of not voiding the manufacturer's warranty by adhering to the installations instructions. The material replacement cost is of little importance when structure and contents damage occurs from piping failure caused by poor hanger installations.

In the Field

Plumbing professionals must exercise caution in the acceptance of hanger products. The codes do not have conformance standards to address all the various hanger products. Independent acceptance decisions are necessary by purchasers and code officials alike.

HANGERS AND SUPPORTS

The proper support of plumbing piping and its contents is one of the important aspects of plumbing that is often overlooked. Piping systems are composed of several different materials, each with its own purpose, characteristics, and strength that are affected by hanger distances. Another aspect is that drainage lines must have the proper slope without interruptions commonly known as pockets. These pockets hinder the flow velocity and promote obstructions, resulting in inoperable conditions.

The supports for piping systems must be compatible with the pipe material they support. The majority of old support systems was metallic and required corrosion resistance considerations. Recent shifts to nonmetallic piping have reduced technicians' awareness of the importance of proper hanger materials. For example, in the past, galvanic action could occur between different pipe-to-hanger materials that might not be understood by technicians accustomed to nonmetallic systems. Another important factor is that hanger manufacturers have developed numerous new superior products that have resulted in cost and installation-labor savings. The nonmetallic piping also has its own unique support needs, including bending support and close scrutiny to avoid chafing the piping material. Chafing on sharp edges is increased by the higher expansion and contraction rate of plastic piping. All plumbing codes address hanger support systems for the piping products identified in their code. Technicians and code officials are strongly advised to refer to the manufacturer's installation instructions, which will have greater installation detail and differ from one manufacturer to another.

The installation of hangers is dependent on distances established by code with consideration to pipe size. The support is not only a consideration of pipe weight but also pipe contents, discussed as load calculations. For a better understanding of support considerations, analyze the following example of a full, 4-inch schedule 40, galvanized steel, horizontal water line based on basic plumbing principles. This example could also apply to a horizontal storm line that is flowing full. The facts are provided without the normal accompanying formulas as a reminder.

The codes require a hanger every 12 feet for a horizontal steel pipe.

1 gallon of water = 8.33 pounds
1 cubic foot of water = 7.48 gallons
1 cubic foot of water = 62.48 pounds
1 foot of 4-inch pipe contains about 5.516 pounds of water; *12 feet equals 66.192 pounds*
1 foot of 4-inch schedule 40, galvanized steel pipe = 10.8 pounds; *12 feet equals 129.6 pounds*

The twelve-foot piece of pipe weighing 129.6 pounds plus the 66.192 pounds of water would equal 195.792 pounds.

Other hanger considerations shall include seismic and sway supports. The building code for the jurisdiction will dictate if seismic consideration must be addressed. When earthquake loads are applicable, the same building codes should provide sufficient installation detail. Contractors and designers who have worked in the same area for a period of time are aware of these details. Sway bracing is required for lines larger than 4 inches in diameter at changes in direction greater than 45°. Waste discharge flows provide lateral forces, which can be significant and damage the system. Sway bracing is different from the thrust-blocking systems on large water mains and hydrants most technicians consider. Another important consideration is that normal hanger and fitting takeoffs are not recognized as a substitute for required sway bracing. Large projects with the involvement of a registered designer and specifications to address swaying concerns are a great help for technicians.

Stress and strain on piping systems must also be considered. Provisions are provided to reduce stress on systems caused by expansion and contraction to manageable levels. Those provisions may be the use of expansion loops or mechanical expansion joints for the particular material. Size, location, material, and manufacturer's recommendations are the factors that dictate the installation requirements.

Hangers

Several codes have a chart to help quickly address the allowable support intervals of various piping materials. Table 36-1 provides a brief overview of some of the major materials. Consult your local code for greater detail when necessary.

These are the maximum distances; it may be wise to use additional hangers. For example, when hanging several cast iron fittings in a series, additional hangers may be necessary. Additionally some codes address larger pipe sizes with tables that specify the size of the threaded rod, necessary for larger pipes that have excessive weights. Installing technicians often only consider $\frac{3}{8}$ inch threaded rod when installing larger pipe sizes or hanger systems (a threaded rod on each end of a supporting bar or unistrut) to support several pipes in a parallel, but such installation may require $\frac{1}{2}$ inch or larger threaded rod.

Underground Support

When installing conduction underground piping, consider that the successful operation of later piping systems is heavily dependent on proper support. Very old plumbing literature often provided illustrations of how to install rigid piping on supports. The concept was to brace uneven, over-excavated trenches with supports and later fill in the beds by hand compaction. This method is no longer acceptable, and modern materials that require firm, even beds with sidewall support will never perform under such conditions. The *National Standard Plumbing Code* sums up the requirements with, "firm, stable, uniform bedding shall be provided under the pipe for continuous support."

Manufacturer's installation instructions and the applicable plumbing code must be referred to for greater detail. The codes will list ASTM standards, which provide guidance for underground plastic piping. Another consideration when installing underground piping is that the backfill must be conducted on layers with proper compaction in addition to the initial side backfill.

Piping Protection

Piping protection has always been an important part of the code and commonly refers to terms such as *relieving arch* (masonry terms for sleeves) and *pipe sleeves*. Code consideration is necessary because plumbing systems for drainage are low in

Table 36-1 Hangers and Supports

Piping Material	Maximum Horizontal Spacing in Feet	Maximum Vertical Spacing in Feet
ABS pipe	4	10
Cast iron pipe	5	15
Copper tubing, $1\frac{1}{4}''$ and smaller	6	10
Copper tubing, $1\frac{1}{2}''$ and larger	10	10
PEX pipe	2.67 (32 inches)	10
CPVC pipe, 1″ and smaller	3	10
CPVC pipe, copper tubing, $1\frac{1}{4}''$ and larger	4	10
Steel pipe	12	15
PVC pipe	4	10
Stainless steel DWV pipe	10	10

buildings and commonly run through exterior concrete and outer concrete walls and footings to outer sewers (see Figure 36–1). Water lines and services are also included in these concerns. The protrusions must be sealed and the sealing process cannot be detrimental to the pipe. Technicians often make the mistake of simply sealing the lines with concrete or embedding the risers in concrete to stop any leakage around the lines. This is a very serious error because concrete and other sealers have a corrosive effect on the plumbing lines. Corrosion shall also be addressed with underground lines in certain soil conditions. Exercise caution by referring to the codes and other details.

Cutting and notching, and the penetration of fire-ratted assemblies are two of the more recent concerns involved with pipe sleeves and openings in structures. The various code-enforcement agencies will have a building code to address cutting and notching issues because these agencies are concerned with the integrity of structures. Some plumbing codes publish the required building code sections. Conformance is the technician's responsibility. Advance planning that considers line changes to accommodate the building's structure is part of a skilled technician's responsibilities. Excessive cutting or changes to the structure's engineered system, for example, cutting laminated or factory-engineered floor trusses, are expensive. Penetration of fire-rated assemblies is another shared concern of the building codes. Contractual issues dictate who performs the fire stopping. However, your installation practices will affect the fire-stopping material costs.

Pipe penetrations in wood-frame structures require their own protection against physical damage by means of nailing plates. The nailing plates cover nonmetallic pipe that penetrates studs or ceiling and floor plates. The nailing shield plates are metal of a specific thickness to stop nail protrusions (see Figure 36–2). When shield plates are used for floor and ceiling plates, they shall extend an additional 2 inches above and below the wooden plates to accommodate floor trim or ceiling decorative trim that may be nailed over wall coverings such as drywall.

The importance of proper supports and penetration is often overlooked in the plumbing system, perhaps because in new construction the technicians do not have to maintain the building systems and live with the installation methods, like an owner must. For example, only a fire would prove the importance of fire penetrations, which address smoke issues or poor cutting, and notching practices, which manifest as wall or ceiling cracks and continually require decorating repairs. Poor installation of newer piping materials is also prone to producing objectionable noise in the building for its occupants. Several contractors have experienced service calls from occupants hearing water leaks that were a result of plastic drain line creep against wood framing.

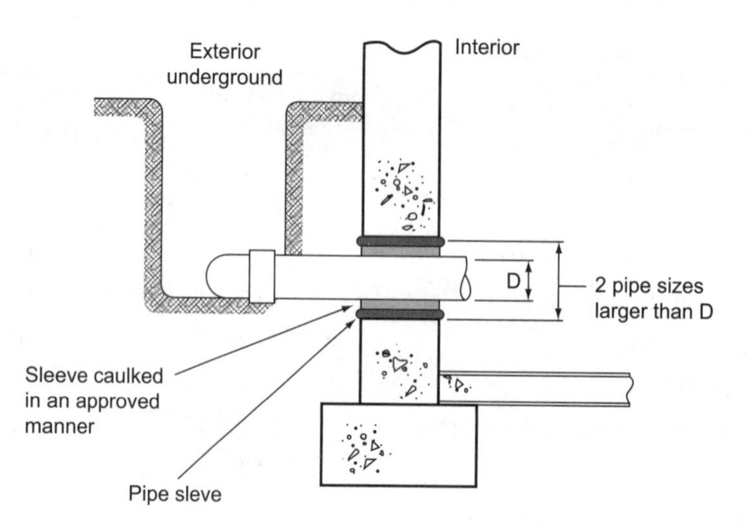

Figure 36–1
Pipe passing through concrete walls to the exterior shall be protected and sealed.

INDIRECT WASTE PIPING

Indirect waste piping connections are important to ensure separation of fixtures, appliances, appurtenances, and other devices from a source of contamination. This safety is provided by an indirect connection. These indirect connections have different levels of protections and are qualified by their names: an air gap and an air break. An air gap (a connection above) and an air break (a connection below) describe how the exit pipe goes into the drain line.

An air gap provides superior safety to an air break. The separation of clear air between the drain outlet and the flood-level rim of the receiver/receptor shall not be less than twice the diameter of the incoming pipe (the effective opening). See Figure 36–3.

Air breaks are allowed to terminate below the flood-level rim of the receptacle but must be above the top of the trap seal. The concept is that the receptor may spill over the floor to relieve itself. It is further understood the all of the piping exposed to atmospheric pressure will not allow any contaminant to flow back into the discharge pipe.

Indirect Waste Applications

Indirect connections are required where increased safety concerns exists, such as food preparation and food storage areas or where clear water from water distribution lines is discharged to the sewer, such as relief valves. Table 36–2 provides a brief overview of these requirements for your consideration.

Receptors

The plumbing code, in general terms, calls the device that receives the air gap or air break discharge a *receptor*. There are several different kinds of receptors, and they shall be constructed of materials that meet the plumbing code. They shall be trapped to be isolated from the sanitary system and shall conform to the code's venting requirements. Receptors shall be designed to reduce splashing from the indirect waste as much as possible. Specific types of receptors are floor sinks, standpipes, open hub drains, some floor drains, or at times other fixtures such as service sinks.

When the receptors receive discharge other than clear water waste, they shall have a strainer or removable basket. Some codes vary on this and state that if the lip of the floor sink is raised above the floor by 1 inch, a strainer may not be

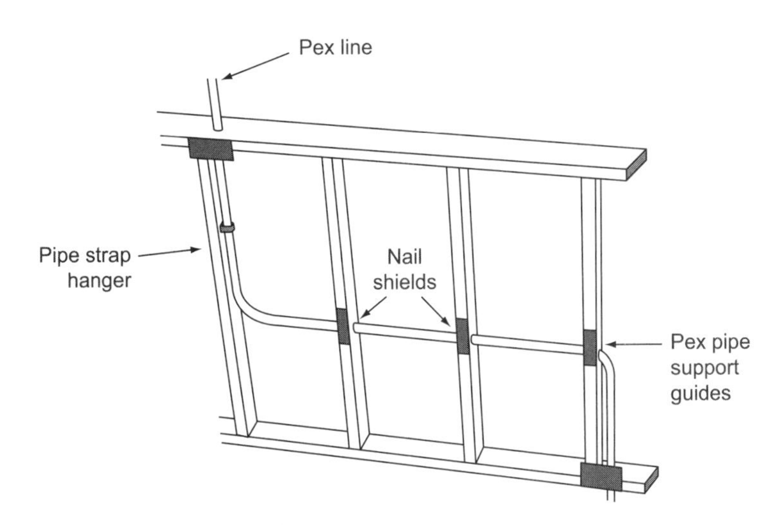

Figure 36–2
The pipe support may be required by the pipe manufacturers for the bend radius, and the code requires pipe nailing protection in specific cases.

Table 36-2 Indirect Waste Piping Connections

Discharge or device	Air Gap (above)	Alternative	Air Break (below)
Automatic clothes washer, (standpipe)			x
Backflow prevention device overflow, (relief port of RPZ)	x		
Clear water waste	x		
Coffee urn	x		
Condensate discharge, (no back pressure excepted)	x	or	x
Culinary sinks for food preparation	x		
Dishwashers, commercial	x	or	x
Food storage areas	x		
Glass-washing bar sinks	x	or	x
Healthcare steamers	x		
Ice machines	x		
Ice storage bins	x		
Non-potable, clear water, waste filters			x
Potable, clear water waste	x		
Refrigerated cases	x		
Relief valve discharges	x		
Sterilizers	x		
Swimming pools and related equipment (filter backwash)	x		
Steam kettles	x		
Walk-in coolers, floor drains	x	or	x
Water-treatment equipment	x		

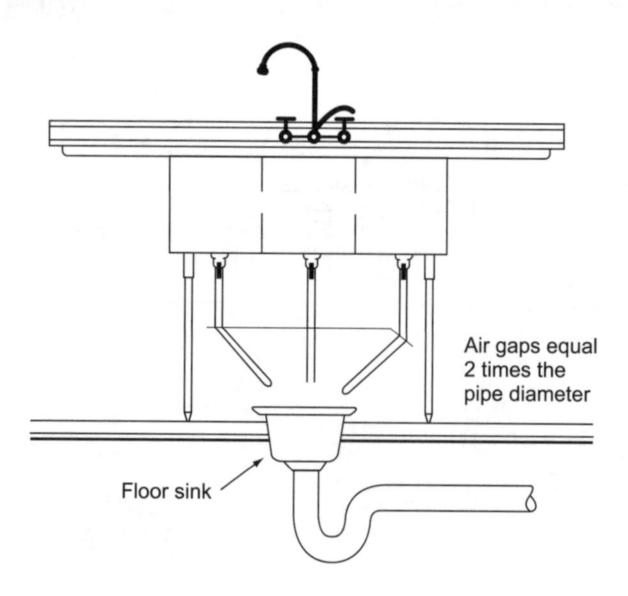

Figure 36–3
The simple connection diagram illustrates the indirect connection's ability to stop back siphonage of drain contents into a sanitary device such as a food preparation sink.

necessary. The idea is that a raised mini-sidewall will not allow someone sweeping the floor to push food scraps, such as french fries, on the floor into the floor sink. This last discussion implies correctly that most receptors are used in areas of food service. Open hub drains installed above the floor by 2 inches are recognized as clear water waste receptors. They may be found in mechanical rooms as well. All codes point out that the receptors shall not be in restrooms or inaccessible places.

Plumbing codes provide clear sizing information for receptors. For example, standpipes are receptors receiving drains, primarily automatic washer discharges, by way of air breaks. The code states they shall be at least 2-inch pipe size, no smaller than 18 inches and no larger than 48 inches, which eliminates splashing. Code also requires the receptor to be at least one pipe size larger than the line discharging into the receptor to reduce splashing. It is very common for the receptor, such as a floor sink, to receive more than one discharge line. For example, a single kitchen floor sink could receive an ice cream dipper overflow, an icemaker discharge, a beverage tray overflow, and a food preparation sink discharge. Codes address this concern by requiring the receptor discharge line to be no smaller than the aggregate cross-sectional area of the indirect waste discharge pipes.

SPECIAL WASTE

Special waste receptor discharges account for miscellaneous products utilizing air breaks or air gaps. Several of these, including high temperature, condensate, and chemical receptors, have been discussed elsewhere in this book. High temperature discharges are a serious threat to the integrity of sanitary systems. Occasional dripping from a relief valve air gap may not be considered a problem, but boiler blow offs would have a receiver. The maximum temperature of waste is 140°F.

The major codes vary on condensate connections and some may allow small discharges into tailpieces. Check the applicable code in your jurisdiction to be sure.

A wise closing statement related to indirect waste and special wastes is that the materials, connections, and installation methods shall meet code requirement to achieve proper sanitation practices.

REVIEW QUESTIONS

1. What is the result of hanger failures on drainage piping installations?
2. Why is sway bracing required for lines more than 4 inches in diameter at changes in direction greater than 45°?
3. What should take place when a receptor such as a floor sink receives more than one discharge line?
4. Indirect connections with different levels of protections are qualified by two different names. What are those names?
5. What is the name of a hanger system using a threaded rod on each end of a supporting bar or unistrut that supports several pipes in a parallel?

CHAPTER 37

Water Supply and Distribution

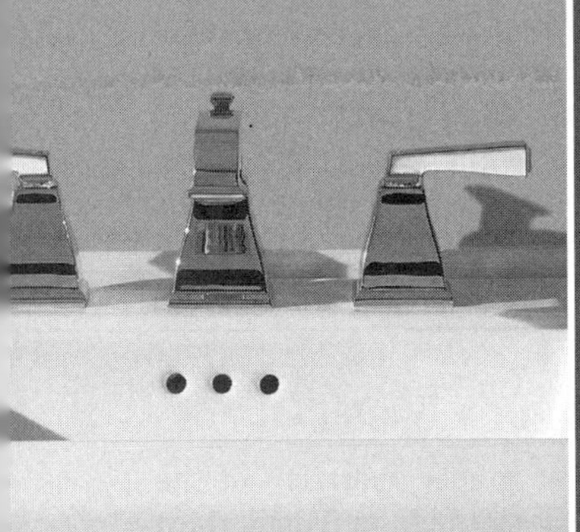

LEARNING OBJECTIVES

The student will:

- Identify the necessary marking requirements for the identification of non-potable water systems.
- Identify the temperatures and pressures in a water supply system necessary for its proper operation.
- Identify the components of a water supply system.
- Discuss water system treatment methods identified in the codes.

SYSTEM REQUIREMENTS

Plumbing codes are of great importance to ensure that safe water is provided for drinking, bathing, food preparation, and proper functioning of all devices utilizing water for their operation. The safe operations of water systems also ensure that other structures using the same water supply will have equal levels of safety. In order for systems to operate safely, the code establishes requirements such as cross-connection protection, pipe sizing, and hot water distribution safety. The *National Standard Plumbing Code* best describes those concerns by stating, "Only potable water shall be supplied to plumbing fixtures used for drinking, bathing, culinary use or the processing of food, medical or pharmaceutical products." Water in sufficient amounts and at the correct operation pressures is also of major concern.

Identification of Water Supply

The identification of piping systems is necessary to ensure that individuals will be able to identify and not use non-potable water. Identification reduces the chances that individuals involved in changing pipes will not connect into the wrong system. Codes require non-potable systems to be identified by color markings, tags, or other methods established by the authority having jurisdiction. The outlets shall be marked with signs identifying the water as being dangerous and unsafe.

Technological improvements and water conservation changes have a bearing on these systems also. New technology in commercial systems has required ultra-pure piping systems, often supplied by reverse osmosis devices. These systems should have labeling requirements also. Other system changes are a result of conservation concerns, which have led to the development and use of gray water recycling systems. Gray water systems shall never be connected to the water distribution system regardless of the degree of treatment. Many codes will not even allow the connection of a public potable water supply with a private potable well supply, even when separated by a valve.

Minimum Requirements for Water Distribution System

The design of the water piping system is critical to the proper operation of the water system. There are several accepted engineering methods and important considerations. The systems shall operate under conditions of peak demand. Peak demand is not all outlets operating at the same time but rather predictions of the maximum amount of water flowing and required at any given time for the particular structure's use. Sizing methods account for this by considering the frequency of use, amount and rate of water required, and the duration of use. Codes state that at peak demand, the water amounts (capacities) shall not be less than their referenced flow and pressure rates listed in the code tables. The codes will also have a table reference to clarify what the minimum supply line size shall be to listed fixtures. The *National Standard Plumbing Code* also requires the water distribution system to be sized for a maximum velocity of 8 feet per second at the design flow rate. Chapter 13 of this book provided details of system sizing. Refer to the specific code related to your licensing examination jurisdiction to resolve the many different, minute areas related to sizing.

All water pipe sizing methods consider the supply pressure in the system's development. Pressures will vary from the supply and there can be instances where the pressures are excessive or inadequate. Codes state that when system pressures vary, the design shall be based on the minimum supply available. This compensation will allow the system to operate correctly. Codes will also state that the maximum allowable pressure in a structure shall be limited to 80 psi under static conditions, no flow. Higher pressures are allowed at a sillcock. The acceptable method to reduce the pressure is a pressure-reducing valve. Some pressure-reducing valves have a built-in bypass that allows excess built-up pressures in a system, such as those created by the water heater, to escape through the valve to the supply. Without that bypass, an expansion device would have to be installed to prevent the water heater relief valve from discharging. Inadequate water pressure conditions are dealt with by using a water pressure booster system, as discussed later in this chapter.

Hot Water

Hot water is required in fixtures and equipment designed for bathing, washing, culinary, cleansing, laundry, or building maintenance purposes. Hot water is defined in some codes as being equal to or greater than 110°F. Strangely, those same codes do not address the high-end limit of hot water with a temperature reference. This has been a very important matter in legal decisions. Some codes require the supply of tempered water to public hand washing sinks, which is between 85°F to 110°F. Exercise caution in how you provide the tempered water; it must not be done by adjusting water heater controls. Several local or state codes require these "accessible" lavatories to be supplied by an ASSE 1070 device. An ASSE 1070 device provides scald protection (from burns), whereas an ASSE 1016 device required in the last code cycle protected against scalds and thermal shock.

The codes for many years have been energy conscious related to the demand for hot water in fixtures. Rather than opening a hot water valve and letting the water run until hot water occurs, the codes have set distance limits of 100 feet. The *National Standard Plumbing Code* describes the concern by stating, "in buildings where developed length of heated water piping from the source of the heated water to the farthest fixture exceeds 100 feet shall maintain heated water temperature in all supply piping to within 25 feet of any heated water outlet." This means a circulation system must be provided for distribution systems longer than 10 feet. Codes also allow an approved heat tracing system. Figure 37–1 is provided from the *National Standard Plumbing Code* with their clarification notes.

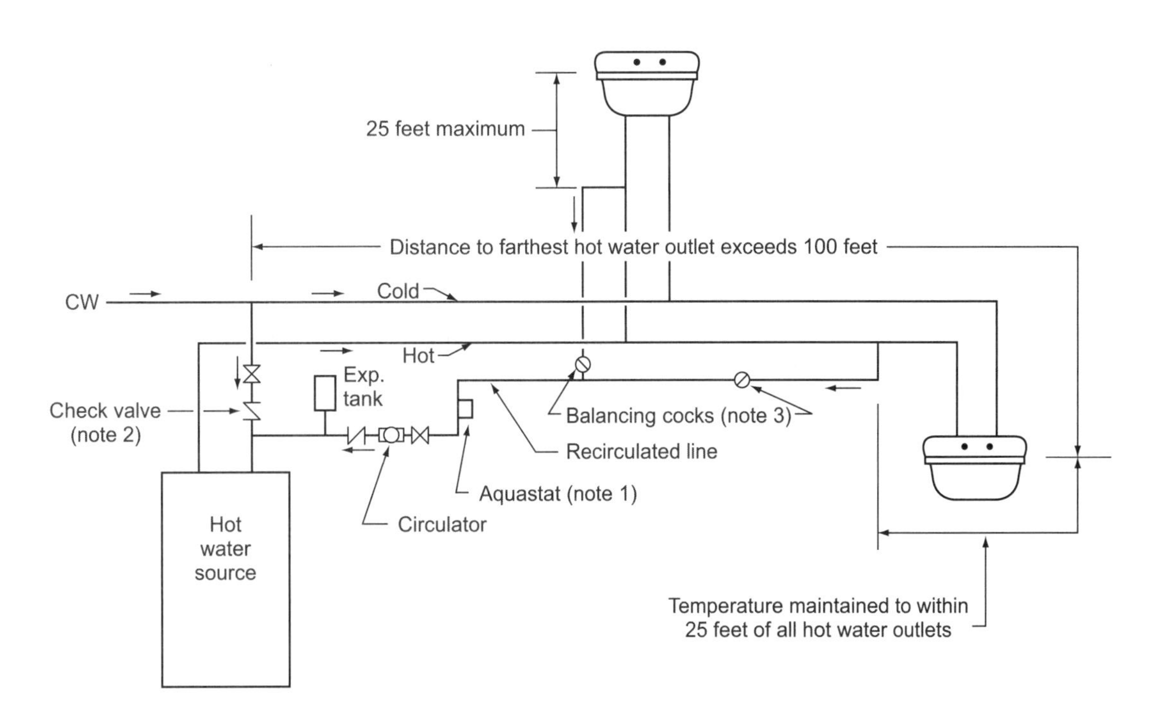

Figure 37–1
Example of a situation requiring a circulation system. Courtesy of Plumbing-Heating-Cooling-Contractors—National Association.

When codes provide hot water discussions, they will also discuss the methods of controlling hot water to the fixtures, incorporated in the hot water section.

In the Field

When water heaters stand alone in a separate chapter, mechanical code officials may move the language out of the plumbing code and into mechanical codes. This is unfortunate because water heaters have been within the expertise of individuals licensed in the plumbing profession. Many jurisdictions with newer mechanical licensing laws tend to license the mechanical business owner rather than the technicians.

WATER HEATERS

Water heaters and related subjects are discussed in the *National Standard Plumbing Code* along with the hot water requirements in Chapter 10, Water Supply and Distribution. The *International Plumbing Code* and the *Uniform Plumbing Code*, using the common code format, address water heaters in Chapter 5, Water Heaters.

Plumbing codes do not list specific sizing criteria for water heaters. The conformance standards or other codes, such as the fuel gas code, might be referenced for electric, gas, or oil units but not for sizing directives. The concept has long been debated and a consensus has yet to be reached. Newer products like tankless and on-demand units have increased the commitment to not listing sizing criteria. The *National Standard Plumbing Code* clearly states the sizing goal in the following two statements:

> Water heaters and storage tanks shall be sized to provide sufficient hot water to provide both daily requirements and hourly peak loads of the occupants of the building.
>
> Hot water storage tanks shall be adequate in size, when combined with the BTUH input of the water heating equipment to provide the rise in temperature necessary.

The *International Mechanical Code* states, "Potable water heaters and hot water storage tanks shall be listed and labeled and installed in accordance with the manufacturer's installation instructions." When the unit is also used for space heating, it shall be listed for such use and its potable temperature shall be limited to 140°F. This limit provides safety to users when the installing technician or owner later would be tempted to increase the water heater operating temperature in order to improve the space heating capability.

Hot water storage tanks, whether incorporated into the heater or separate, must be capable of being drained and have information identifying their maximum allowable working pressure. The piping to water heaters is a concern and the code having jurisdiction must be referred to for its unique requirements. Plastic piping, which is affected by higher temperatures to a greater degree than metallic piping, must be distanced from the heaters by specifically stated distances found in the codes or manufacturers installation recommendations. Working pressures and temperature protection for water heaters have been addressed in earlier portions of this book and will be addressed for the entire water system later in this chapter. The use of plastic piping also has an effect on the selection of the relief valve setting. Refer to the *National Standard Plumbing Code*, Section 10.15.8 for greater detail.

Water heaters and storage tanks installed in locations where leakage could cause structural damage are required to have drip pans. The purpose is to collect the leakage and conduct this water to a safe place for disposal. For example, a pan would not be required for a unit on concrete floor with a nearby floor drain. A pan would be required for a unit on a wood floor or mounted above the ceiling of an office restroom. The pans must be corrosion resistant, not less than $1\frac{1}{2}$ inches in depth, and of sufficient size to hold the heater and not disrupt its operation. Plastic pans may be approved if they meet specific requirements. Most codes will not allow a relief valve to discharge into these pans. The pan drain, with a gravity drain and minimum size of $\frac{3}{4}$ inch, will not provide sufficient drainage. This is a controversial matter because some codes will not state that the relief valve discharge shall be provided with a drain. Installers will maintain that the pan at least provides an acceptable drainage means. The pan drain line shall be extended to an approved point of discharge.

PROTECTION OF THE WATER SYSTEM

Protection of the water system is generally divided into two areas of consideration: how the system is installed and isolated from physical problems, and the use of backflow prevention devices to isolate contaminants from the water system. For example, with regard to the installation of a system, waste piping installed in close proximity to a water service could increase the chances of contaminating the water supply.

Protection of Potable Water Supply

The codes state that water systems must be installed, designed, and maintained to prevent contamination by non-potable gases and liquids. However, technicians often do not consider the gases because of their reduced contamination capabilities. In order for gases to be introduced in a system, the system's liquid would have to be diminished to provide room for the gases. The system has to have an absence of pressure to pull the gases back into the piping system. Technically speaking, normal habitable air in an occupancy space is not acceptable because it contains harmful bacteria, but the gases most often considered are chemical fumes present in laboratories or manufacturing facilities. Water supplied around or in fume hoods used in laboratories is the classic example.

The underground and out-of-sight location of a water service may pose more problems than normal water distribution piping. Earlier plumbing materials had far more joints and connection in their piping than current materials, which increased the likelihood of developing leaks (both drainage and water lines). For example, clay tile sewers had more joints than the new 10- or 20-foot lengths of pipe today. Water services in the past could have had threaded connections every 20 feet. Increased connections increased the probability of leaks. The sewer leaks could have contaminated the surrounding soil, whereby the leaks in the water service during loss of pressure could have allowed contaminates to be siphoned back into the water supply. The codes to address these issues require various distance separations between the sewer and water lines. Consult your code for details because there is a great deal of difference in the numbers. Some codes also address the separation of elevation; for example, the water service shall be on a shelf of undisturbed earth 1 foot above the sewer. Local health codes having jurisdiction also list the separation distance between well water supplies and sewer and drain lines. The water service shall be sized properly, generally using the same sizing method as the water distribution system. The minimum code size is $\frac{3}{4}$ inch nominal.

Healthcare concerns are also expressed in the code related to sustaining the healthcare environment. Some plumbing codes require two different water services to hospitals. If one of the water services becomes inoperable, for example, if underground construction hits a service and causes it to be shut down, the other service line would be able to supply the facility. It is understood that the lines should be installed at opposite ends of the hospital property.

WATER SYSTEM COMPONENTS

A structure's water supply is composed of many different areas that serve several functions. The functions vary from having adequate supply pressures to controlling system portions. Often they are dependent upon each other. For example, low-pressure supplies will require increased pressure from booster systems and the booster systems will require protections using backflow preventers; they will both require the use of valves. The following clarifications apply to code requirements for different system components.

Water Pumping and Storage

Water pumping and storage equipment must be protected from contamination. This protection is typically the use of backflow preventers in accordance with the code as discussed in Chapter 17. Many older systems used gravity storage tanks at elevated locations in the building, usually at much higher floor locations, to develop additional pressure through their discharge. Those gravity tanks were not always sealed. For that reason, codes will reference the old occurrences by stating that gravity tanks shall not be under soil or waste piping. Plumbing codes continue to address these gravity systems with specifics on the overflow outlet sizes and discharge locations, the covers, the filling provisions (ballcock type of float-fill devices), and the tank draining methods. The overflow sizing provisions are important because they provide a type of air gap protection for the supply. Today's backflow concepts recognize that proper backflow protection can be provided by a device rather than a tank and overflow. Recent licensing examinations do not cover these systems on their tests; however, Figure 37–2, a helpful diagram contained in the *National Standard Plumbing Code*, provides an excellent visual summary for the systems.

Some codes clarify that there shall be ready access in or to pumps to provide service or repairs to the entire pump and its components. Other pump details may include a statement that well pits are prohibited or may require pumps to be elevated above floor levels to reduce the contamination possibilities.

Water Pressure Booster Pumps

Water pressure booster systems designed to increase system pressures are effective when inadequate system supply pressures are present. However, the pumping does present unique concerns related to backflow protections. Pumps will elevate system velocities and, in the process, draw water at a rapid rate from the supply, which

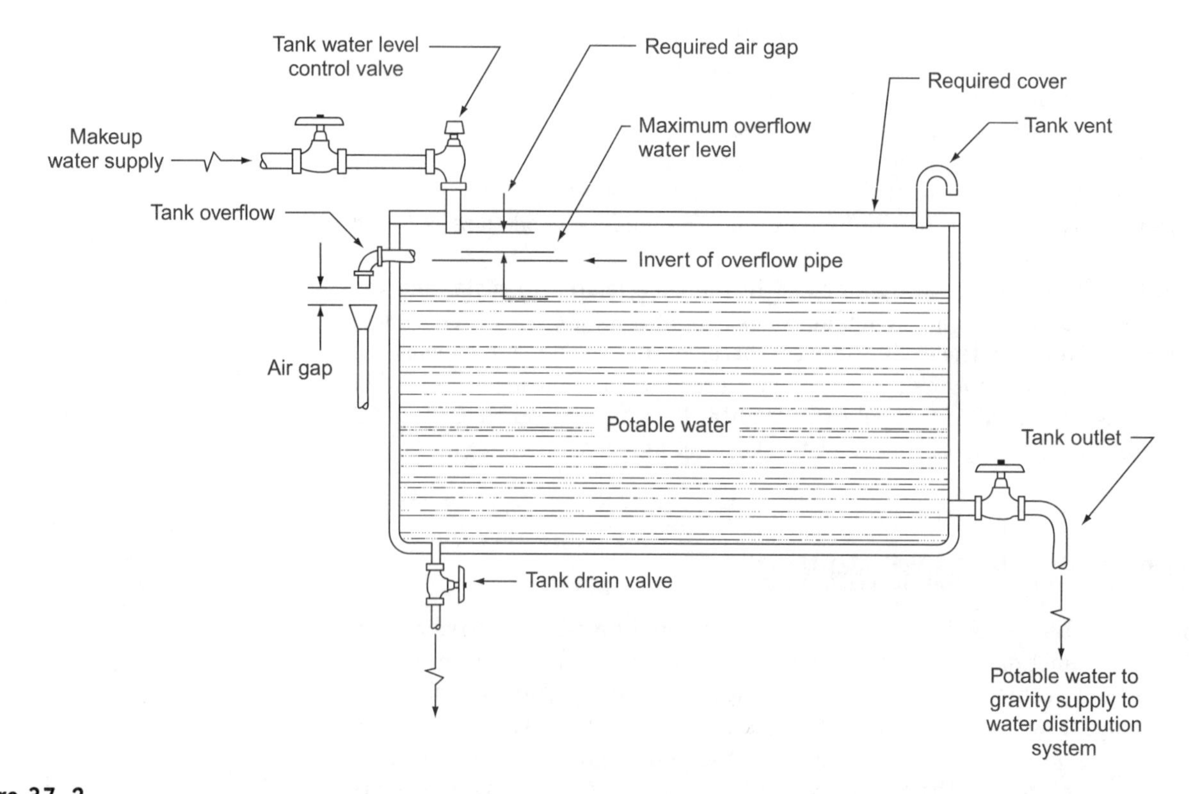

Figure 37–2
A gravity supply system in an elevated location has many requirements for protection of the water supply. Courtesy of Plumbing-Heating-Cooling-Contractors—National Association.

creates a negative pressure or a vacuum on the pump's supply side. This negative pressure could draw contents from other lines connected to the supply side. There is an increased possibility that one of the lines may be connected to a contaminant such as a garden hose in a puddle of ground water. To protect against this low-pressure suction, a low-pressure cutoff is installed. These low-pressure cutoffs by some codes are set to shut off the pump when pressures reach 10 psi or lower.

Water Supply Safety and Control Valves

Valves in the plumbing system are devices that regulate flow. The flow can be of potable water fluids, excessive pressures or temperatures, or air into or out of the plumbing system. Technicians commonly use many flow control valves. Code discussions consider these and other specialized valves, such as temperature and pressure relief valves discussed earlier and vacuum relief valves on storage tanks. All valves, to some extent, are involved in the safety of the system or its occupants by control or a relief processes.

Safety

Valves are heavily involved in tanks, including pressure tanks, storage tanks, and water heater tanks. Pressure tanks require pressure relief valves to displace or bleed off excessive pressures for safety purposes. These valves are generally installed on the supply side of the tank and are rated for the maximum design pressure of the tank. Vacuum relief valves must also be installed on pressure tanks. The vacuum relief valves protect the tank by stopping a vacuum when the loss of pressure in a tank could subject the tank to damage, even to the point of collapse. Tanks installed at higher elevations with bottom-supplied inlets are more susceptible to this condition. Vacuum relief valves do not have large pipe sizes and have a cap over their inlet to prevent debris from falling into the opening port. The valve, which is basically a check device, shall be rated at a maximum pressure of 200 psi and a maximum temperature of 200°F. Storage tanks are generally associated with hot water proving systems; they must meet the same requirements as the pressure tank because they are under pressure, and they must be insulated. Another consideration relates to the basic valve use: the tanks shall have control valves.

Water heaters require pressure and temperature relief valves and incorporate several valves in their operation. Several codes state that when water heaters are replaced, the relief valve shall not be reused. In addition to the pressure and temperature relief valves, water heaters may require vacuum relief valves dependent upon the piping arrangements.

Control

In addition to safety matters, plumbing systems require numerous types of valves in order to operate efficiently. Discussions in this area should start at the beginning with the water service.

Curb valves, often referred to as curb cocks or plug valves, are located where the codes usually begin their jurisdiction, at the private property line. Access shall be provided by a valve box, which allows a long, handled extension to open or close the valve (a street key). These valves shall not have a weep hole for self-draining purposes because a weep hole would form a cross connection.

Building service valves are located at the termination of the water service and provide isolation safety and maintenance capabilities for a structure. Some codes require the valve or take-off fitting to have bleeding capabilities (pressure removal and draining) at this location. Structures with a private well supply source shall have a full open valve after the pressure tank, defined as a water supply tank valve, as required by code.

The discharge side of the meter is the specified location for meter valves. The building service valve is before the meter and many times both these valves are incorporated into the water meter bar, a factory-built assemblage that incorporates

In the Field

Full open valves provide full flow capabilities without restricting the water volume in ways other than friction loss. They are referred to as gate, ball, or butterfly valves and are installed at the curb, structure entrance, dwelling units, meters, risers, and tanks (includes water heaters).

the valves and enables a meter or a jumper line to be installed between the valves. Some meter bar assemblies may be provided with an isolation double check device built into the assemblage.

A dwelling unit isolation valve is required to facilitate maintenance within a unit. Dwelling units include an apartment or condominium in a multi-occupancy structure. Exceptions can be considered when all the fixtures (outlets) have their own control valves. Verify these details by the appropriate code; they all vary in small details.

Riser valves are required when water lines pass through more than one floor. These full open valves are required regardless of the number of individual dwelling units or fixture control valves. An additional consideration is that valves shall be accessible, which may require an access panel in common corridors.

Tanks and water heater control valves shall be provided on the water heater supply. The code does not prohibit a control valve on the hot side to enable technicians to shut off the hot and cold lines to facilitate easier tank removal. Valves on water heaters should remind technicians that valves between tanks and their safety devices are never permitted. Further, a valve is never acceptable in a relief valve discharge line.

Shutoff valves fit a broad definition and provide installation access for service and repair of appliances, fixtures, mechanical equipment, and sillcocks. They shall be accessible and identified when necessary.

There are several devices considered as valves, ranging from temperature control products to yard hydrants. Care must be taken not to sacrifice the potable water system's integrity for proper function. For example, a yard hydrant would not be allowed to have a weep hole, and stop and waste valves shall not be installed in underground locations.

Flexible Connectors

Flexible connectors have been an important part of plumbing installations for many years and ironically were somewhat inflexible. Those connectors were referred to as lav supplies and closet tubes. Today's connectors are far more forgiving in flexible abilities for technicians. Although all of them must meet toxicological concerns of NSF 61, they are not constructed with the same materials covered in piping standards. For that reason, ASME A112.18.6, a national consensus recognition standard, was developed to provide a conformance standard. Flexible connectors may also apply to automatic washers, dishwashers, and water heaters.

WATER SYSTEM TREATMENT

Water system treatment encompasses several things, from system treatment itself for cleanliness (potability) to the refinement of its supply for use, and other system needs such as reverse osmosis purification. Prior to use, the potable water system's potability is ensured by the public supplier or health department having jurisdiction and is then flushed and disinfected. The next treatment areas are point-of-entry treatment (softeners and filters) and point-of-use treatment, where extra purification aids drinking or equipment functions.

Flushing and Disinfecting

New water systems or those being repaired require two steps for cleaning prior to being placed in service. The concern is to have a safe system that has addressed the bacteriological and contamination levels. The first step is flushing, which requires the system to be brought up to pressure and then water is run through each outlet to flush out foreign matter from the line. Care should be taken to remove any screens or aerators first, which could be damaged during this process.

The second step is to fill the system with a water and chlorine solution, whose concentration is measured in parts per million of chlorine content, at each outlet. The solution must be present at each outlet for equal system dispersion, meaning everything will be cleaned. Different parts per million content can be considered for specific times. The general idea is that higher levels work more effectively. A word of caution: codes and processes were developed based on metallic piping rather than newer plastic systems. High levels of chlorine are known to have a detrimental effect on piping systems. This has been pointed out in past litigation concerning polybutylene tubing and fitting failures. Check the product manufacturer's disinfecting recommendations. Following treatment the system must be purged, measured for potability, and placed in service.

In the Field

Plumbers installing large water mains are familiar with system disinfection. Unfortunately, structure installers are not practicing disinfecting on a regular basis, and increased sensitivity should be practiced prior to local health department mandates for the monitoring of disinfections.

Residential Water Softeners

Water softeners, also referred to as conditioners, are devices used to remove dissolved materials and objectionable odors. The process has the commonly stated purpose of making the water "more usable." The washing capabilities are improved, and the odor associated with high-iron content is reduced. Softeners do not remove higher iron content levels; the iron-removal treatment package and filter are more expensive devices. They all require periodic attention and maintenance. Plumbing codes do not mandate the use of conditioning equipment; however; some codes address sizing requirements. The *National Standard Plumbing Code* outlines the following residential sizes (see Table 37-1).

In the Field

Some plumbing codes will reference the following standards for potable water system disinfection: AWWA C651 Disinfecting Water Mains and AWWA C652 Disinfection of Water-storage Facilities.

Drinking Water Treatment Units

Drinking water treatment units are designed for point-of-use supply, processes such as high-purity laboratory use, reverse osmosis (RO) units, or sterilizers. The reverse osmosis units are briefly referenced because of their new popularity and availability to homeowners. The systems remove the dissolved solids by the use of pressure to force water through a filter membrane. Many of the units will use plastic tubing, which shall conform to NSF 14, NSF 61, and NSF 58.

Care must be taken for all treatment devices to connect the drains by an air gap through an indirect waste connection to the sanitary system. The drains are a result of backwashes designed to purge the filters, membranes, and minerals from objectionable materials and dissolved solids. Some of the treatment devices will have overflows. The standards, which some codes reference for the devices, are listed as follows:

1. NSF 42, Drinking Water Treatment Units: Health Effects
2. NSF 44, Residential Cation Exchange Water Softeners
3. NSF 53, Reverse Osmosis Drinking Water Treatment Systems
4. NSF 62, Drinking Water Distillation Systems

Table 37-1 Sizing of Residential Softeners

Required Size of Softener Connection	Number of Bathroom Groups Served
$\frac{3}{4}$ inch	Up to 2
1 inch	Up to 4

For more than four bathroom groups, the softener shall be engineered for the specific installation.

REVIEW QUESTIONS

1. What is the minimum temperature water must reach to be considered hot water?
2. What is the maximum allowable pressure in a structure under static conditions?
3. Are yard hydrant valves with freeze protection provided by a weep hole in a gravel bed acceptable in the code?
4. When a water heater is used for space heating, what is the maximum permissible operating temperature according to code?
5. Codes require non-potable systems to be identified by markings, tags, or other methods. Is there a requirement to identify the outlets?

CHAPTER 38

DWV and Storm Drain Systems

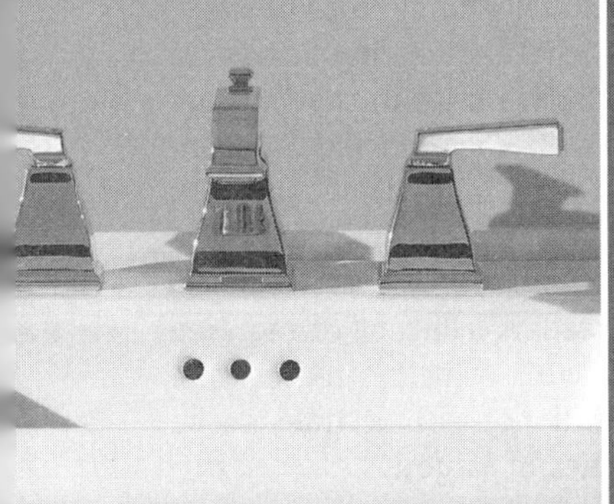

LEARNING OBJECTIVES

The student will:

- Describe the necessity and benefit of using code-defined slope and drainage fixture unit values.
- Interpret the drain- and stack-sizing requirements in the code in order to prepare for examinations or improve installation skills.
- Compare the various vent methods found in the national codes.
- Discuss the concerns involved in roof drainage systems.

SANITARY DRAINAGE

The goal of this chapter is to provide plumbing code information and additional insights into the second most important area of the code. The first area was the supply of safe, potable water to the structure. This chapter will address the drainage, waste, and venting (DWV) of the sanitary system, followed by storm systems.

The code chapters on drainage and venting provide the technician with the best of plumbing because the technician uses the technical code requirements to turn their installation skills into craft. *Webster's New World Dictionary* describes *craft* as a special skill or art. Being a part of a skilled trade, using your mechanical abilities with longstanding hydraulic principles and physics is rewarding. In an overview of the drainage, waste, and venting areas, drainage is a factual matter where slope produces velocity and loads in drainage fixture unit (DFU) values. The DFU values are accumulated into established, engineered line size for drains and stacks. Then venting applies air movement and pressures in time-proven methods, resulting in operable systems.

Slope

Pipe slope creates force within the line to wash out waste from the drainage system. The desirable force is a velocity of 2 feet per second. This rate produces somewhat of a self-cleaning flow called a scouring effect. Past codes identified the desirable velocity for grease kitchen waste as 4 feet per second. In order to achieve 3 feet per second velocities, the codes require lines 2 inches and smaller to be installed at no less than $\frac{1}{4}$ inch per foot. Lines 3 inches and larger may be installed at $\frac{1}{8}$ inch per foot of slope. Very large lines on the structure's exterior often fall under civil engineering principles; any might be installed at $\frac{1}{16}$ inch per foot. Horizontal drainage lines in the code are sized as being $\frac{1}{2}$ inch to conduct waste and air in systems. The *National Standard Plumbing Code* in Appendix K provides valuable gallon-per-minute flow rates and velocities for the various pipe sizes, considering the pipe slope. The appendix considers rough-in smooth wall lines, which affect the velocity discharges.

Table 38–1 is provided as a visual to enhance the readers' concept of required pipe slopes. A 5-inch line at $\frac{1}{8}$ inch per foot slope will have a velocity greater than 2 feet per second. The *International Plumbing Code* will not permit lines to be installed at $\frac{1}{16}$ inch per foot until the line size is 8 inches or larger.

Drainage Fixture Unit Values

The next step in sanitary drainage sizing is calculating loads in the system based on individual fixtures and drains with assigned values. The value development was discussed previously and has become a sophisticated method that changes from time to time to adapt to low-flow fixtures. For example, the *International Plumbing Code* recently assigned a drainage fixture unit (DFU) value of 0.5 for waterless urinals. Most codes have the same values; however, they may group fixtures together based on the building's use-group designation. The *National Standard Plumbing Code* recognizes individual dwellings, three or more dwelling units, other-than-dwelling units, and heavy-use assembly. This is wise because

Table 38-1 Minimum Allowable Slope of Horizontal Drainage Piping

Size	Minimum Slope
2 inches or less	$\frac{1}{4}$ inch per foot
3 inches or larger	$\frac{1}{8}$ inch per foot

Note: See the applicable code edition for acceptable $\frac{1}{16}$ inch per foot installation size

commercial structures or hospitals have a much higher degree of use and the different consideration provides a built-in operation-function factor that has been time proven.

A very simplified DFU table is provided in Table 38–2 as a visual example, which will be used in our sizing discussion later.

Line Size

The drainage pipe sizing is not only based on the amount of drainage fixture units connected to the line but also on where the line is in the system. The pipe for the building sewer, building drain, and branches to the building drain have different considerations than the short stacks, taller stacks, and branches connected to the stacks. These adjustments are necessary when considering there is greater airflow and less stress in lower horizontal building drain and building sewer applications than in stack connections. The author uses the term *stress* to challenge the readers' understanding that stacks, with their branches, are more susceptible to increased turbulence and decreased flow capacities than lower portions of the sewer system. Another basic fact about drainage line size is that as more DFUs are added to a line, the line size (in the direction of flow) may be required to increase based upon the code tables. Further drainage lines shall not be reduced in the direction of flow, and lines shall not be smaller than the traps they serve.

Building Drains and Sewers

Building drains and building sewers are the lowest portion of drainage system and are sized in accordance with code-provided tables, in this case The *National Standard Plumbing Code* table (see Table 38–3). Fixture branch connections, defined as those lines serving two or more fixtures discharging into another line, are also addressed. A very evident fact in reviewing the table is that increased slope offers the technician

Table 38-2 Drainage Fixture Unit (DFU) Values

Type of Fixture	DFU Value
Automatic clothes washer, domestic, (2-inch standpipe)	3
Drinking fountain	$\frac{1}{2}$
Floor drains, auxiliary (emergency, only serving spills)	0
Service sink, with one 3-inch trap	3
Kitchen sink, domestic, with food waste grinder	2
Kitchen sink, domestic, with dishwasher $1\frac{1}{2}$-inch trap	3
Lavatory with $1\frac{1}{4}$-inch waste	1
Laundry tray (one or two compartments)	2
Shower stall, domestic (2-inch trap)	2
Urinal, 1.0 GPF	4
Wash sink, fountain (2-inch trap)	3
Water closet, 1.6 GPF (gravity, pressure tank, or flushometer valve)	4
Water closet, 3.5 GPF	6
Trap size $1\frac{1}{4}$ inches or less	1
Trap size $1\frac{1}{2}$ inches	2
Trap size 2 inches	3
Trap size 3 inches	5

Note: See the applicable code edition for factual information. Minimum trap size requirements will also be provided in some plumbing code tables. Where a discharge rate in gallons per minute is provided in the NSPC, 1 gpm shall equal 2 DFUs.

Table 38-3 Maximum Number of Drainage Fixture Units (DFUs) Connected to the Building Drain or Sewer. Courtesy of Plumbing–Heating–Cooling–Contractors—National Association.

	Slope per Foot			
Diameter of Pipe (Inches)	$\frac{1}{16}$ Inch	$\frac{1}{8}$ Inch	$\frac{1}{4}$ Inch	$\frac{1}{2}$ Inch
$1\frac{1}{2}$[a]	—	—	3	—
2	—	—	21	26
3	—	[b]	42	50
4	—	180	216	250
5	—	390	480	575
6	—	700	840	1,000
8	1,400	1,600	1,920	2,300
10	2,500	2,900	3,500	4,200
12	3,900	4,600	5,600	6,700
15	7,000	8,300	10,000	12,00

[a] The $1\frac{1}{2}$ inch branch limit is taken from the NSPC table addressing horizontal fixture branches and stacks. This NSPC table (Building Drains and Building Sewers) does not include a reference to $1\frac{1}{2}$ inch lines, because the minimum size underground is 2 inches. However, a building drain could be installed overhead (hung from the ceiling) in a basement rather than underground, which would result in a branch to the building drain $1\frac{1}{2}$ inches in size.

[b] The *National Standard Plumbing Code* and the *Uniform Plumbing Code* have several restrictions for water closets on 3-inch lines, which must be referred to for detail. The *International Plumbing Code* does not restrict the number of water closets on a 3-inch line but rather uses the number of assigned DFU values as the limit. The first two code limits are no doubt based on the fact that a 3-inch line at $\frac{1}{8}$ inch per foot slope has a velocity of 1.60 feet per second in a rough-walled pipe. This is below the desired limit of 2 feet per second and considers drain line carrying concerns of water closet discharges.

the opportunity to conduct a greater amount of DFUs, translated to more fixtures. Consider that these are the line sizes and that venting provisions will be acknowledged later.

In the Field

It is extremely important when applying code requirements that technicians apply all the code principles in new text portions. For example, drainage line sizing mandates a line shall apply when venting requirements address line-location requirements. Requirements accumulate and work together rather than independently. Line size, distance, slope, vents, and other factors provide proper performance.

Stacks with Connecting Horizontal Branches

Stacks, the vertical portion of sanitary piping, have their own sizing table for several practical reasons. Significant discharges in the stack from its branches may cause a backup from the branches connected to the stack; therefore, stacks are sized for flow discharges of $\frac{7}{24}$ full. This compensates and reduces the turbulence in the incoming discharges.

The varying turbulence in the stack changes pressures, which have negative effects on the trap seals. The fixtures, branches, and stack require venting to protect the trap seals, which is addressed later in this chapter.

Discharges in stacks tend to flow in a circular motion down the stack, at times accumulating and acting as a piston with air in the middle. The air might break through, resulting in pulsations. You can often view stack-pressure changes by trap seal movement up and down in fixtures like water closets, and even more so in urinal trap seals. Smaller discharges generally fall down the center of the stack. Distance also affects stack flow movements, which are discussed as terminal velocity, gravitational force, and maximum speed achieved (normally 15 feet of fall distance). The flow is described as a maximum because it does not increase from air and pipe-wall friction. The lack of increase allows stacks of greater height to accept larger discharges. The discharge velocity in stacks may be in the range of 15 to 20 feet per second.

These facts require four final considerations when using the code-provided table. First, the total number of fixtures discharged into each branch interval is limited by the table column as the first safeguard against overloading pressure or volume. The second consideration is the number of branch intervals on the stack. Refer to the definition of branch intervals and consider that floors without fixture discharges are not considered branch intervals. The third consideration is the total number of fixtures discharged into the stack. A provided table column uses this as the bottom line for fixtures allowed in stacks over three branch intervals. Finally, the fourth table column, which is listed first in the table, clarifies the branch size required for lines into the stack.

Stack Offset Sizing

A drainage stack with an offset greater than 45° from the horizontal shall be provided with additional protective sizing requirements because the stack's flow is subjected to increased pressure and turbulent flows. These increases are greater than those already factored into the stack sizing limitations discussed earlier.

Three adjustments are required to address concerns about the stack's increased pressure and turbulence. First, the horizontal offset shall be sized as a building drain (see Table 38-3). This will more than likely increase the pipe size. Second, the stack below the offset shall be increased to the same size as the offset or increased based on the stack table's total number of fixtures draining into the stack, whichever is greater. Third, relief vents shall be installed as stated in the code to equalize pressure differentials. The major codes have minor differences here and must be consulted for conformance.

In the Field

Designers may place offsets in a system as a means of compensation for stack movement developed by expansion and contraction of piping materials.

Additional Stack Connections

There are additional areas that must be considered briefly to ensure the drainage system operates effectively. Horizontal drain connections shall not be made within 10 pipe diameters of the stack base owing to hydraulic-jump conditions. This is explained by our previous discussion of stack waste discharging at rates up to 15 feet per second, which meets a horizontal system designed at 2 feet per second. Horizontal branch drain connections to horizontal stack offsets are also addressed by the codes. Discussion details of 10 pipe diameters and prohibition of connections above and below 2 feet of the offsets are detailed in the individual codes.

The *National Standard Plumbing Code* and the *Uniform Plumbing Code* contain criteria to ensure proper drainage in stacks with suds-producing fixtures. Suds-producing fixtures include kitchen sinks, laundry sinks, automatic washers, and others that could discharge sudsy detergents. These suds are objectionable because they may come into other fixtures or worse hinder the draining and venting of other fixtures. Refer to the codes for detail.

Sumps and Subdrains

The location of sewers or treatment systems is not always below the level of the structure's drainage system. When this occurs, the lines must discharge into a sump pit and be lifted or pumped by automatic pumping equipment. All other drains shall be discharged by gravity to avoid unnecessary utility costs and reduce problems such as pump failure that can render fixtures inoperable. Backwater valves, commonly referred to as checks in the vertical position, and full-way shutoff valves for servicing are required. Most codes have minimum sump and pump sizing criteria, which greatly affect the number of pump operations. Alternatives to sumps serving more than one fixture are small dual-purpose sumps and pumps for a single fixture; refer to your jurisdictions code for additional details.

VENTING PRINCIPLES

Vents are required to ensure that plumbing traps do not lose their trap seal protection mandated by the code. Codes state that systems shall be installed where a pneumatic-pressure differential shall not be more than 1 inch of water column (wc) pressure. One inch of wc equals 0.0357 pounds per square inch.

Proper venting is extremely important and not difficult to understand once the basics are pointed out. The venting requirements will be listed alphabetically with a brief general discussion providing insights; these will be followed by method explanations and illustrations.

1. Basic premise: Every fixture shall be vented. This does not mean that every fixture shall have its own individual vent but that every fixture will receive some type of venting.
2. Crown venting: Crown vents are vents that are within 2 pipe diameters of the trap weirs and they are prohibited. They increase trap seal evaporation and promote clogging.
3. Drainage stack venting:
   A. A vent stack connected to the stack base is required for stacks of five branch intervals or more. The purpose of the vent is to ensure adequate amounts of vent air are provided to the building's drain fixtures for proper operation. This was a critical concern when older plumbing systems allowed independent fixtures such as floor drains to run off the building drain without additional vents. Fixtures on the stacks obtain their venting by other means.
   B. Stack vent extensions on top of the stack and extend up through the roof or connect to other vent headers. This vent may be the only vent for individual, common, or waste-stack vented fixtures. Large soil stacks may have a stack vent or it may be omitted because the fixtures would be vented in their various branches.
4. Future fixtures: The codes are different on this provision and may be very strict. If there is a basement in new construction, future vents may be required. Check your local code.
5. Frost closure: Vent terminals subject to freezing shall be increased in size above and below the roof to guard against frost closure. This is very common in colder climates and can come and go, often resulting in confusing service calls when proper drainage is not taking place.
6. Headers: Common vent headers are vents connected together to reduce unnecessary piping and roof penetrations. A common vent off the header will extend to the open air.
7. Offset vents: Vents shall be provided when there are horizontal and vertical offsets in drainage stacks having five or more branch intervals with stack offsets. The purpose is to relieve back pressure from the hydraulic effect of the stack's offset.
8. Relief vents: Relief vents shall be provided when stacks extend more than ten branch intervals and shall be provided every ten branch intervals from the top down. The lower end shall be piped as a yoke vent and connect below its tenth branch interval. The upper end shall be connected to its vent stack at least three feet above the floor served. Its purpose is to relive pressure surges from tall stacks.
9. Side inlets: Side inlet closet bends are permitted where the side inlet is vented and washed.
10. Slopes and connections: Vents shall be free from drops and sags and slope to the drainage system. Moisture from condensation could form in sagging lines, which would stop the free movement of air necessary for proper venting.
11. Terminals: Vents through the roof shall extend in accordance with code and shall have a waterproof flashing. Heights will vary from one code to the next and can range from a minimum of 6 inches to 7 feet when the roof is used for other purposes. The vent terminal shall not be located where its vapors can enter the structure, whether through windows or ventilation openings. Refer to the code having jurisdiction for other details. Sidewall venting is permitted in some circumstances. Technicians and inspectors do

not readily embrace sidewall vents, but they do have benefits with metal roofs because sheets of ice or snow can skid down the roof, shearing off conventional terminals.

12. Vertical rise: All vents shall rise vertically 6 inches above the flood-level rim of the highest trap or trapped fixture being vented. Some codes allow exceptions to this concept for bathtubs or floor drains and discuss rising above the centerline. Check your local code for deviations that might require items such as extra cleanouts and slope.
13. Water closet vents: Check the applicable code for these requirements. The *International Plumbing Code* is very liberal in that it allows a closet line to travel an unlimited length from its vent because this code operates under the assumption that the trap seal will be replenished at flushing. The *National Standard Plumbing Code* is clear in stating the distance between the fixture outlet to its vent connection shall not exceed 3 feet vertically and 9 feet horizontally.

Trap Arm Distance

The distance from each trap to its vent is limited to reduce the trap's possibility of being self-siphoned. The *International Plumbing Code* has a very descriptive phrase in its Code and Commentary, which states, "The intent is to locate the vent within the hydraulic gradient of the fixture drain so as to prevent the opening to the vent from being occluded by waste flow." The meaning of *occluded* is to close, shut, or block. These distances are very important and are provided for consideration in Table 38-4.

Various Methods

Table 38-5 represents the accepted methods and venting applications permitted by the three major codes, with applicable code section numbers for ease of location. Use caution because terminology varies from one code to another.

VENTING OVERVIEW

General insights are provided below for the different venting methods shown in Table 38-5 to increase your application understanding, although all codes vary somewhat in detail.

Individual Vent

Individual venting (see Figure 38-1) is the simplest form of venting because every fixture is vented. The vent is connected to the individual drain of the trap or trapped fixture being vented. This is not a cost-effective method because other methods could eliminate number of vents installed, thereby reducing the expense.

Table 38-4 Maximum Length of Trap Arm. Courtesy of Plumbing-Heating-Cooling-Contractors—National Association.

Diameter of Trap Arm in Inches	Length: Trap to Vent	Slope: Inches per Foot
$1\frac{1}{4}$	3″6″	$\frac{1}{4}$
$1\frac{1}{2}$	5′	$\frac{1}{4}$
2	8′	$\frac{1}{4}$
3	10′	$\frac{1}{4}$
4	12′	$\frac{1}{8}$

Table 38-5 Nationally Accepted Venting Methods (including 2006 codes)

Method	IPC	NSPC	UPC
Individual vent	Yes, Section 907	Yes, 12.8	Yes, 901
Common vent	Yes, 908	Yes, 12.9	Yes, 905.6
Vertical leg	No, was in residential code	Yes, 12.8 exception	No
Island fixture vent	Yes, 913	Yes, 12.18	Yes, 909
Wet vent, horizontal	Yes, 909	Yes, 12.10	Yes, 908.4
Wet vent, vertical	Yes, 909	Yes, 12.11 (stack venting)	Yes, 908
Circuit vent	Yes, 911	Yes, 12.13	Yes, L 7 alternate
Combination drain and vent system	Yes, 912	Yes, 12.17 (waste and vent)	Yes, 910 (waste and vent)
Waste stack	Yes, 910	Yes, for dwelling units 12.10.5	No
Air admittance valves	Yes, 917	Yes, E.8	No
Suds pressure	No	Yes, 12.15	Yes, 711
Engineered vent systems	Yes, 918	By authority having jurisdiction E.3	Yes, L 2 alternate systems
Computer vent design	Yes, 919	By authority having jurisdiction E.3	No

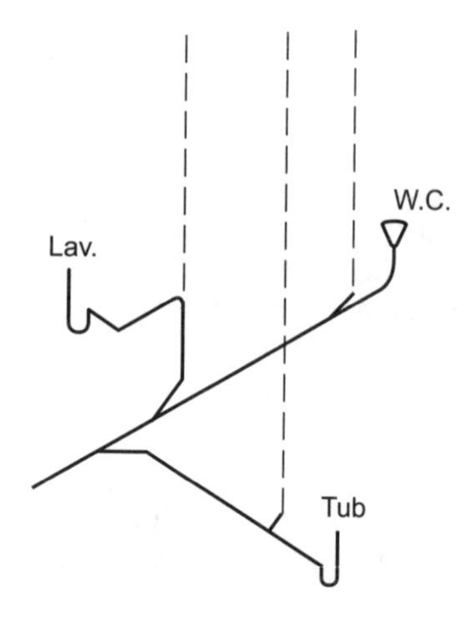

Figure 38–1
An individual vent.

Common Vent

A common vent (see Figure 38–2) serves two fixtures, which may be connected at the same level with a double sanitary tee or at two different levels with sanitary tees stacked. A common vent could also be installed between two fixtures of a horizontal branch. A vent sizing table is provided to address the line between the fixtures because the line carries the top waste and, at the same time, vents the lower fixture. The most common application would be two lavatories side by side or a laundry tub connection on top of an automatic washer takeoff. The latter would be perfect for connections at different levels because the takeoffs commonly have different rough-in heights.

Vertical Leg Vent

Vertical leg vents (see Figure 38–3) are recognized by exception in the *National Standard Plumbing Code* as venting extensions. A vertical leg vent is made by over sizing the portions of its risers. Individuals worried about S-trap prohibitions have been successful in having it removed from several codes. It has been a holdover from

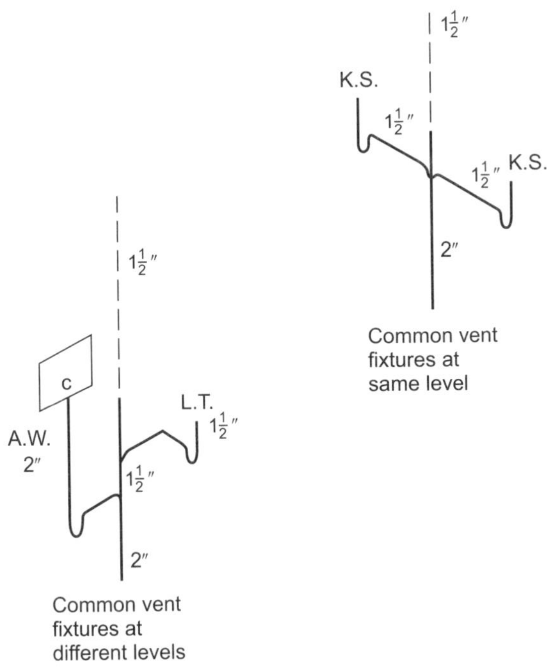

Figure 38–2
A common vent.

very early codes such as the Hoover Code and works very well. Older plumbers quote the former section concept by heart, "Single fixtures within 8 foot of developed length, not over 3 foot vertical drop, do not require additional venting." The CABO (Council of American Building Officials) code recognized vertical leg venting where the vertical drop was increased from the trap arm and then the next horizontal leg was increased, which went into a vented system. This method was perfect for lavatories off a horizontal and tub traps that were tucked between floor joists that ran to a line beside a wall perpendicular to and underneath the joist in a single-family ranch home.

Wet Vent, Horizontal

Horizontal wet vents (see Figure 38–4) can serve up to two bathroom groups connected to a horizontal run. The restrooms can be installed in series or back to back. The individual vent from the system is installed in front of the last fixture and allows air to travel down the system to all the fixtures served, while providing drainage to its fixtures. Applicable vent sizing tables provide protection, and several other considerations provide safeguards. For example, fixtures shall be connected independently and fixtures with a vertical rise, such as lavatories, would be considered an S-trap and need individual or common vents. Other fixtures not belonging to the restrooms could not discharge through the system; for example, a kitchen sink or laundry would have to discharge with its own vents downstream of the horizontal wet vent system.

Wet Vent, Vertical (Stack Vent)

Vertical wet vents (see Figure 38–5) can serve up to two bathroom groups connected to the top of a stack riser. A vertical wet vent has very similar requirements to those of a horizontal wet vent. The vent for the system is an extension of the stack vent and all the fixtures shall connect independently to the vertical wet vent. Applicable vent sizing tables provide protection, and several other considerations provide safeguards. For example, fixtures shall be connected independently and fixtures with a vertical rise would be considered an S-trap and need individual or common vents. S-traps should not be present because the stack would be centrally located and, more than

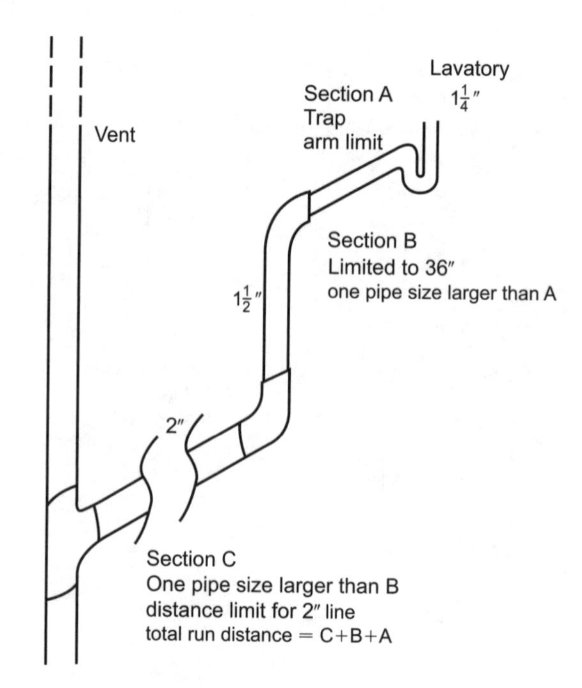

Figure 38–3
Vertical leg vents.

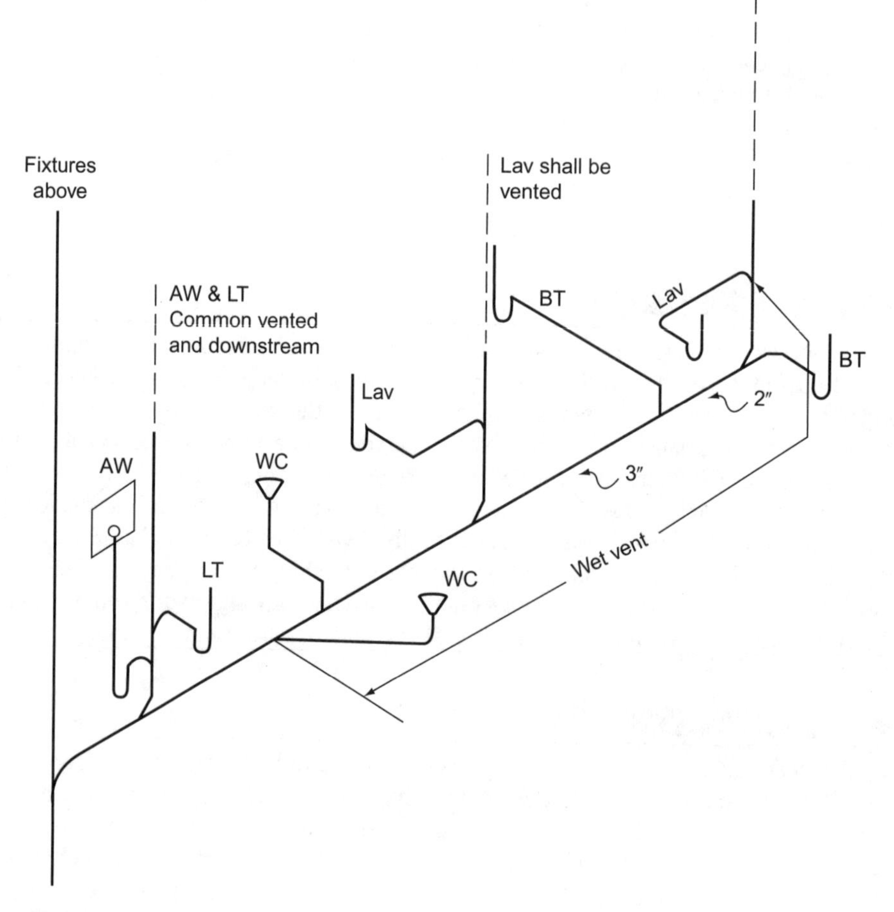

Figure 38–4
A horizontal wet vent.

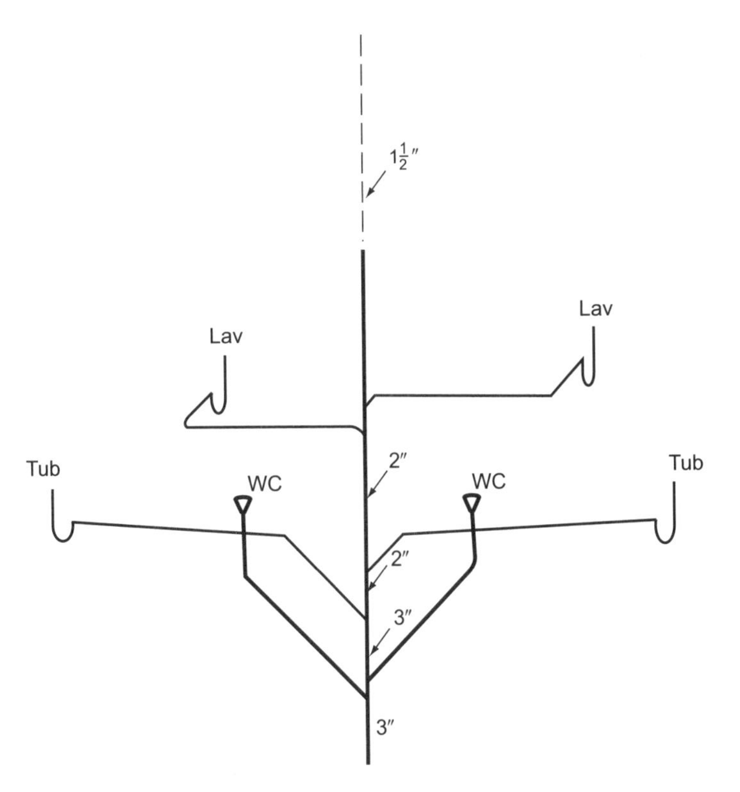

Figure 38–5
A vertical wet vent.

> **In the Field**
>
> The plumbing profession is unique in that past proven practices should not be abandoned and new products and methods should be viewed with a very open mind. These new methods and products have been of extreme benefit to technicians and consumers. An example might be that at one time professionals questioned the need for reduced-pressure zone (RPZ), backflow preventers.

likely, lavatories would be a run directly off the stack. Water closets must be the lowest connection in the vertical wet vent and connect at the same elevation. Other fixtures not belonging to the restrooms could not discharge in the system; for example, a lower-floor laundry tub would have to discharge below the wet-vented group with its own vents.

Waste Stack

A waste stack is a stack with waste fixtures connected to it with limitations to their amounts. See Figure 38–6. The main venting concept is that fixtures shall connect individually to an oversize stack, which will allow the fixtures to drain while, at the same time, allowing air to flow down the stack to protect the trap seals. It is an excellent method to use in stacked apartments where the same waste fixtures are located directly above each other. There are several safeguards built into the code language; for example, no offsets are permitted and sizing tables are provided. Consult your code for details. Waste fixtures are defined in the code and are considered lower volume in nature. There is another stack concept for soil and waste fixtures that has been used for many years, called the Philadelphia System. Many very old codes used the concept with limited application. This system has gained recognition lately due to ASPE advocates at code hearings in the code-development cycles.

Circuit Vent (Battery Venting)

Circuit vents (see Figure 38–7) serve individually connected fixtures in series (battery) in groups of as many as eight fixtures. Other fixtures may connect to the system, such as urinals or lavatories, but they must have their own vent. The considered fixtures for this system are water closets and floor drains, usually in a commercial-structure setting. The system obtains its venting from a circuit vent installed in front of the last fixture on the line. Exercise caution; the *National*

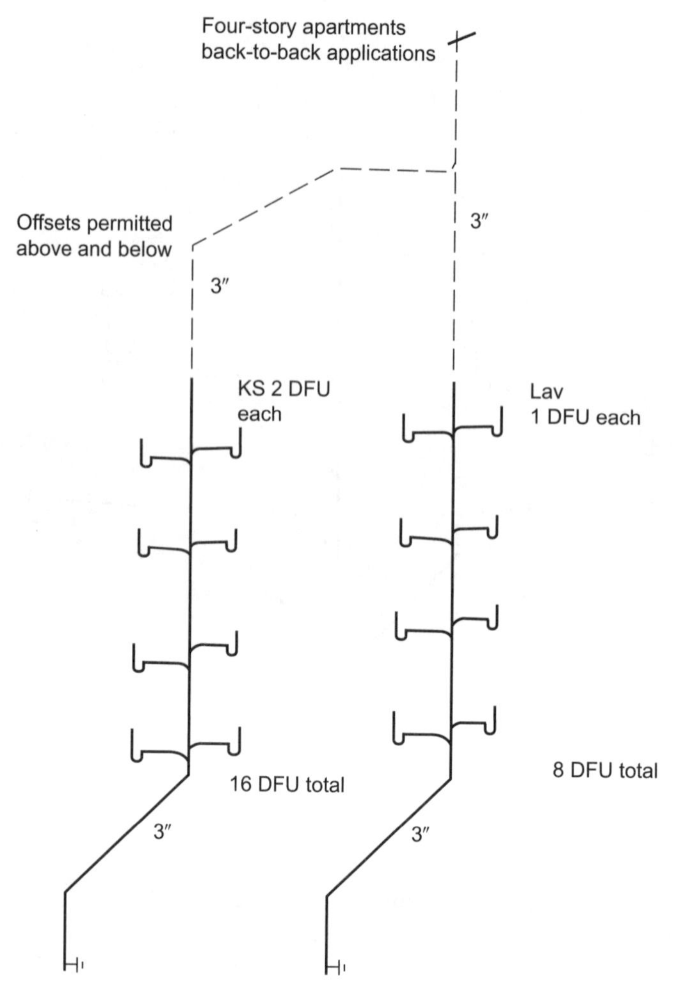

Figure 38–6
A waste stack vent.

Standard Plumbing Code would allow you to wash the vertical portion of the vent but the *International Plumbing Code* will not allow any fixtures into this protected vent. The *International Plumbing Code* concept is that the vent is washed at its heel by the last fixture. Multiple-floor applications will require a relief vent under certain conditions when batteries are installed on stacks above. Consider the allowable grade and that the groups of up to eight fixtures shall maintain the line size in each group. In other words, if you are running a 4-inch line through these eight fixtures, you cannot reduce to 3 inches simply because you feel the DFU value has reduced to the point that a 3-inch line is acceptable.

Combination Drain and Vent (Combination Waste and Vent)

Combination drain and vent systems (see Figure 38–8) are for low-volume waste fixtures and limit the system slope. The concept is that these fixtures have a great deal of distance and vertical rise because the system drain line is far oversized. The design should allow only a quarter of the main line size for waste flow and three-quarters for air circulation. One vent will be allowed to service the system. A word of caution here: many inspectors recognize these as systems with special conditions. A growing number of individuals believe the concept allows them to indiscriminately use combination drain and vent systems. These individuals place single fixtures in a branch or building drain without the necessary vent and call the fixture a combination drain and vent. That is one of the reasons why the vertical-leg venting method was removed from the

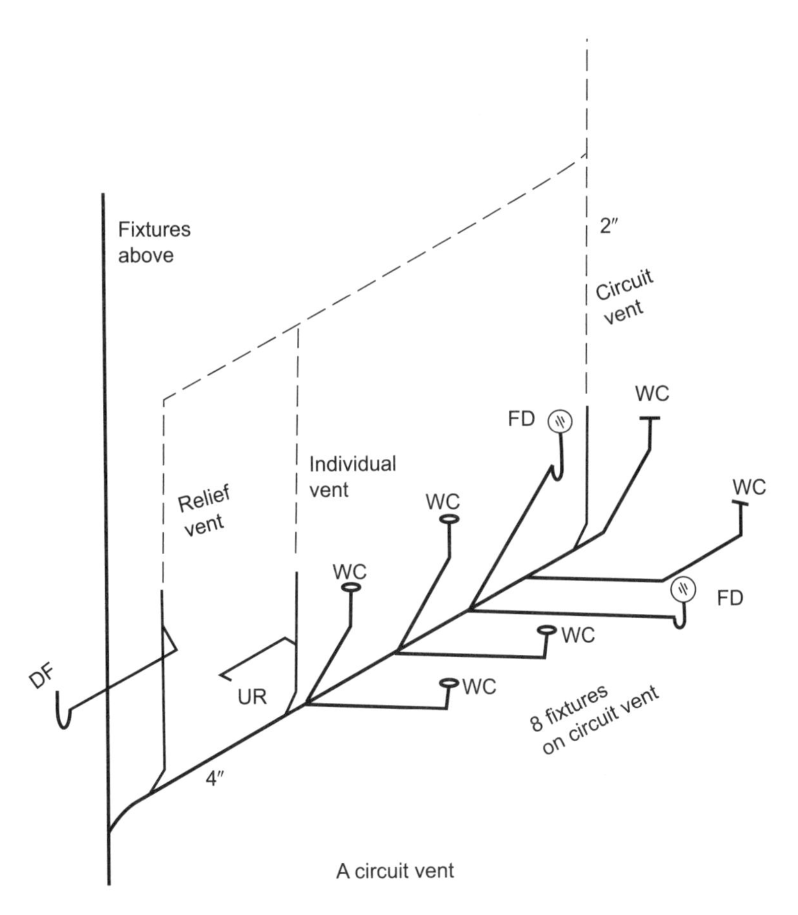

Figure 38–7
A circuit vent.

International Residential Code. The *International Residential Code* for one- and two-family dwellings includes several chapters applicable to plumbing and is used in several areas of the county.

Island Fixture Vent (Barbershop Loop)

The island fixture vent (see Figure 38–9) is used for single waste fixtures where the conventional vertical rise of a vent is impractical. An example would be a kitchen sink in a log home where the installation technician could not drill the back wall to accommodate a vent. The island vent loops up as high as possible, serving the fixture, and travels back or over to an appropriate vent connection. Cleanouts are required to maintain the vent and the vent shall have slope to care for condensation matters.

Air Admittance Valves

Air admittance valves have applicable ASSE standards for conformance and the manufacturer's installation instructions must be used. At least one vent in a drainage system shall extend outdoors. The log home discussed above would be an excellent application for the device.

Refer to your local code for enforcement details.

Venting of Building Subdrain Systems

Fixtures in drainage systems that drain by gravity and those served by sump pits shall be vented like other systems and may connect to the venting system serving the building. Sumps shall be sealed and vented and the vent may connect to the venting system serving the building.

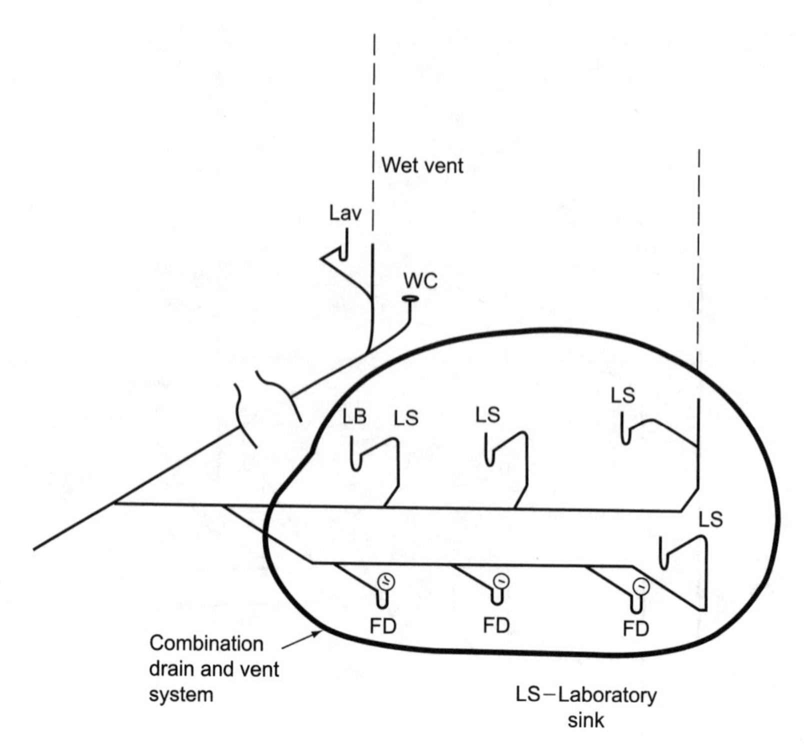

Figure 38–8
A combination drain and vent system.

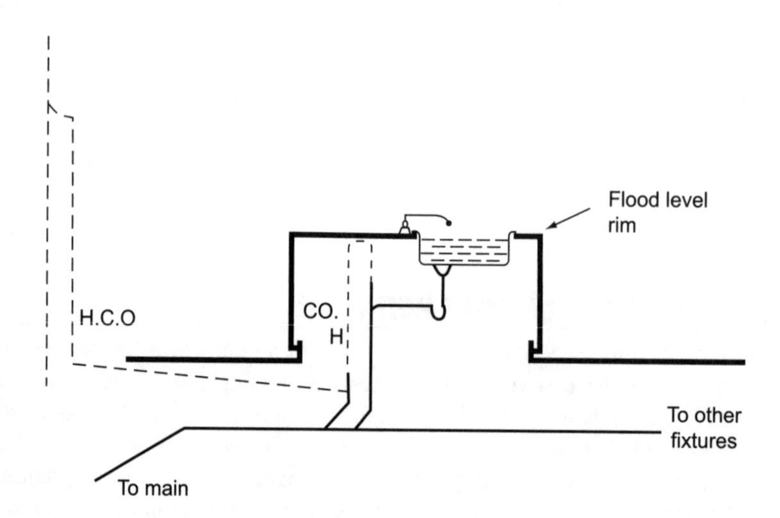

Figure 38–9
An island fixture vent.

Vent Size and Length

Additional discussions related to vent size and length tables may be found in Table 11-4 and the accompanying discussion.

STORM WATER DRAINAGE

Plumbing includes storm drainage, and all structures are required to have separate or combined systems that discharge to approved locations. Storm water shall not be drained into sewers designed for sanitary sewage only. Jurisdictions, which had allowed combined systems, are doing their best to change to separated systems to address proper treatment. Vast portions of the storm drainage portions of the code are similar to the code for sanitary systems, such as piping, connections, changes in direction, and many others.

Several factors enter into consideration for technicians, such as rainfall rates and the structural integrity of the roof. For example, some roofs are designed to carry additional loads for things such as ponding. This type of roof design is called controlled roof drainage. Other designs require additional drainage backup, known as emergency secondary systems. These are necessary when the main system could fail. The secondary systems are only required when ponding could take place because parapet walls surround the roof's edge. Rainfall rate information will be provided in the appropriate plumbing code. Sizing methods may vary and they will all address the conductors, building drains, and building sewers. Information is provided to address sizing for other concerns, such as ejectors providing pumped discharges for subsoil systems and semi-continuous discharges from clear water waste equipment. Many codes will include requirements for exterior roof gutter systems, which serve pitched roofs. Other codes will amend such requirements for removal because their licensing program does not include these technicians.

Please refer to Chapter 12 for greater detail related to storm drainage.

REVIEW QUESTIONS

1. What is a common vent?
2. What is the purpose of slope in a drainage pipe system?
3. What is the advantage in increasing drainage pipe slope?
4. Where are some roof vent terminals increased to a larger size?
5. Why is it that storm water shall not be drained into sewers designed for sanitary sewage only?

CHAPTER 39

Medical Care Facilities Plumbing Equipment

LEARNING OBJECTIVES

The student will:

- Discuss why healthcare facilities are given additional life-safety importance in the code.
- Compare the different local vent requirements for clinical sinks, bedpan washers, and sterilizers.
- Identify the three major groups affected by ASSE 6000 series standard requirements.
- Discuss the various supply areas affected by NFPA 99C.

FACILITIES

Healthcare facilities are unique in their requirements and seek fail-safe systems because of their life-safety concerns. Insights in this chapter will focus on code conformance to the *National Standard Plumbing Code* and medical gas piping. NFPA 99C discussion will provide an overview of medical gas piping systems for those individuals considering certification training.

All areas of healthcare facilities are of concern for life safety and begin with the water service by requiring dual water service lines to maintain a water supply in the event of a water main failure. This allows a single water main break to be isolated and repaired without shutting off all water service to the facility. The code moves on to address other portions of the facility, equipment such as specialty fixtures, and patient medical gas piping systems. Icemakers or ice storage chests shall not be located in a soiled utility room or similar areas where they are subject to possible contamination. Protrusions from walls are prohibited and drinking fountains, cleanout covers, and other devices such as medical gas controls shall be fully recessed in corridors and other areas where patients may be transported. Mental patient rooms and correctional facilities shall have piping, trap drains, fixtures, fittings, and pipes concealed and fixtures shall be vandal-proof to protect the individuals who are institutionalized.

EQUIPMENT

Healthcare facility equipment is far different from code fixtures in that it does not have published conformance standards to guide manufacturers and plumbing professionals. For that reason, plumbing codes have listed specific guidelines for fixtures such as sterilizers, bedpan washers, and aspirators. The basic code provisions such as backflow protection and drainage still apply for all these products. Several codes state that vacuum breakers for bedpan washer hoses and laboratory areas shall be 5 feet and 6 feet above the floor, respectively.

Clinical Sinks and Bedpan Washers

Clinical sinks, bedpan washers, and flushing-rim service sinks shall be installed similar to water closets. A non-flushing service sink shall not be used to clean bedpans, and clinical sinks shall not be used as a substitute for service sinks. See Figure 39–1 for examples of bedpan washers.

There are some bedpan washers and clinical sinks with local vent connections to remove odors and excessive humidity. Local vents are separate systems extended to the outdoors, above the roof by themselves, apart from the sanitary vent. These vents could connect to other bedpan washer vents and are not permitted to connect to sterilizer vents or other local vents. Further, the vent materials shall conform to code requirements for sanitary vents. The local vent serving a single clinical sink or bedpan washer shall not be less than 2-inch pipe size. Several of these local vented sinks could connect together in accordance to specific sizing criteria found in the code. The connections to the local vent stack shall use sanitary tee or tee-wye fittings oriented for upward flow from the branch.

The condensate drainage from the vapors in the local vent may drain back to the fixture served. The base of a local vent stack serving more than one fixture shall be directly connected to a trapped and vented waste branch of the sanitary drainage system. The trap shall be primed by the water supply of a clinical sink or bedpan washer at each floor level on the local vent stack. The priming line, no smaller than $\frac{1}{4}$-inch OD size, shall be extended from the discharge or fixture-side of the vacuum breaker protecting the fixture to the local vent stack. These lines shall prime the trap at the base of the local vent stack each time the fixture is flushed. The trap seal depth shall be no smaller than 3 inches, and the trap and branch shall be the same size as the local vent stack. Finally, the vent for the waste branch shall be at least $1\frac{1}{4}$ inches but not less than half the size of the waste branch.

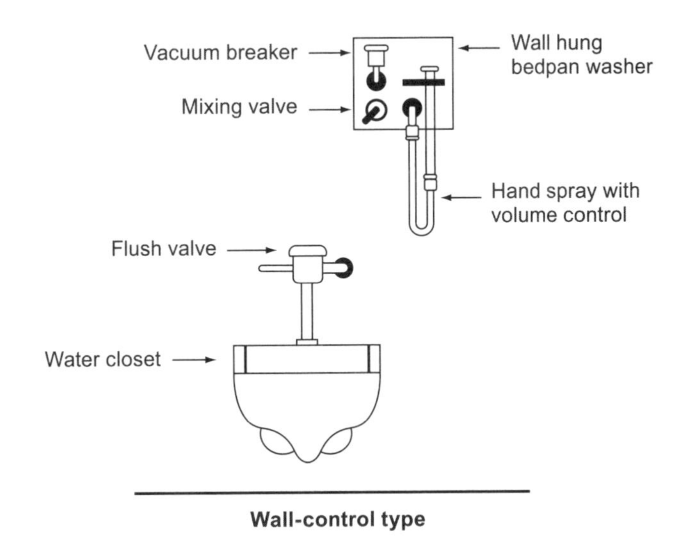

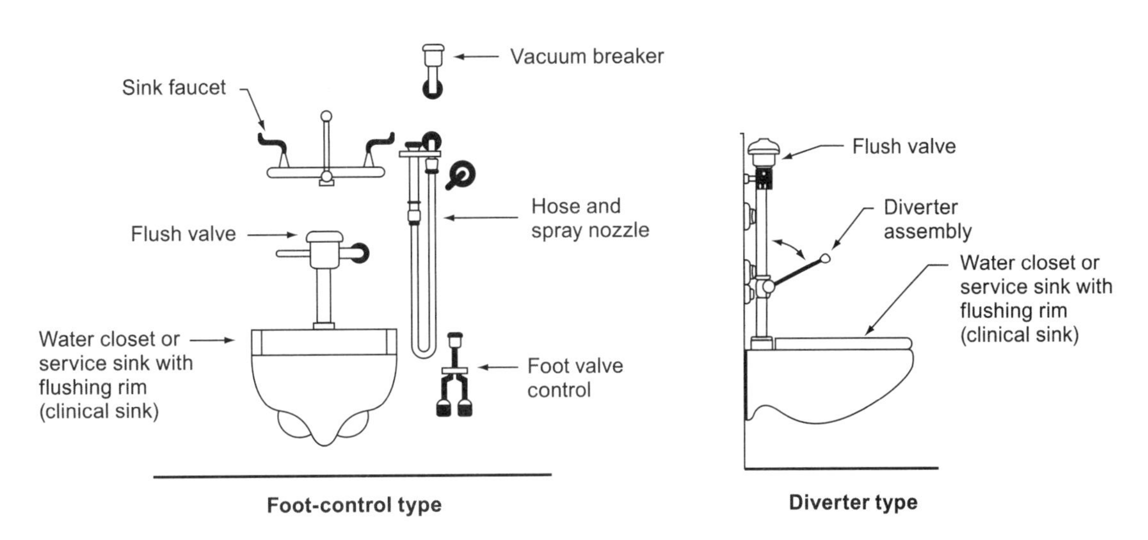

Figure 39–1
The newer, conventional types of bedpan washers are illustrated from the 2006 *National Standard Plumbing Code, Illustrated.*

Sterilizers

Sterilizers and bedpan steamers may be located at different locations in healthcare facilities to properly clean bedpans and similar devices at high temperatures and steam to ensure proper sanitation. The equipment shall be installed in accordance with the code and the manufacturer's instructions.

Drains

All drainage waste from sterilizer and bedpan steamers shall be indirectly connected to the sanitary drainage system through an air gap. In addition, an isolation trap with a minimum seal of 3 inches shall be provided in the indirect waste line located between the fixture and the air gap at the indirect waste receptor. Each fixture shall have a separate waste line, except up to three sterilizers may share a common indirect waste pipe if the developed length does not exceed 8 feet. The size of such common indirect waste pipes shall be no smaller than the aggregate cross-sectional area of the individual sterilizer drain connections.

A trapped and vented floor drain shall be provided in each room or space where recessed or concealed portions of sterilizers are located. The floor drain, no smaller than 3-inch pipe size, shall drain the entire floor area and shall receive the indirect waste from at least one sterilizer. Where required by the sterilizer manufacturer, a floor drain shall be located directly beneath the sterilizer within the area of its base. The waste drainage from condensers or steam traps shall be cooled below 140°F before being discharged indirectly to the sanitary drainage system.

Local Vents

Where sterilizers have provisions for a vapor/local vent and required by the manufacturer, a vapor vent shall be extended to the outdoors above the roof. These sterilizer vapor vents shall terminate with the same requirements as sanitary vent terminals. As stated above, they shall not be connected to local vents for clinical sinks or bedpan washers or to any drainage system vent. The vent piping for sterilizer vapor shall be of a material acceptable for sanitary vents.

The sterilizer vents and stacks for individual sterilizers shall be no less than the size of the sterilizer vent connection and no less $1\frac{1}{2}$-inch pipe size. Where vapor vent stacks serve more than one sterilizer, the cross-sectional area of the stack shall be no less than the aggregate cross-sectional areas of the vents for all of the sterilizers served. The vent connections to the stack shall be made using sanitary tee or tee-wye fittings oriented for upward flow. Provisions are provided for the drainage of condensation within vent piping. Finally, the base of stacks shall drain indirectly through an air gap to a trapped and vented waste receptor connected to the sanitary drainage system.

Central Vacuum Systems

Gravity drainage from medical, surgical, dental, and similar central vacuum systems shall be directly connected to the drainage system in accordance with NFPA 99C, Gas and Vacuum Systems. The direct connection eliminates the chance of vapors being aspirated by occupants in the same room as the vacuum system drainage.

Holding tanks for the gravity drains of dental vacuum tanks shall extend from the vacuum check valve on the waste outlet of the tank and be connected directly to the sanitary system through a deep-seal trap. The trap shall be conventionally vented within the plumbing system. In addition, a vent shall be installed between the vacuum check valve and the drain trap, on the inlet side of the trap, to seal the check valve when the holding tank is operating under vacuum and collecting waste. The trap and drain size shall be at least two pipe sizes larger than the waste outlet and vacuum check valve, but no smaller than 2-inch pipe size. Finally, the trap shall be no shallower than 4 inches deep. This vent shall be connected to the plumbing system vents. The vent for the vacuum check valve shall not be less than the size of the check valve. The trap vent shall not be less than half the size of the trap and drain branch. Both vents shall extend vertically to no fewer than 6 inches above the top of the holding tank before making any horizontal turns.

Positive-pressure drainage from air/waste separators in dental vacuum systems shall be directly connected to the sanitary drainage system through a deep-seal trap that is conventionally vented within the plumbing system. The trap vent shall extend vertically to no fewer than 6 inches above the top of the separator before making any horizontal turns. The vacuum exhaust airflow from the separator shall be separately vented outside as required by NFPA 99C. The trap and drain branch size shall be at least two pipe sizes larger than the waste pipe, but no less than $1\frac{1}{2}$-inch pipe size. The trap seal shall be at least two times the exhaust backpressure in the separator, but no fewer than 4 inches deep. The vent shall be the full size of the trap and drain.

An overflow floor drain used as a visual indication of a drainage problem shall be provided for gravity and pressure discharges from a dental vacuum air/waste separator or waste-holding tank to the sanitary drainage system. The overflow will prevent the backup from reaching the level of the trap for the air/waste separator or the drain check valve for the waste holding tank. The trap of the floor drain or receptor shall have a trap seal primer.

Aspirators

Provisions for aspirators or other water-supplied suction devices shall be installed only with the specific approval of the authority having jurisdiction. Where aspirators are used for removing body fluids, they shall include a collection bottle or similar fluid collection trap. Aspirators shall indirectly discharge to the sanitary drainage system through an air gap and the potable water supply to an aspirator shall be protected by the appropriate backflow preventer.

MEDICAL GAS AND VACUUM PIPING SYSTEMS

Plumbing codes state that the installation of medical gas and vacuum piping systems shall be in accordance with the requirements of either NFPA 99, Standard for Health Care Facilities or NFPA 99C, Gas and Vacuum Systems. NFPA 99C deals specifically with the medical-gas piping systems, where as NFPA 99 addresses the entire facility including electrical. Many plumbing codes only reference NFPA 99C to save inspectors' costs when purchasing standards. NFPA 99C references ANSI/ASSE Series 6000, Professional Qualification Standard for Medical Gas Systems Installers, Inspectors, and Verifiers, and some plumbing codes specifically reference the Series 6000 as a conformance requirement. This standard sets the criteria for installers (including brazers), inspectors, and verifiers of medical gas and vacuums. NFPA 99C and ASSE Series 6000 are documents that every individual involved in healthcare facilities installation and inspection should have. The United Association of Journeymen and Apprentices of the Plumbing and Pipe Fitting Industry of the United States and Canada (UA) and many private training programs provide excellent training in the certification of installers and inspectors.

Medical gas piping concerns developed because there were serious problems with flammable gas in medical facilities. Later developments and improvements in piping systems clarified the need for clean systems that did not have cross connections. The cross connections were simply incorrectly supplied products or vacuums from the wrong line. Early flammability problems were addressed by the individual in control (an anesthesiologist), who directly adjusted the product in manageable amounts. Non-flammable gases such as oxygen have many uses in medical facilities including inhalation in the anesthetic processes and treatment. Oxygen is an oxidizing gas and creates an environment that supports combustion. The safety issue is clearly understood when you recall your early safety training related to greasy or oily gloves used near oxygen tanks, which can best be described as very dangerous.

The NFPA 99C standard contains the criteria for the design, installation, system components, materials, testing, and safety requirements for medical gas systems. Medical gas systems include vacuum systems, and the major discussion of facilities covers level-one healthcare facilities, level-two facilities (such as freestanding birth centers), and level-three facilities (such as dental facilities). Level one is the most stringent classification and provides the most protection. Discussions pertain to systems that provide services to human beings; for that reason, a veterinary clinic is not considered in these discussions.

New construction and remodeling of healthcare facilities include four main components: designer, installer, inspector, and verifier. The designers, with their expertise and experience, are covered in their professional registration. Installers must be qualified in brazing (the joining process for the pipelines) and shall have the training (generally 32 hours) and understanding to properly install the systems. Inspectors are charged with inspecting the installation to ensure conformance to the proper codes and standards; their training is generally 24 hours and they have more than likely been installers. Verifiers are the final line of safety assurance with their testing verification. All of the latter three are involved in system testing.

In the Field

CAUTION: Information will be provided in the discussions of NFPA 99C and medical gas systems as a very general review to direct the reader to proper training programs, where accurate details can be provided. These discussions are not adequate for such a serious life-safety matter. However, there is a great need for individuals to have a basic understanding of where resources are available in these matters. Some of the information in text or graphics may be from older editions of the standard.

Document Resources

There are several helpful resources related to medical gas installations, which are listed for your convenience.

1. American Society of Mechanical Engineers (ASME), ASME Section 10, Qualification Standard for Welding and Brazing Procedures.
2. American Society of Sanitary Engineering (ASSE), Series 6000.
3. American Welding Society (AWS), AWS B2.2-91, Standard for Brazing Procedure and Performance Qualifications.
4. Compressed Gas Association (CGA), CGA G-4.1, Cleaning Equipment for Oxygen Service.
5. Copper Development Association (CDA), the Copper Tube Handbook.
6. National Fire Protection Association (NFPA), NFPA 99C, Health Care Facilities, and Health Care Facilities Handbook.
7. Plumbing Code Commentaries.

Explanation of System Details for Level One

The design, installation, storage, handling, and use of nonflammable medical gas systems including inhalation anesthetic and vacuum piping systems are identified to reduce the potential fire and explosion hazards associated with medical gas piping systems. The term *medical gas* applies to all patient systems; however, the standard may name a specific gas with specific requirements. The systems commonly consist of nitrous oxide, nitrogen, medical compressed air, medical vacuum and evacuation, oxygen, carbon dioxide, helium, and various mixtures of gases. Labeling and identification of the systems are critical, and there is an emphasis on isolation of connections. The systems begin at the supply source and proceed up to the mains, branches, risers, feeders, service outlets, and alarms to the patient's point of use. The goals are to supply safe, adequate amounts of pure product with uninterrupted flow and, most importantly, not to have cross connections. System hazards include explosive gases, infectious diseases, and the buildup of oxidizing gases. See Figure 39–2.

The systems are best understood by viewing them in basic groups: the source and equipment, piping, and outlets. The sources are composed of cylinders, bulk supplies, air compressors, and vacuum pumps.

Cylinders (Source)

The cylinders are high-pressure storage containers that contain gases at pressures of 2,000 pounds per square inch. Non-medical gas installers generally equate them with cutting-torch setups, oxygen and acetylene. Low-pressure containers of cryogenic liquids are also included in the cylinder group. Cylinder systems are divided into two different categories. There are systems with reserve and systems without reserve.

Cylinder systems with reserve (see Figure 39–3) contain a minimum of two banks. Each bank shall contain a minimum of two cylinders per bank. The reserve bank shall contain a minimum of an average day's use or three cylinders, which is for emergency use only. The main banks can be low-pressure cryogenic. The main and reserve have different safety storage requirements, which are provided by addressing check valve and alarm locations. The banks are not required to have a pigtail check valve. The reserve shall have a check valve in the pigtail or an alarm to indicate a one-day supply balance.

Cylinder systems without reserve (see Figure 39–4) contain a minimum of two banks. Each bank shall contain a minimum of two cylinders per bank or an average day's supply, whichever is greater. However, the reserve bank shall not be required to ensure ample supply. The pigtails shall contain check valves, and the banks shall have an automatic change-over function.

Medical supply system
Nurse's station
9
3rd Floor
Delivery room
6 7 Recovery room
8
Patient room
5
2nd floor
Operating room
Recovery room
Patient room
Patient room
10
1st floor
1 Main shutoff
2 Master alarm sensor
3 Gauge
4 Riser valve
5 Area alarm sensor
6 Zone valve
7 Gauge
8 Station outlet
9 Area alarm panel
10 Master alarm panel
4
3 2 1
Separate systems
Evacuation - Purple
Carbon dioxide - Gray
Helium - Brown
Medical air - Yellow
Nitrogen - Black
Nitrous oxide - Blue
Oxygen - Green
Vacuum - White
Others

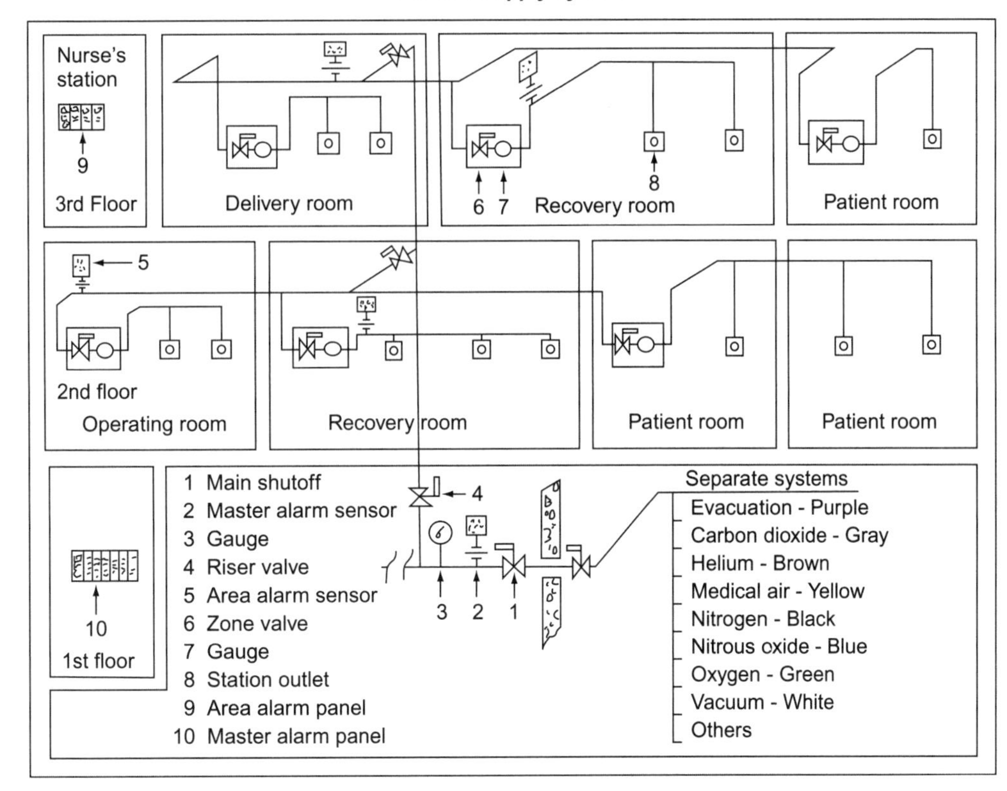

Figure 39–2
A medical supply system.

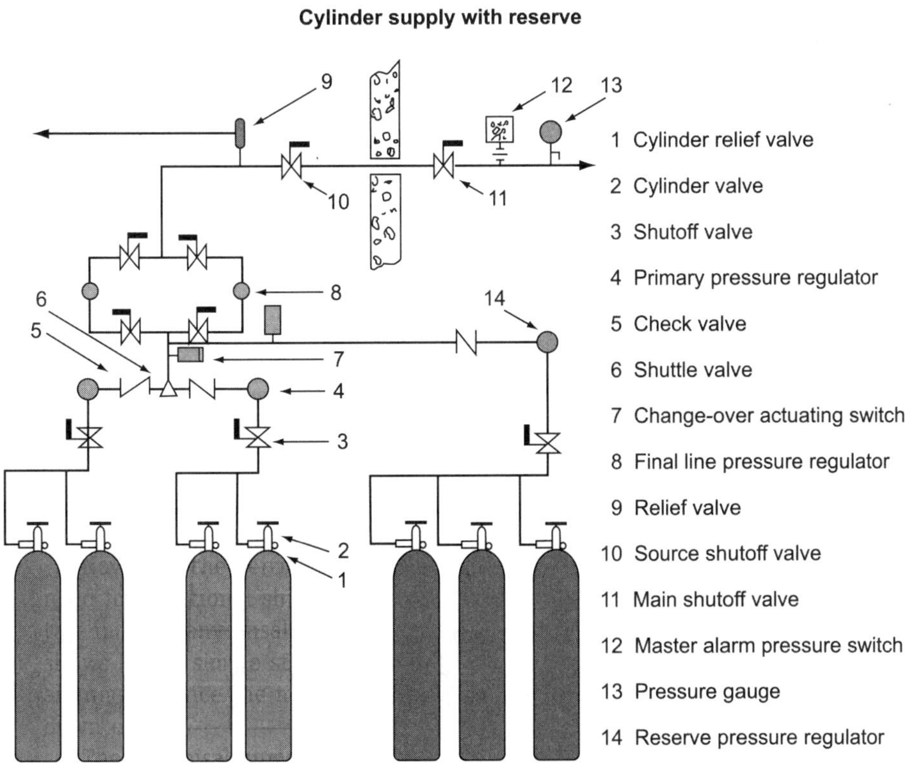

Cylinder supply with reserve
9
12
13
10
11
6
5
8
14
7
4
3
2
1
1 Cylinder relief valve
2 Cylinder valve
3 Shutoff valve
4 Primary pressure regulator
5 Check valve
6 Shuttle valve
7 Change-over actuating switch
8 Final line pressure regulator
9 Relief valve
10 Source shutoff valve
11 Main shutoff valve
12 Master alarm pressure switch
13 Pressure gauge
14 Reserve pressure regulator

Figure 39–3
A cylinder supply with reserve.

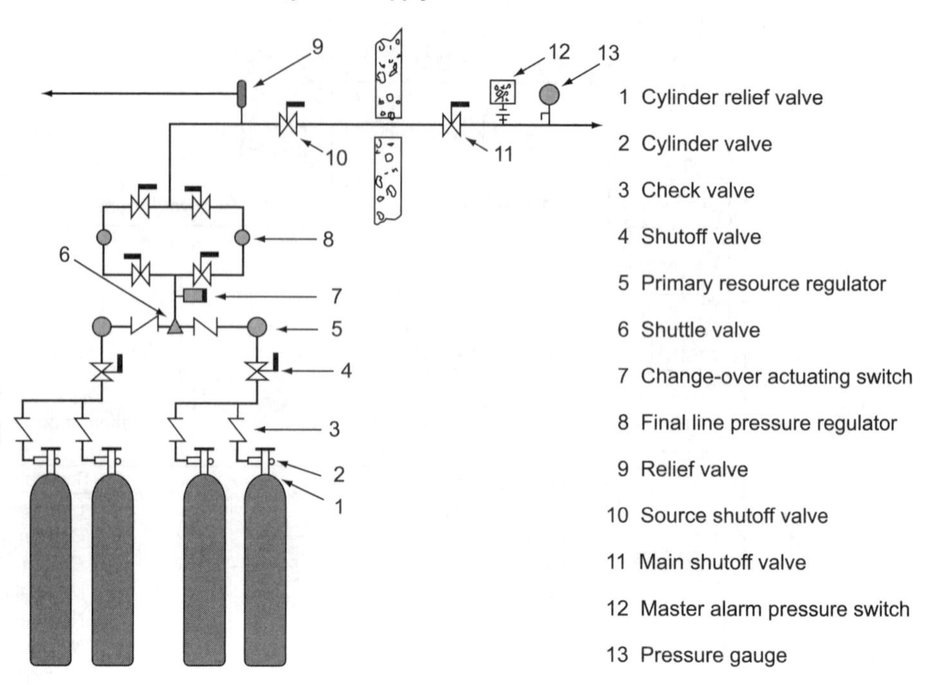

Figure 39–4
A cylinder supply without reserve.

Bulk (Source)

Bulk containers are highly insulated and contain cryogenic liquids in various volumes at high and low pressures. The cryogenic liquid allows the facility to store larger amounts in a smaller area. There are two types of bulk supply systems, alternating and continuous supply sources. Bulk systems shall have a reserve (emergency) supply, one of three options. The first option requires at least three high-pressure cylinders with check valves in the pigtail. The second option requires at least three high-pressure cylinders with a master alarm signal when the cylinders do not have check valves in the pigtails. The third option requires master alarm signals indicating a one-day supply at sufficient pressure levels with the use of cryogenic liquid storage.

Bulk oxygen systems shall also have added supply assurance in the form of an emergency low-pressure oxygen inlet. The female connection shall be located on the outside of the structure and tie in downstream of the main shutoff valve. Its presence requires two additional check valves and an additional relief valve; check valves should not isolate alarm sensors.

Bulk systems as detailed in NFPA 99C also are addressed by NFPA 50, Bulk Oxygen Systems at Consumer Sites and NFPA 51, Design and Installation of Oxygen-Fuel Gas Systems for Welding, Cutting, and Allied Processes.

Source Storage

The requirements for storage areas are critical to ensure safety to the building's occupants prior to the dispensing of the gas products. The standard identifies larger amounts of product with the clarification of bulk systems. Bulk designations are applied to amounts larger than 20,000 cubic feet for oxygen and 28,000 cubic feet for nitrous oxide. Further clarification is provided when the amount of product is greater or less than 3,000 cubic feet. Bulk storage will be located outside the structure. It is helpful to visualize one large cylinder containing 333 cubic feet, and the nine cylinders as containing 3,000 cubic feet.

Air Compressors (Sources)

Medical air systems are used to supply respiratory air (used for breathing) and are addressed extensively in the standard. The system will generally consist of air

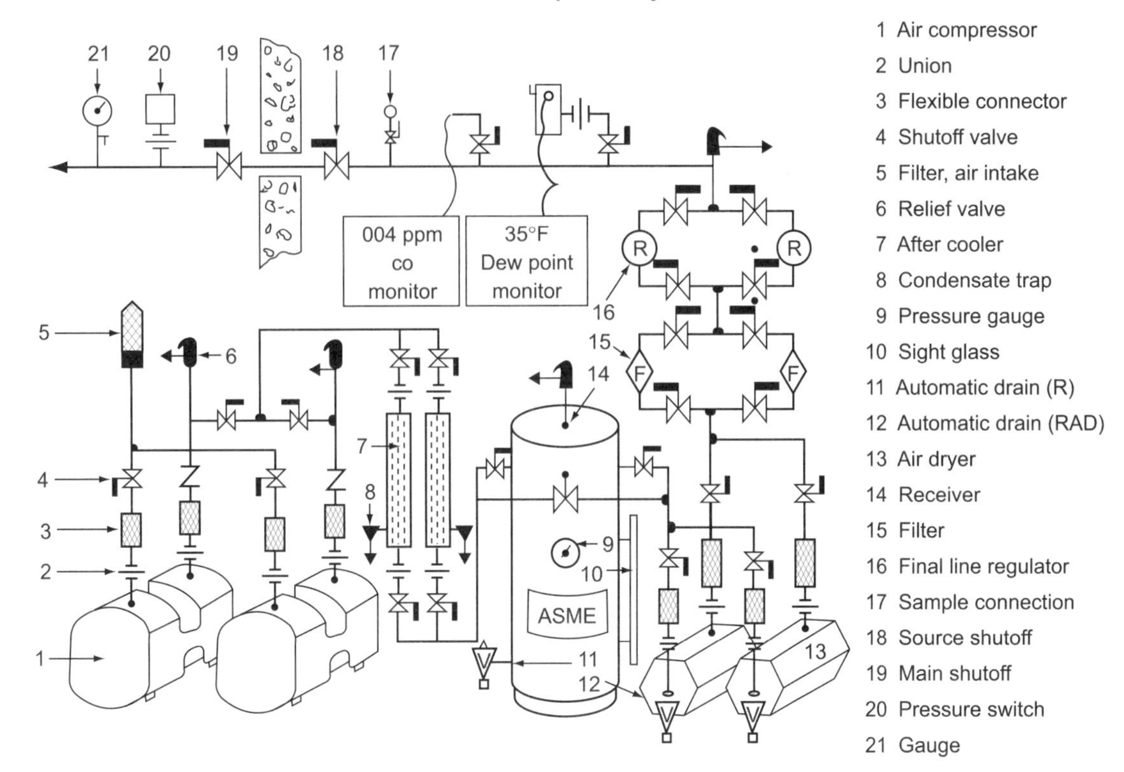

Figure 39–5
A medical air-compression system.

compressors, after coolers, a receiver, dryers, filters, regulators, and supply piping with various alarm indicators. See Figure 39–5.

A minimum of two compressors shall be provided, as with most medical gas systems, to ensure backup. The two products will always be piped parallel to each other. The three types of acceptable compressors are oil less (including liquid ring), oil free, and rotating element (rotary screw). The standard has several conditions to ensure adequate clean air is available for the compressor intake.

Two after coolers are provided to ensure that the air temperatures do not exceed 100°F. The medical air is supplied to a single receiver, which requires a three-valve bypass around the receiver from the after coolers. The receiver is a storage device, which shall have the following operating accessories:

1. A pressure gauge to indicate the storage pressure.
2. A pressure relief valve to eliminate excessive pressures.
3. A sight glass to provide a visual indication of the moisture level in the receiver.
4. A receiver tank that meets ASME pressure vessel standards as indicated by an ASME label.
5. An automatic drain to purge the receiver of excessive liquid.

Air is supplied to at least two desiccant-type dryers from the single required receiver. The two dryers, with automatic draining ability, supply air to the two filters (98 percent at 1 micron with visual indication) piped in parallel with one on standby. They must be designed to deliver 32°F air at 50 pounds per square inch. The filtered air is then provided to two final line regulators piped in parallel, with one on standby. The air exits the regulators to the main, which contains a relief valve set at 50 percent above line pressure.

The supply main located at the source has several monitoring capabilities to ensure suitability. A dew-point monitor with alarm capabilities to ensure moisture-free air at 35–39 degrees dew point is located adjacent to a carbon-dioxide monitor with alarm

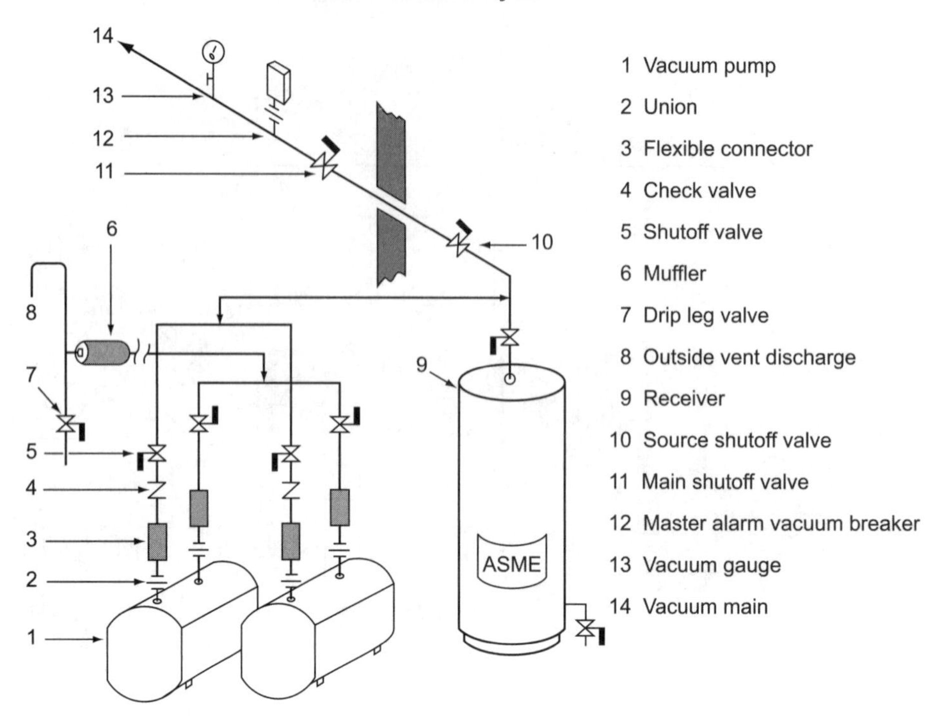

Figure 39–6
A medical air-vacuum system.

capabilities set at 4–10 parts per million. The source main will also contain a sample port valve used at verification time and a source shutoff valve. The medical air main within the structure shall have a main shutoff valve followed by a pressure alarm switch and a pressure gauge. Alarm functions are downstream of the source valve or main valve.

Vacuum Pumps (Sources)

Vacuum systems (see Figure 39–6) are commonly used to remove fluids from patients during surgery. A different type of vacuum system, called Waste Anesthetic Gas Disposal system (WAGD), is designed to remove flammable anesthetic gases from patients during surgery. This system is separate from common vacuum systems and far more critical because it might be removing anesthetic gases that are detrimental to the staff performing surgery. Surgical vacuum systems consist of vacuum pumps with isolation devices, receiver, shutoff controls, and supply piping with various alarm indicators. A minimum of two sources shall be provided, as with all medical gas systems, to ensure backup. Receiver drains are connected directly to the sanitary system to avoid contamination of the surrounding atmosphere. The exhaust discharge shall also be connected to avoid problems.

SYSTEM PIPING

The installation shall be made by qualified, competent technicians experienced in making such installations. Materials and methods require compliance with the manufacturer's installation instructions. Several inspections are to be performed by the installer, with a discussion of the inspector and detailed requirements for the verifier.

Cross Connections

The standard states, "no two medical gas pipelines shall be interconnected at any time." It further clarifies that systems shall not even be connected commonly during testing. For ease of understanding, cross connections can be understood to be DEADLY!

While cross connections are a major concern, the quality of the product cannot be compromised, a process commonly referred to as being contaminated. To contaminate is to make impure by mixture, taint, or pollutant. The three most common contaminants are particulate matter (dust, fillings, dirt), moisture, and oxides (formed by brazing temperatures when oxygen is present). Oil-free nitrogen purging is mandated to avoid the oxidation scale formed during brazing.

Line Pressure

The delivery of consistent design pressures is extremely important to medical gas systems. Normal operating pressure for oxygen, for example, is 50–55 psig. Normal operating pressure for nitrogen is 160 psig. Systems above these pressures have added material, labeling, and certification conditions.

Labels

Identification of the system and its components is necessary at all times for both the installer and the structure owner. Correct identification during installation reduces the risk of interconnection of different systems. Several system components are gas specific and have multiple parts to ensure separation. The outlets are one example. Proper identification is also critical to the owner during maintenance and system expansion. The obvious reason is to avoid cross connections, but the most important reason is to avoid cutting into an incorrect line, resulting in downtime during medical procedures.

The most common pipe used in medical gas systems is copper with brazed joints for positive-pressure gases. The copper pipe shall be identified for medical use and be of specific thickness and sealed to avoid contamination. The lines shall be labeled at least every 20 feet with the gas content, at least in every room, at each floor, and pressures shall be identified.

Color-coding and applications for the most common gases are as follows:

1. Oxygen, O_2, (breathing) is green.
2. Medical air (breathing) is yellow.
3. Nitrogen, N, (power tools) is black.
4. Nitrous oxide, N_2O, (anesthetic) is blue.
5. Helium, He, (breathing, burns) is brown.
6. Carbon dioxide, CO_2, (surgery) is gray.
7. Vacuum is white.
8. Evacuation (WAGD) is purple.

Mixtures shall color-coded with the primary gas being the primary color and the wording label being the color of the lesser gas.

Materials

Pipes and connections shall have sufficient strength and cleanliness to deliver and remove medical gases in sufficient quantities to achieve the design goals. Copper tubing shall be hard-drawn Type L or K that meets ASTM B819 and shall be cleaned and capped. When the pressure is above 185 psi and the line size is larger than 3 inches NPS, only Type K shall be utilized. The mains and branches shall not be less than $\frac{1}{2}$ inch NPS for all pressure piping and a minimum of $\frac{3}{4}$ inches for all vacuum piping. The gauge run out may be $\frac{1}{4}$ inch nominal. All directional changes shall be made with fittings.

Brazing procedures for medical gas piping are developed and followed to ensure leak-proof assemblies. Classes and testing verification by independent, third-party agencies are available to meet the standard's certification for brazers. The procedures address the materials, cleanliness, work area, cutting and deburring, assembly, continuous purge, and filler materials used.

Outlets

Outlets that incorporate product-specific connections and are not interchangeable ensure that instruments and devices are provided with their specific products. The outlets require primary and secondary check valves and each shall be legibly identified with the gas product. The color code, as identified previously, is established by the Compressed Gas Association (CGA). The outlets should be color-coded and identify the gas; they shall have primary and secondary check valves; and shall be properly spaced to allow equipment use. The outlets may be quick-connect or DISS style but must have gas-specific connections.

Joints

The most common method for joining medical gas pipe systems is brazed joints installed by certified individuals. The certification must be in accordance with qualification procedures under AWS B 2.2 or Section 8 of the ASME Boiler and Pressure Vessel Code, as modified by NFPA 99C. The brazing papers address procedures in the practice of purge-gas flow rate, cleaning, joint clearance, overlap, and filler metal. The socket-type fittings shall be brazed using copper-phosphorus (BCuP) brazing alloys for copper-to-copper metals without flux. All materials shall be cleaned and supported by documentation. The pipe shall be sealed and fittings shall be bagged to prevent contamination.

Cleaning the joint is of utmost importance to ensure capillary action in the proper brazing techniques. The pipe shall be cut straight and deburred with all metal shavings removed. Before the brazing process begins, the system shall be purged to prevent oxidation or flaking of the products, which contaminates the system. The purging shall use clean, oil-free nitrogen throughout the system, generally at 15 liters per minute, and will vary on the pipe being brazed. However, there is a requirement for the interior of the piping to contain less that 1 percent oxygen and be measured with an oxygen analyzer. Insufficient flows will result in oxidation and excessive flow will blow out the filler in the brazing process.

Full-flow shutoff valves shall be provided to isolate medical gas piping sections for maintenance or testing. The valves shall be three-piece construction and all valves, except zone valves, must have lockable handles or be located in secure areas. All valves shall be placed in a secure area for authorized personnel only or locked open when not in identifiable valve boxes. A public valve on control-valve boxes shall have a frangible cover and be clearly identified with the gas and area it controls. New or replacement valves shall be quarter-turn ball valves with an indicating handle that has extensions for brazing purposes and shall be open port. Medical gas shutoff valves are location specific and identified as source valves, main line valves, riser valves, zone valves, or in-line control valves.

Gauges

Pressure and vacuum gauges are important assets in the performance monitoring of all medical gas vacuum systems. Pressure gauges should be labeled in psi, and vacuum gauges should be labeled in inches of mercury. A pressure and/or vacuum gauge shall be installed in the main line adjacent to the actuating switch, shall be labeled, and shall be readily visible from a standing position. A pressure and/or vacuum gauge shall also be installed at the alarm panel of each area, downstream of each zone valve, and at each receiver.

Alarms

Alarms with actuator switches are required to provide notification of operational deficiencies in medical gas piping. Alarms include master alarms, main and equipment source alarms, area alarms for gas pressure at specific areas, and local alarms for conditions of equipment. Master alarms shall consist of at least two panels wired in parallel and located separately. The primary monitor shall be in an area where 24-hour surveillance is provided. Area alarms are required for critical-care areas and anesthetizing areas indicating 20 percent above or 20 percent below operating pressure. The alarms are usually found at the nurses' station. Local alarms that monitor the condition of equipment in a source room can be singular and shall send a signal to the master alarm panel.

Dental Air

Dental air requirements are different from level-one systems because some areas do not require duplexed components.

MEDICAL SYSTEM SAFETY

Maintaining conformance to NFPA 99C in a medical gas system is extremely important to safeguard patients in a medical setting. Consider that patient could very well be you or a family member.

REVIEW QUESTIONS

1. What are the concerns in healthcare structures?
2. What is the minimum size of a sterilizer local vent pipe?
3. Can clinical sinks be used as a substitute for service sinks?
4. What does NFPA 99C specifically address in the plumbing code?
5. Why are two medical air compressors provided in level-one medical gas systems?

CHAPTER

40 Plumbing System Tests and Individual Sewage Disposal Systems

LEARNING OBJECTIVES

The student will:

- Identify the different portions of the plumbing system that require inspection tests.
- Describe how to perform the different system tests and what pressures are necessary for the individual tests.
- Summarize the various components of a private sewage disposal system.

PLUMBING SYSTEM TESTS

The testing of plumbing systems is extremely important. The code requires systems be left open and tested to verify that they comply with the code and operate in a safe manner as designed. Inspectors first visually inspect the system, whether rough in, finish, or other system portions, and then the inspectors witnesses applicable testing. Inspectors never conduct the testing; that is the responsibility of the permit holder and/or technician and they must furnish the equipment, material, and labor required for testing.

Rough Plumbing Tests

Following their completion, plumbing drainage and venting systems shall be tested by water or air and proved watertight. The authority having jurisdiction (hereafter referred to as the inspector) may require the removal of any cleanout plugs for assurance that the pressure is located in all parts of the system under test. Either a water test or an air pressure test may be used.

Water testing may be applied to the entire system or portions of the system depending on the size and height of the system being tested. The openings in the piping shall be closed, except the highest opening, and the system filled with water to point of overflow. If the system is tested in sections, each opening shall be plugged except the highest opening of the section under test, and each section shall be filled with water. All portions of the systems under test shall have at least a 10-foot head of water applied to the connections and pipe. The water shall be kept in the system and held for at least 15 minutes to ensure the system is sealed.

Air tests shall be made by attaching an air-compressor testing apparatus to any suitable opening and after closing all other inlets and outlets to the system, forcing air into the system until there is a uniform gauge pressure of 5 pounds per square inch or sufficient to balance a column of mercury 10 inches in height. This pressure shall be held without introduction of additional air for a period of at least 15 minutes in order to ensure the system does not have any leaks. See Figure 40–1. Plumbers often select the air test because of the inability to obtain water on construction sites and because in freezing climates the water used in a test would have to be completely removed to protect the system from breakage due to freezing.

In the Field

Important consideration should be given to the gauges used in systems testing. For example when testing a DWV system with air (5 psi), you should not be using a 100-pound gauge. A small amount of pressure loss would not be visible to the technician conducting the test. Some plumbing codes specify the gauge shall reflect the pressure in increments in accordance with the test pressures, such as:

1. Up to 10 psi shall have increments of $\frac{1}{10}$ pound per square inch.
2. Tests between 10 and 100 psi shall have increments of 1 pound per square inch.
3. Tests greater than 100 psi shall use increments of 2 pound per square inch.

Finished Plumbing Tests

Completed plumbing systems may be required to demonstrate that the final fixture connections to the drainage system are gas-tight following visual and operating systems testing. Gas-tight testing is conducted after the plumbing fixtures have been set and their traps filled with water. A final smoke or peppermint test is the method used to ensure connections are gas-tight.

A smoke test is conducted by filling the entire system with a thick, pungent smoke produced by smoke machines. When the smoke appears at stack openings on the roof, they are closed and a pressure equivalent to a 1-inch water column shall be developed and maintained for the inspection.

When the smoke test is not conducted due to practical difficulties, a peppermint test may be substituted. Peppermint tests are conducted by inserting 2 ounces of oil of peppermint into the roof terminal of every line or stack to be tested. Ten quarts of 140°F water is then placed into each stack terminal and sealed. A test failure, which reveals leakage, shall be the detection of the odor of peppermint at any trap or other point on the system. Individuals who have placed the materials on the roof shall not be the same individuals monitoring the connections.

Sewer Tests

The building sewers, whether storm or sanitary, shall be tested by insertion of a test plug at the point of connection with the public sewer, private sewer, individual sewer disposal, or other point of disposal. The sewer is then filled with water to a head no less than 10 feet. The water level at the test head shall not drop for at least 15 minutes. There are occasions when air tests might be used. Occasionally a sewer is installed in wet locations, which may involve soil and trenching dewatering. The lines are then compacted and filled in except for the joint connections. If water is present around those connections, air can be a very useful method because it shows leaks by the presence of bubbles. Some codes state that where the final connection of the building sewer cannot reasonably be subjected to a hydrostatic test, it shall be visually inspected. Pressurized sewers, commonly smaller in size than normal sewers and often called force mains, are tested by 5 psi above the working pump rating.

In the Field

Several projects may have extensive lengths of sewer and may be very large. It is common in those cases to test from manhole to manhole. A great deal of force can accumulate in those cases and individuals must be careful when removing the system thrust blocks while removing line plugs in order to drain the line after head testing. Research into construction accidents shows test flanges or plug blow offs in manholes or large meter pits will result in loss of life.

Example

A rough, approximate example regarding force will be the consideration of a 4-inch ID pipe, and an 8-inch ID pipe with a pressure of 5 psi established by water head. You may complete the math to verify the numbers if you wish.

The formulas for consideration are:

Area = 3.14 (π) $\times$ radius square
Thrust (the force) = pressure (5) $\times$ area
A 4-inch line will have 62.8 pounds of thrust
An 8-inch line will have 251.2 pounds of thrust

Water Supply Systems Test (Test Gauges Soapy Water)

Water distributions, commonly referred to as the water supply system test, shall be conducted by the technician following completion of the system to prove the system is watertight with no leaks. It shall utilize a water test with pressure no less than the system's operating pressure or 80 pounds per square inch, whichever is greater. When a clean, potable water source is not available, an air pressure test for metallic pipe may be used to meet the water system pressures in accordance with the manufacturer's recommendations. Testing by compressed gas or air pressure is prohibited in plastic pipe. Piping shall be disinfected after testing prior to in-service applications.

In the Field

Air tests are not permitted for plastic piping because they are too dangerous and are discussed in the manufacturers' recommendations.

Existing Systems Tests

The *National Standard Plumbing Code* provides important clarifications for plumbing and drainage systems in existing structures by stating that they shall be maintained at all times in compliance with the provisions of the code, and that existing plumbing installed under prior regulations may remain unchanged unless hazards are evident. While the language addresses performance concerns, the codes are understood to apply to new construction or remodeling. This concept is challenged by other codes, such as the *International Plumbing Code*, when they list the frequency of inspections for backflow preventers. Most authorities having jurisdiction will abide by the safe drinking water requirements of the local water purveyor's cross-connection control program.

INDIVIDUAL SEWAGE DISPOSAL SYSTEMS

Individual sewage disposal systems are required in the absence of a public sewer system. These sewage disposal systems are generally for a single structure or small group of buildings under the same ownership control. Often the local health department has control over these disposal systems. The *National Standard Plumbing Code*, Chapter 16, Regulations Governing Individual Sewage Disposal Systems for Homes and Other Establishments Where Public Sewage Systems Are Not Available,

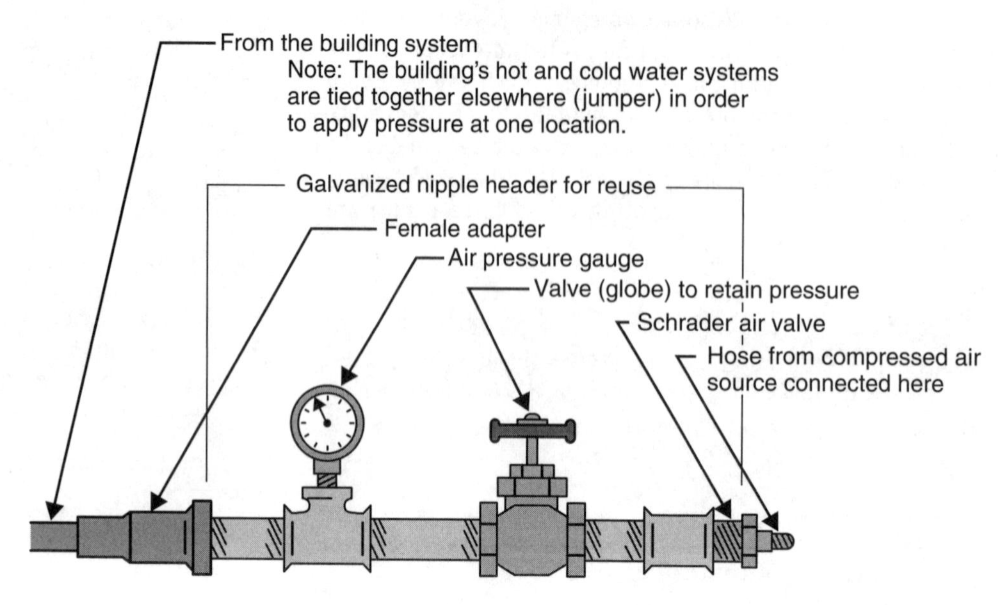

Figure 40–1
An air-pressure test assembly may be used in winter conditions where a water test in an unheated building may be impractical. Courtesy of the International Code Council.

provides excellent direction for use and maintenance of private sewer disposal systems. The code chapter is applicable in the absence of state or other local laws. The authority having jurisdiction generally adopts the codes by reference through legislative processes. The remainder of this chapter will outline those requirements in the orderly manner used in the code. Other resources are local county health departments and the *International Private Sewage Disposal Code*.

The code defines sewage disposal as all private methods of collecting and disposing of sewage, including septic tanks. Sewage shall be disposed of by an approved method of collection, treatment, and discharge. Sewage or sewage effluent shall not be disposed in a manner that will cause pollution. Further, water-carried sewage from structures shall pass through a septic or other approved sedimentation tank prior to its discharge into the soil or into a sand filter. When underground disposal or sand filtration is not feasible, considerations can be made for special methods such as a pump-and-haul system. The code frequently references domestic sewage and does consider other establishments.

System Design

The designer of an individual sewage disposal system shall consider the location with respect to wells, water table, soil characteristics, available area, and occupancy of the building. There are several different types of systems based on soil permeability. Systems consist of a septic tank discharging into either a subsurface disposal field, (commonly called a conventional absorption system), one or more seepage pits (commonly called cesspools), or into a combination of both. Elevated pressurized drain fields (commonly called pressure distribution systems or mound systems) have been recently utilized in some cases. Clear water waste from subsoil or roofs shall not be directed to disposal systems. Important considerations here will also include water treatment discharges, which may contain brine. There is a great difference of opinion in this matter; check with your local authorities prior to design and installation.

The code contains a design criteria table, which lists several types of structures with projected-use discharges per day. A code table assists in the design consideration for sewage flows. For example, an assembly hall is listed at 2 gallons per seat in the code table.

Location of Systems

The code provides a table that lists the minimum required design distances for various components of the disposal system. For example, a septic tank shall not be closer than 50 feet to a deep well and components of a disposal system shall not be within 10 feet of the property line. General principles state that systems should be located at the lowest point on the premises, consistent with the general topography and surroundings; disposal facilities shall not be located on any watershed for a public water supply system.

> **In the Field**
>
> Topography elevation problems may be overcome by sewage pumps, which can pump effluents in a system to different elevations than others.

Percolation Test

Standard percolation tests are one of the most important steps when determining the acceptability of soil conditions for a drain field's absorption capabilities. The test shall be performed prior to the system's design.

The following percolation test method is provided for your consideration:

1. A test pit shall be prepared 2 feet square and at least 1 foot deep. A hole 1 foot square and 1 foot deep shall be prepared in the test pit.
2. The hole shall be filled with water to a depth of 7 inches for pre-wetting purposes, the water level shall be allowed to drop 6 inches before recording is started.
3. Water is added to the hole, raising the depth back to 6 inches. and the amount of time it takes to be absorbed in the hole is measured. The time required for the water level to drop 1 inch in depth, from 6 inches to 5 inches, shall be noted.
4. The recorded time shall be used in factoring the amount of drainage pipe to be used, along with other considerations such as the type of piping material and gravel bed information. The code further states that in no case shall fewer than 100 feet of tile be installed when 1-foot trenches are used.

Seepage pits (often called dry wells) are also an option, although they have fallen out of favor recently because of safety concerns. The code contains a method for determining a percolation test for seepage pits, which is very helpful. To begin, prepare a 5-foot pit, then perform a percolation test at the bottom of the pit, and finalize the pit diameter and effective square footage of absorption area for the pit based on the test results.

Capacity of Septic Tanks

Septic tank capacities are identified in the code by consideration of the type of structure and load based on the number of occupants, use group, bathrooms, and fixtures served. The code then provides sizing details; for example, when tanks have more than one compartment, the inlet compartment shall have a capacity no smaller than two-thirds of the total tank capacity. The tank's construction materials are also clarified. For example, steel tanks are no longer accepted because they have a short life expectancy because of rusting.

Concrete tanks shall have a 24-inch manhole with a handle or be cast in three or four sections cemented in place. The wall thickness shall be at least $2\frac{3}{4}$ inches and the top and bottom shall be 4 inches thick. All tank walls and bottoms shall be reinforced with approved materials and be watertight.

The effluent discharge of septic tanks shall be disposed of underground by subsurface irrigation or seepage pits or both. A distribution box to disperse the effluent is required when more than one line of subsurface irrigation or more than one seepage pit is used. The code has greater details on distribution chambers. Very large systems use dozing chambers, which have large bell caps over large traps serving different drain fields. The systems operate off a principle of self-siphoning to automatically cycle between the different fields.

Seepage Pits and Sand Filters

Seepage pits may be used either to supplement the subsurface disposal field or in lieu of such a field where conditions favor the operation of seepage pits, if they are accepted by the authority having jurisdiction. Seepage pits shall not penetrate the water table.

A sand filter consists of a bed of clean, graded sand on which septic tank effluent is distributed by means of a siphon and pipe, and the effluent percolates through the bed to a series of under drains through which it passes to the point of disposal. The filter size is determined on the basis of 1.15 gallons per square foot per day if covered, and 2.3 gallons per square foot per day if an open filter is used. The code offers construction details for both these methods.

Absorption Trenches

Absorption trenches shall be designed and constructed on the basis of the required effective percolation area. The following direct quote is taken from the *National Standard Plumbing Code* in section 16.9.2, Filter Material,

> The filter material shall cover the tile and extend the full width of the trench and shall be not less than 6 inches deep beneath the bottom of the tile, and 2 inches above the top of the tile. The filter material may be washed gravel, crushed stone, slag, or clean bank-run gravel ranging in size from $\frac{1}{2}$ to $2\frac{1}{2}$ inches. The filter material shall be covered with burlap, filter cloth, 2 inches of straw, or equivalent permeable material prior to backfilling the excavation.

The size and minimum spacing layout requirements for absorption fields are detailed in the code with a clarification that the length shall not exceed 100 feet. Excessive length will increase the opportunity for one particular area to fill with infiltrated sand, for example, and render a large extent of the line inoperable. Limiting sorter runs will increase the system's operability. The absorption lines shall be constructed of 4-inch, open-jointed or perforated pipe composed of several different materials. Finally, the trench bottom shall be uniformly graded to slope from a minimum of 2 inches to a maximum of 4 inches per 100 feet.

An alternative to the elevated pressurized system with sand filters is the tee trench. Applications of tee trenches require the nonporous soil in trench layouts to be removed so there is only acceptable, porous soil. The trench is then backfilled with stone, gravel, or sand up to the drain field piping.

In the Field

Technicians must always consider sewage treatment systems while designing drain exits from a structure. The basic consideration is to keep the sewer line from the structure as high as possible, which enables the drain field to be closer to the terrain surface. Drainage water not only is absorbed by the soil for dispersion into the ground but some of the absorbed water is evaporated through the surface. This surface evaporation reduces the load on the ground's expected absorption.

REVIEW QUESTIONS

1. What is a percolation test?
2. What is the alternative when public sewers are not available?
3. When do the inspectors conduct inspection tests with their test equipment?
4. Why is it important to keep the drain exit from a structure as high as possible?
5. Why are traps of completed plumbing systems required to be filled with water when conducting smoke tests?

CHAPTER

41

Potable Water Supply Systems

LEARNING OBJECTIVES

The student will:

- Contrast the different well-drilling methods.
- Identify the different pumping devices employed by the code.
- Describe the various code requirements for the different components of a well-supplied water system.

POTABLE WATER FROM PRIVATE SUPPLY SYSTEMS

The potable water supply service and distribution systems provided by public/municipal sources were addressed in Chapter 39. This chapter will address private water supply systems serving one or more structures used for human occupancy. Those systems are generally lumped into one category, well supply systems. Local health departments generally exercise control over well supply systems because they were in place long before the local department of public works (DPW) had public systems and other programs dictated by the Safe Drinking Water Act. This, however, is not true in large metropolitan cities. The following information from the *National Standard Plumbing Code* is provided where the authority having jurisdiction has not established their own requirements nor amended the code in this matter. The authority having jurisdiction generally adopts the codes by reference through their legislative process.

Wells are typically bored, drilled, or driven depending on the soil conditions above the water aquifer to the water table. Bored wells are made by the use of an auger and have well casings added after completion. Drilled wells are required where the earth is rock or too hard to be augured. Driven wells are pushed down (driven down) using force, generally in shallow well cases. The depth is highly dependent upon the area. Figure 41–1 depicts the hydrological cycle, which is critical for private water supply systems. Also see Figure 41–2, which shows how a well system functions in a private residential setting.

Pumps for wells with their various components are our primary concern, but some locations have pumps installed in springs or cisterns that comply with applicable rules and regulations, because of the water table and earth conditions around the well. The term *earth conditions* is used rather than *soil* because it encompasses rock formations. Accumulation cisterns are used in these conditions and should not be confused with rainwater cisterns.

Quantity of Water Required

Regardless of the source, the codes mandate that safe potable water shall be supplied in proper amounts for its occupants. The code clarifies what amounts of water are proper. The minimum capacity in gallons per minute for a single dwelling shall equal the number of fixtures installed. Structures other than a single dwelling shall be designed to meet code sizing tables for water distribution that deal with water supply fixture units (WSFU) and the table converting WSFU values to gallons per minute (NSPC Chapter 10). The supply for structures other than single dwellings shall be capable of supplying the maximum demand according to usage and for no less than a minimum period of 30 minutes.

The requirements listed above are impractical for a well pump on its own, which leads the code to a practical amendment: when the available primary source of water does not meet the minimum requirement, additional water storage facilities shall be provided. Three methods are listed: a pressure tank of sufficient size, a gravity tank, and a two-pump system. The two-pump system is a combination of the first and second methods. The supply pressure shall deliver adequate flow rates for fixture operations.

In the Field

Some code agencies are involved in the licensing of plumbers. There are several jurisdictions where licensing programs do not include licensed well drillers and pump installers. When that is the case, jurisdictions may amend their code to exclude wells and the associated piping. In such instances, jurisdictions generally start their authority at the pressure-tank valve and classify piping as water-distribution piping.

Piping Materials

Piping from the well to the inside of the dwelling shall meet the requirements for water service piping. The code lists the acceptable standards for metallic and nonmetallic water service products. Plastic piping shall meet NSF 14, and all materials shall be water-pressure rated at no less than 160 psi at 73°F.

Storage Tanks

Storage tanks are an important part of well supply equipment and shall be certified under Water Systems Council standards for size and pressure rating. All tanks shall be coated or made of material that resists corrosion, and tanks shall be constructed

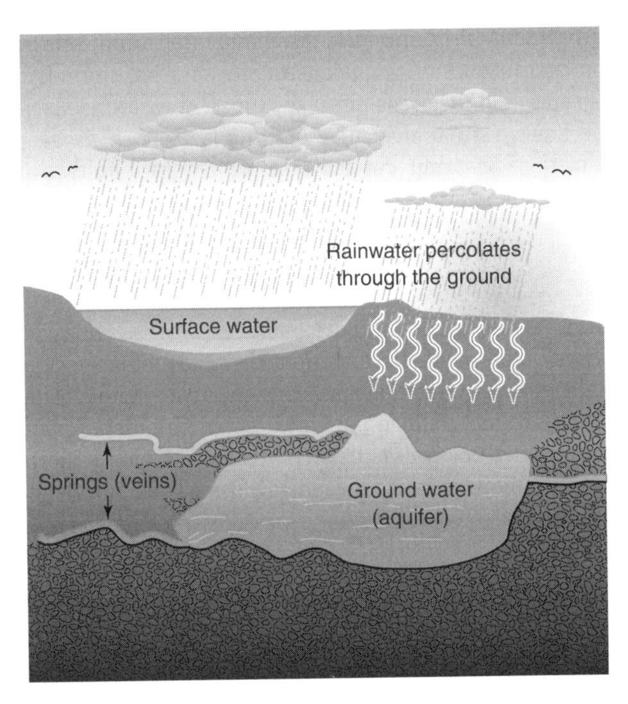

Figure 41–1
Groundwater is replenished by water that cycles through the environment in a process referred to as the hydrological cycle.

of materials and/or coatings that are non-toxic. The hydropneumatic tanks shall have a working pressure rating in excess of the maximum required system pressure. The tanks shall be provided or piped with a method for draining.

PUMPS

The pumps shall also be certified under Water Systems Council rating and rating standards. They shall be installed in accordance with the manufacturer's recommendations, and the installing technician must provide assurance that the pumping equipment is correctly installed to prevent the entrance of contamination or objectionable material either into the well or into the water that is being pumped. Pumps shall be located properly to provide clearance for maintenance and repair, including overhead clearance for removal of the drop pipe and other accessories. Care shall be taken in mounting pumps to avoid objectionable vibration and noise and to prevent damage to pumping equipment.

Well pumps include shallow-well jet pumps, deep-well jet pumps, and submersible pumps. Jet pumps circulate water through an ejector fitting that creates a vacuum and draws water from the well. Shallow-well jet pumps are used for wells fewer than 25 feet deep, and the ejector fitting is located outside the well. Deep-well jet pumps can be used for wells up to 75 feet deep, and the ejector fitting is located within the well casing. Some jet pumps are convertible and can be used for installations of either one-pipe shallow wells or two-pipe deep wells, although their rate capacity is reduced for deep wells.

The design and operating principles of each type of power-driven pump determine where each may be located with respect to a well. The location selected for the pump determines what factors must be considered to make an acceptable installation.

Pumps themselves also have code-addressed concerns. For example, power-driven pumps with bearings that require lubrication by water or oil shall be protected so the quality of water pumped is not adversely affected. To protect pumps, a separate structure housing the water supply and pumping equipment

In the Field

Some authorities having jurisdiction will not accept shallow-well installations due to established well-depth criteria intended to protect against the increased likelihood of groundwater contamination.

shall have an impervious floor and rain-tight walls and roof. Several agencies now prohibit installations in outside well houses, the small shelters with a waterproof roof just above grade. The concerns primarily arose in resort areas near lakes where owners may not have a great deal of storage space, such as garages or storage sheds. Owners were not using common sense safety habits and often stored gasoline or insecticide in outside well houses to keep containers out of the elements.

Controls

The code addresses pump controls and accessories by stating that they shall be protected from the weather. It then lists the controls and accessories including a pressure switch, thermal-overload switch, pressure-relief valve on positive-displacement pumps, and low-water cutoff switch where the pump capacity exceeds the source of water.

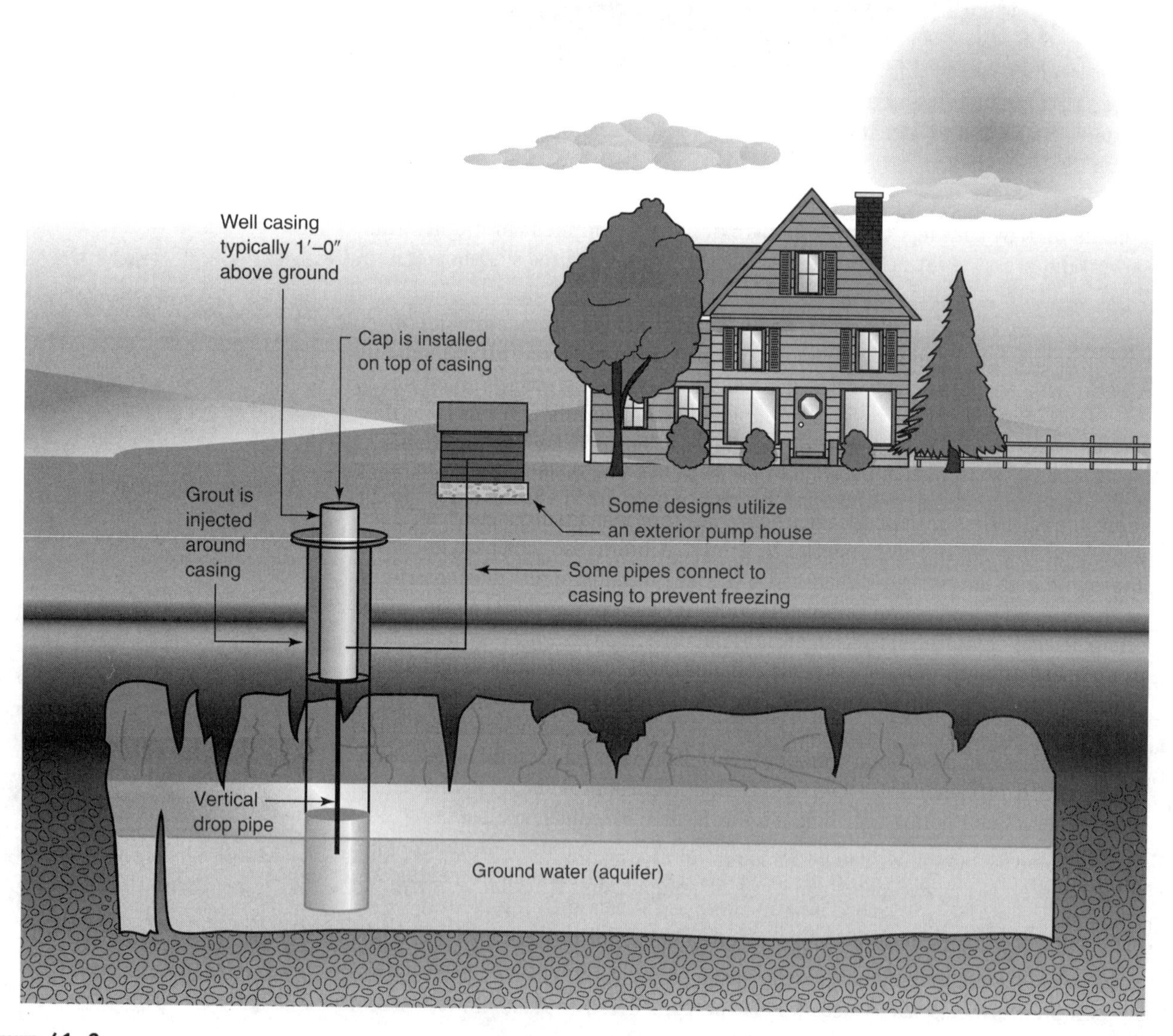

Figure 41–2
A well is an opening in the ground whereby piping can be inserted and pumps can be used to extract the groundwater.

WELL TERMINAL

Well casings, curbs, and pitless adapters shall terminate no fewer than 8 inches above the finished ground surface or pump house floor and at least 24 inches above the maximum high-water level where flooding occurs. Casing shall be cut off or cut in below ground level, except when installing a pitless adapter. Pitless adapters designed to replace a section of well casing or for attachment to the exterior of a well casing shall be constructed of materials that provide strength and durability equal to the well casing. The connection installation shall be by threaded, welded, or compression-gasketed to a cutoff casing and shall be watertight. Adapter units designed to replace a section of the well casing shall extend above the finished ground surface, and the top of the adapter unit shall be capped with a cover that has a downward flange to overlap the edge. The cover shall be securely fastened to the unit and be vermin proof and watertight for electrical cables and vent line, if installed.

In the Field

Pitless adapters allow a connection from the submersible pump's drop pipe to the line supplying the water-pressure storage tank.

Consideration of all pump devices includes hand pumps. Hand pumps shall be force-type equipped with a packing gland around the pump rod, a closed delivery spout directed downward, and a one-piece bell-type base that is part of the pump stand or is attached to the pump column in a watertight manner. The bell base of the pump shall be bolted with a gasket to a flange that is securely attached to the casing or pipe sleeve.

Any power-driven pump located over a well shall be mounted on the well casing, pipe sleeve, pump foundation, or pump stand to allow a watertight closure to be made for the open-end of the casing or sleeve. The pump base, which is bolted with a neoprene or rubber gasket or an equivalent watertight seal to a foundation or plate, provides an acceptable seal unless the pump makes insufficient (not watertight) contact with the gasket. When pumps are not located over the casing and suction pipes emerge from the top, a watertight rubber seal or the equivalent shall be installed between the well casing and piping to provide a watertight closure.

Underground discharge lines serving submersible pumps shall have a sealed pitless adapter and a check valve shall be installed in the discharge line above the pump in the well. Submersible pumps with a discharge line that leaves the well at the top of the casing shall be sealed watertight. When the discharge line remains above ground with a slope draining back to the well, the check valve can be located beyond the well.

Pumps offset from the well, if not located in an above ground pump house or other building, may be located in an approved basement, provided the pump and all suction pipes are elevated at least 12 inches above the floor. These buried lines below the ground surface between the well and the pump shall be located the same minimum distance from sources of contamination as prescribed by code regulation tables for sewage systems. If the distances cannot be obtained, the suction line shall be enclosed in a protective pipe sealed at both ends.

VENTS

Systems, which have vents, shall be of adequate size to allow equalization of air pressure in the well and shall not be less than $\frac{1}{2}$ inch in diameter. Vent and pump locations shall be out of the vicinity of flammable or toxic materials. If either of these types of materials is present, the vents shall be extended to discharge outside of the building, where they will not be a hazard. Vent terminals shall end in a downward direction covered with no fewer than sixteen mesh screen wires.

CROSS CONNECTION

Cross connection between an individual water supply system and other individual or public water supply systems are prohibited.

CONSTANT-PRESSURE SUBMERSIBLES

A new product, constant-pressure submersibles, deserves some attention. Systems employing these products begin with a variable-speed, constant-pressure submersible pump, which claims to provide greater volumes of water at a steady pressure. The constant pressure is achieved by varying the speed of the pump motor. When there is more demand for water, the pump speed will increase, and when there is less demand, the pump speed will decrease. A pressure transducer wired to a controller that sends a signal to the pump maintains the constant pressure. A 2-gallon storage tank is mounted on the wall with the control, rather than a much larger freestanding storage tank.

Advocates of the technology state the advantage of the system is the reduced wear and tear on the electric pump motor in the submersible pump. The motor will not be subjected to high-amperage startup torque demands, which will extend its life. Average users can relate to this simply by standing in a mechanical room and observing a conventional pump at its storage tank and pressure-switch unit.

REVIEW QUESTIONS

1. Define a pitless adapter.
2. What are the three main types of pumps?
3. Name the three methods typically used to obtain or make a well from the surface to the aquifer water table.

CHAPTER 42

Mobile Home and Travel Trailer Park Plumbing Standards

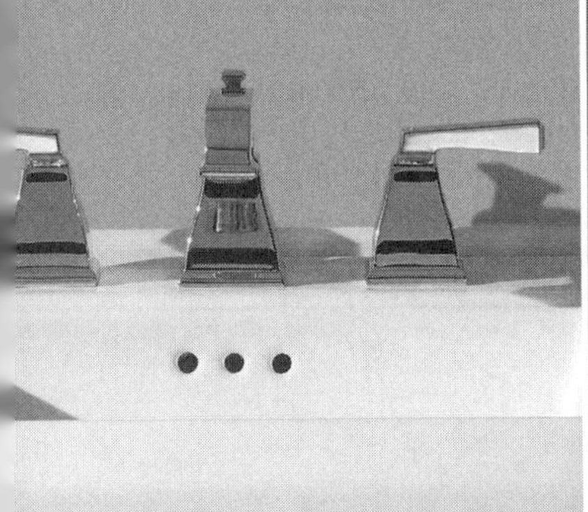

LEARNING OBJECTIVES

The student will:

- Discuss required design distances.
- Identify the type of waste and venting system details used to ensure proper waste flow in the drainage system.
- Interpret where the connections shall be located based on the uniformity requirements for different units.
- Recognize the number of required fixtures for a park's sanitary facilities requirements.

MOBILE HOME AND TRAVEL TRAILER PARK PLUMBING REQUIREMENTS

The requirements for manufactured mobile home and travel trailer parks are provided in the *National Standard Plumbing Code*, Chapter 18, to ensure that sanitary plumbing installations are provided in parks and campgrounds. Travel trailers and recreational vehicles are recognized as temporary dwellings for recreational purposes while factory-built mobile homes are recognized as dwelling units. Factory-built manufactured homes conforming to the Department of Housing and Urban Development (HUD) standards are frequently called mobile homes and can be moved; however, the increased size of most mobile homes and park amenities indicate that moving is not likely. Also, these homes are often located on private property outside a park setting. The authority having jurisdiction governing the establishment and operation of parks varies from state to state. Other homes similar to these HUD homes have standard construction that conforms to conventional building codes and are constructed in manufacturing facilities. They are designed to be set in place on permanent footings and/or walls. These dwellings are referred to as premanufactured units and have different site regulations enforced by the authority having jurisdiction.

STANDARDS/ESTABLISHED RULES

Plumbing systems in parks for premanufactured homes, mobile homes, or travel trailers shall conform to other portions of the code where applicable, and to plumbing code requirements, which are outlined in the following text. Before any plumbing or sewage disposal facilities are installed or altered in any park, duplicate plans and specifications shall be filed, and permits shall be obtained. The plans must include a plot plan of the park, drawn to scale, indicating elevations, property lines, driveways, existing or proposed buildings, and lot sizes. The plans will include piping layouts (both water and waste), which may be former altered systems, and the new site's piping and specifications.

DEFINITIONS

Definitions are necessary to specifically identify the park and its setting. The terms *mobile home* and *trailer* will not be used because the homes may be doublewides, two stories, and so on, and have recently become very large. They are dwellings that can be moved and are constructed on a chassis without a fixed foundation. Recreational units may be vehicles but the category does not include the smaller units previously called trailers. Another term used in place of travel trailer park or RV park is *seasonal mobile home park*. The definitions are provided for your convenience as follows:

1. A *coach* is any camp-car, trailer, or other vehicle with or without motivation power, designed and constructed to travel on public thoroughfares in accordance with provisions of the motor vehicle code and designed or used for human habitation.
2. A *coach, dependent* is a unit that is not equipped with a water closet for sewage disposal.
3. A *coach, independent* is a unit that is equipped with a water closet for sewage disposal.
4. *Coach, left side* is the side farthest from the curb when the unit is being towed, driven, or in transit.
5. *Coach connection fixture* is the connection between a trap and the park drainage system. It receives water, liquid, or other waste discharge from a coach.

6. A *coach drain connection* is the removable extension connecting the coach drainage system to the connection fixture.
7. A *park* is an area or tract of land where space is rented, held for rent, or occupied by two or more units.
8. A *park branch line* is the portion of drainage piping that receives the discharge from no more than two trailer connection fixtures.
9. A *park drainage system* is the entire system of drainage piping used to convey sewage or other waste from a trailer connection fixture to the sewer.
10. A *park sewer system* is the piping that extends from the public or private sewage disposal system to a point where the first drainage system's branch fitting is installed.
11. A *park water service main* is the portion of the water distribution system that extends from the street main, water meter, or other source of supply to the trailer site's water service branch.
12. *Service buildings* are structures that house toilets, laundries, and other facilities that may be required.
13. A *sewer connection* is the portion of the drainage piping that extends as a single terminal under the unit for connection with the park drainage system.
14. A *site water-service branch* is the portion of the water-distributing system that extends from the park's service main to a trailer site, and includes connections, devices, and appurtenances thereto.
15. A *unit site* is the area set out by boundaries on which one trailer can be located.
16. A *water service connection* is the portion of the water supply piping that extends as a single terminal under the trailer coach for connection with the park's water supply system.

SEWER

A park's drainage system must be designed and installed in accordance with the requirements of the code. The drainage system may be designed and installed using a combination waste and vent drainage system, which consists of the installation of waste piping that does not have separately or independently vented traps for one or more unit connections. They are vented through properly sized waste piping to secure free circulation of air. Vents are still required on the end of mains and are called *wet vents* in the venting discussion that follows Table 42–1. Each independent site shall be provided with a trapped connection consisting of a 3-inch-minimum, horizontal, iron, pipe-size threaded connection, installed a minimum of 3 inches and a maximum of 6 inches (from the bottom of the connection) above the finished grade. The vertical connection to the trailer's connection fixture shall be anchored in a concrete slab 4 inches thick, and 18″ × 18″ square. The portion of the plumbing system extending above the ground shall be protected from damage when deemed necessary.

Table 42-1 Drain Pipe Sizing. Courtesy of Plumbing–Heating–Cooling–Contractors—National Association.

Maximum Number of Units, Individually Vented Systems	Maximum Number of Units, Wet-Vented Systems	Size of Drain
2	1	3″
30	10	4″
100	50	6″
400	—	8″
1000	—	10″

Traps shall be located based on the immediate boundary lines of the designated space or area within each site that will be occupied by the unit. Traps shall be located in the rear third-quarter section along the left boundary line of the parking area; be no less than 1 foot and no more than 3 feet from the roadside; and be a minimum of 5 feet from the rear boundary of the trailer site. These specifications may vary by permission of the authority having jurisdiction when unusual conditions are encountered, often dependent on the construction of the mobile home. All the traps, the upper 5 feet of any horizontal vent, and the first 4 feet of any trap branch shall be fabricated from materials approved for use within a building. Cleanouts shall be provided according to requirements of the plumbing code, and cleanouts shall be provided in the vent stacks 1 foot above grade.

Traps at the base of risers that serve the units are discussed by the code and often installed in older units that may not have been trapped and vented. Without trapped risers, the park's sewer gases would directly enter the home. Modern units are constructed with trapped and vented systems, and may not require a system trap. You may encounter several designs in which the registered design specialist for a park does not include traps.

Drain connections for manufactured homes and recreational units shall be made of approved, semi-rigid or flexible, reinforced hose with smooth interior surfaces, no fewer than 3 inches in diameter. Main connections shall be equipped with a standard quick-disconnect, screw- or clamp-type fitting, not smaller than the outlet. These connections shall be gas-tight and no longer than necessary to make the connection between the coach's drain connection and the site's connector fixture. Each unit's connection outlet will have a screw-type plug or cap, and be effectively capped when not in use.

To determine pipe sizes, each unit's connection shall be assigned a waste-loading value of 6 drainge fixture units, and each drainage system in the park shall be sized using Table 42-1. The slope of sewers shall provide a minimum velocity of two feet per second when the pipe is flowing half full, which is the same discharge rate in conventional structures. The discharge of the park's drainage system shall be connected to a public sewer. If a public sewer is not available for use within 300 feet, an individual sewage disposal system of a type that is acceptable and approved by the authority having jurisdiction must be installed.

VENTING

The drainage system shall have wet vents no farther than 15 feet downstream from its upper trap, and long mains shall be provided with additional relief vents at intervals of no more than 100 feet. The minimum size of each vent is identified in Table 42-2.

All vent connections shall be removed above the center line of the horizontal pipe. All vent stacks shall be supported by a 4″ × 4″ redwood post, set in at least two feet of concrete that extends at least 4 inches above the ground, or supported by another approved method, such as steel posts. Vent pipes, which are made of galvanized steel, may extend below the ground vertically and directly connect to the drainage line with an approved fitting, if the entire section around both the drain and the galvanized pipe is encased in concrete to prevent any movement.

Table 42-2 Vent Sizing. Courtesy of Plumbing–Heating–Cooling–Contractors—National Association.

Size of Wet-vented Drain	Minimum Size of Vent
3″	2″
4″	3″
5″	4″
6″	5″

A galvanized-steel pipe encased in concrete shall be coated with bituminous paint or an equivalent protective material.

These vent pipes shall be at least 10 feet above grade and 10 feet from the property line. Vents shall not terminate horizontally within 10 feet of any door, window, or ventilating openings of the building, unless it is 2 feet above the top of the openings. All vent stacks in a wet-vented system shall be 3 inches or more in diameter; however, a 3-inch branch line may be vented by a 2-inch vent.

Unless they are properly vented, 3-inch branch lines shall not exceed 6 feet in length, and 4-inch branch lines shall not exceed 15 feet in length.

In the Field

With regard to mobile homes and travel trailer parks, *wet-vented* means the riser (vent stack) extends directly off the top of the main and is washed by the flow on the bottom of the main, hence the wet vent.

WATER DISTRIBUTION SYSTEM

Each park's water distribution system shall conform to the requirements of the plumbing code for water supply and distribution, with residual pressure of at least 20 psi at each unit site. Each site shall have at least a $\frac{3}{4}$-inch water service branch line. Each trailer outlet on the water distribution system shall be rated at 6 water supply fixture units.

A control valve shall be installed on the water service branch, and a backflow preventer that conforms to ASSE 1024 or CSA B64.6 (a dual check-valve backflow preventer) shall be installed on the discharge side of the control valve. The backflow preventer will have with a pressure relief valve on its discharge side, and a hose connection will be installed on the unit side of the relief valve. The backflow device and relief valve shall be at least 12 inches above grade; the pressure relief valve drain shall be at least 6 inches off the ground with more than 2 feet above grade between the pressure relief valve and the trailer it serves.

The service connection shall not be rigid; flexible metal tubing is permitted. Quick-disconnect fittings that do not require special tools or knowledge to install or remove shall be located at either end. Each unit's water-outlet supply unit shall be located near the center of the left side of each parking spot.

The design of the park's water distribution system shall provide fire-outlet stations (hydrants) for emergencies.

SANITARY PARK FACILITIES

The code requires sanitary public facilities for all parks, but the numbers vary based on the type of units in the park. Separate public restrooms shall be provided with water closets, lavatories, and showers in accordance with the following ratio of unit sites:

1. Parks constructed and operated exclusively for dependent units (a dependent unit is not equipped with a water closet) shall have one water closet, one shower, and one lavatory for each ten sites or fractional part thereof.
2. Parks constructed and operated exclusively for independent units (an independent unit is equipped with a water closet) shall have one water closet, one shower, and one lavatory for each one hundred sites or fractional part thereof.

Some parks may be combined for dependent and independent units, such as those near lakes. The code has detailed sizing criteria for those in the form of a table. See Figure 42–1 for an example.

Showers

Every park shall have shower facilities with hot and cold running water in separate compartments. The compartments shall be provided with a self-closing door or waterproof curtain. The compartment walls can be cement, concrete, metal, tile, or other approved waterproof materials and must extend to a height of at least 6 feet above the floor. Shower floors shall be concrete or other waterproof material and be sloped $\frac{1}{4}$ inch per foot to the drains.

Laundry Facilities

Every park shall be equipped with an accessory utility building containing at least one mechanical washing machine or laundry tray served with hot and cold running water. There shall be one machine or laundry tray for every twenty sites or fractional part thereof.

Maintenance

The owner, operator, or lessee of the park or his designated agent shall be responsible for the maintenance, and all required devices or safeguards shall be maintained in good working order.

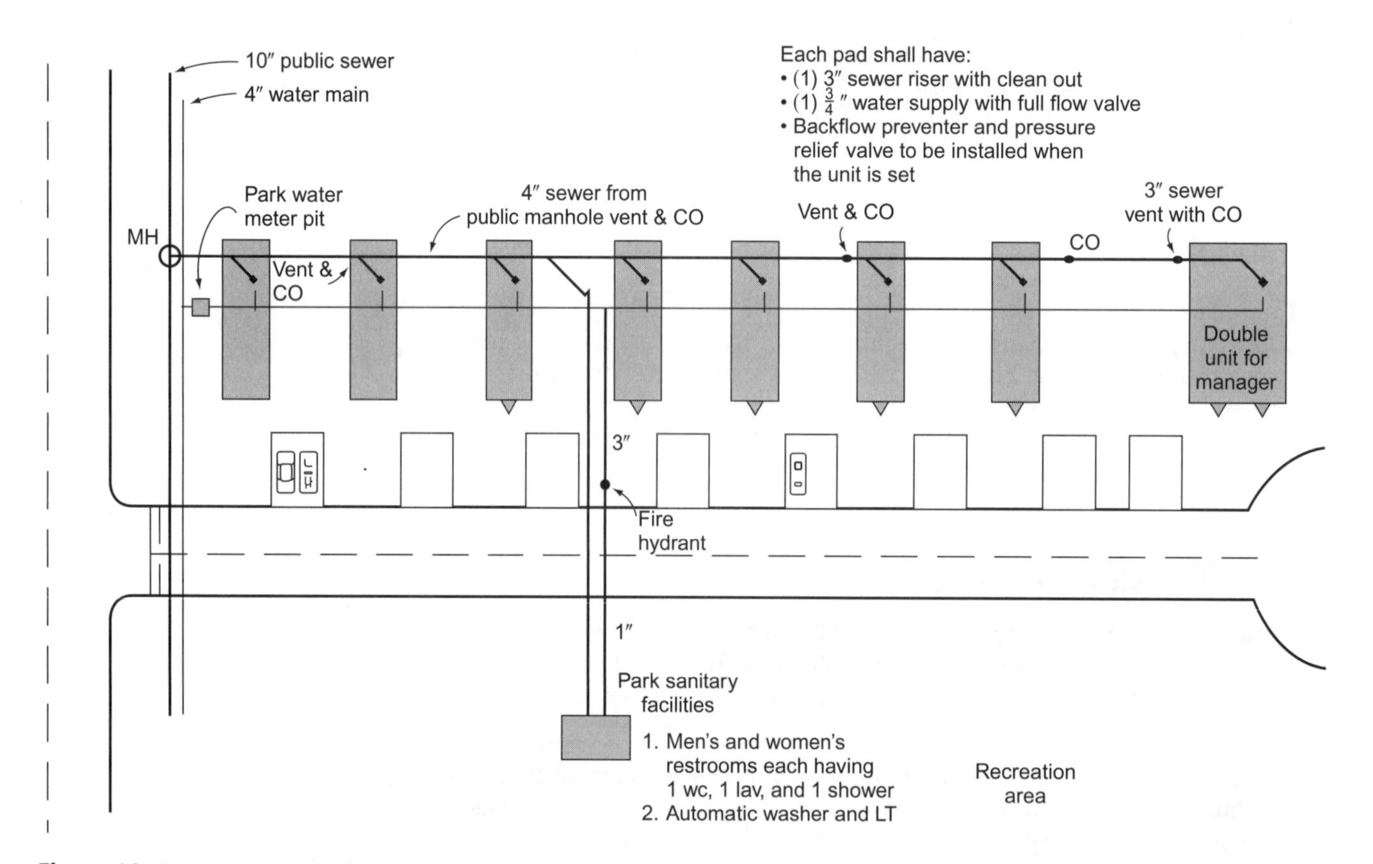

Figure 42–1
Example of requirements for a small eight-unit park for individual living units. Note that a building with sanitary facilities for residents near the public recreation area is provided.

REVIEW QUESTIONS

1. What does the designation of *mobile home* indicate, according to HUD standards?
2. Why are parks required to have sanitary-facility structures with laundry facilities?
3. Are mobile homes always located in parks with rental spaces for the various units?
4. Are all mobile home parks now designed to have sewer system traps at the base of the riser serving the unit?
5. Why is it of concern that the left side is the side farthest from the curb when the unit is being towed, driven, or in transit?

APPENDIX A: ANSWERS TO REVIEW QUESTIONS

CHAPTER 1

1. Customer relations.
2. Respect.
3. Diagnose.
4. Professionalism.
5. Replacement.

CHAPTER 2

1. Diaphragm deterioration.
2. Waterless urinals.
3. You can avoid the use of strong chemicals and increase the flushing capabilities.
4. Compression-type single bibb.
5. It facilitates a quick and easy repair.

CHAPTER 3

1. Water-supplied and drain-supplied trap seal primers.
2. The occurrence of weapons is reduced providing greater safety, and replacement costs are lowered.
3. Reduced touching of handles promotes better sanitation.
4. Safety concerns are paramount because the atmosphere may not support life.
5. Personal cleanliness is practiced during service work to protect the employee's health and safety.

CHAPTER 4

1. Identification enables the technician to make changes in the repair to avoid future problems.
2. Asbestos is a deadly substance. By law, only authorized persons using approved methods may remove, transport, or dispose of asbestos material. There are severe penalties for violation of this law.
3. The extended depth of installation and type of soil, such as sand, may disguise visible water leaks.
4. The cause of a problem could be attributed to a variety of causes. Similar conditions may result later and cause another similar problem, and your report will allow more effective completion of the repair.
5. The water department may insist you use different repair materials and, in extreme cases, may not let you have access to their valves.

CHAPTER 5

1. Bituminized fiber pipe.
2. Provide proper installation methods including supports, use insulation, use cast iron pipe.
3. Rerouting waste piping over food service areas should not be done because of the risk of leaks.
4. Plastic pipe materials may be more prone to ground settlement due to their flexible nature.

CHAPTER 6

1. Bubble testing provides a visual leak indication at the exact location of the leak.
2. Concern over the contamination of the water supply system. Also, other materials are less expensive and easier to work with.
3. No, it is appropriate for modified or newly installed gas piping systems.
4. Serious hazards.

CHAPTER 7

1. It acts as a sacrificial anode to protect the steel in the tank construction.
2. When replacing circulating fluid in a solar system, especially in colder climates.
3. Cold water will mix with the tank contents above the lowest level. The effect will be to reduce the apparent capacity of the heater because the heated water is significantly cooled by cold water dilution at the top of the water heater.
4. Ball valve.
5. Using the relief valve test lever is the primary method to be sure that the valve mechanism is free-working. Manufacturers recommend operating the valve at least once a year and examining by removal every three years.

CHAPTER 8

1. A plunger will clear stoppages up to the vent connection at the trap arm. Using a plunger first will help identify whether the stoppage is located prior to the vent connection or after it.
2. Chemical drain cleaner may have been used and might splash back. Drains in medical labs or hospitals also might contain infectious materials or medical chemicals that form explosive mixtures when exposed to metals and agitated.
3. Not necessarily. Local codes might dictate who has authority to open a cleanout.
4. A sudden stoppage suggests a solid object in the drain; the slowly developing stoppage suggests a gradual accumulation of soap, hair, scum, etc.

CHAPTER 9

1. Water hammer occurs when a rapidly closing valve is struck by water moving at a moderate velocity, which produces a large pressure increase. The water is reflected off the closed valve and reverses direction, which may produce a hammer-like sound.
2. A larger line size reduces the velocity and, therefore, reduces or eliminates water hammer.
3. Most solenoids are quick-closing devices and stop the water flow, which creates water hammer.
4. The loose hangers in a water distribution system will not add to water hammer, but will increase the noise level and might loosen the fittings.
5. Manufactured devices contain air volume sealed into a bellows or a piston device that prevents the air from being absorbed into the water.

CHAPTER 10

1. The purpose of the device is to activate and stop a vacuum when a tank is drained and siphoned. A vacuum relief valve allows air to enter and even out with atmospheric pressure, while a drained tank under a partial vacuum would allow external atmospheric pressure to damage the drained tank or burn out the elements (now dry).
2. Tempered water might be supplied to hand washing facilities to protect the users such as elementary schoolchildren and individuals who may have lost the sense of temperature in their hands, subjecting them to greater burn risks.
3. The electronic flow mechanisms are practical and hygienic and increase water conservation.
4. Pressure-reducing valves without a bypass seal the structure's system under static conditions and do not allow expanded water to migrate back out to the water service, which may cause the water heater relief valve to discharge.

CHAPTER 11

1. No, slope and line size are also considerations.
2. To provide a material take-off for ordering and installation purposes. This layout indicates the fixture connections and fixture unit values.
3. They are sized for greater DFUs.
4. Yes, vents with greater distances or lines with higher DFU values may be increased in size for proper operation.
5. Increasing the pipe diameter reduces the amount of necessary fall.

CHAPTER 12

1. No, it is very common for the jurisdiction having authority to allow other trades to install sloped-roof drainage systems, which commonly discharge on the ground surrounding the structure.
2. If the two systems were tied together below the roof and the primary had an obstruction far downstream, both systems would be inoperable. This could result in a roof failure.
3. Yes, it can carry about twice the load.
4. $80 \times 80 = 6400$ sq. ft. roof
   $4(80)(2) = 640$ sq. ft. parapet
   6400 sq. ft. $+ 1/2(640)$ sq. ft. $= 6720$ sq. ft.
5. No, plumbing code does not specify because storm drainage systems with their appropriate sizes are based upon the rainfall rates experienced at the location of the building and the roof area of the structure being considered.

CHAPTER 13

1. Peak demand is the maximum amount (rate of flow) of water required at any given time during 24 hours of operation.
2. Knowing that water is corrosive, based upon its chemical composition, pH value, scale-forming tendency, and other factors, helps a designer choose a proper piping material when considering the life expectancy of distribution systems. The water quality may warrant a water-conditioning device such as a softener or iron filter.

3. No, during periods of peak demand, inadequately sized lines will increase the likelihood of negative pressures, which contribute to backflow from sources of contamination.
4. No, the major codes have different water-sizing requirement tables that must be considered, even though some do not address specifics.
5. Equivalent length.

CHAPTER 14

1. Category II, III, and IV venting methods are dependent upon the manufacturer's recommended procedures.
2. 75 percent.
3. The amount of gas burned in an hour.
4. No.
5. 1.5.
6. 5 inches.

CHAPTER 15

1. The air gap is the ultimate protection against backflow.
2. Twice the effective diameter of the discharge pipe.
3. No. When a device is subject to a vacuum, an air gap is required.
4. An air break actually will result in less splashing.
5. Overflow connection.
6. Yes, added equipment means more servicing.

CHAPTER 16

1. Yes, interceptors must be located and installed so that they can be serviced readily.
2. To reduce the amount of piping exposed to the undesirable substance discharged and its effect on the system.
3. To prevent accumulation of material in system lines that retard the flow for proper operation and to prevent unwanted materials from interfering with the proper function of the sewage treatment system.
4. No, recovery devices operate on the principle that the material to be intercepted is light and will float on top of water.
5. No, only the fixtures lower than the public sewer's manhole overflow level must be protected.

CHAPTER 17

1. This arrangement makes servicing easier because the water supply is maintained while the unit is isolated for testing or repairs.
2. The protection of the potable water system is of the utmost importance and the first responsibility of the plumbing professional.
3. Air gap.
4. No, only an RPZ or a proper air gap protects against backpressure.
5. Air gap, RPZ, pressure-type vacuum breaker, atmospheric-type vacuum breaker, and double check-valve assembly.

CHAPTER 18

1. a. Air handler coils.
   b. Baseboard that includes electric, forced air, or hydronic fin tube or cast iron.
   c. Convectors.
   d. Radiant panels.
   e. Radiators.
   f. Registers.
   g. Unit heaters.
2. Heating systems consist of a power plant, distribution system, and terminals.
3. No, they would cause the system to hammer.
4. It prevents complete loss of boiler water if a leak develops in a wet return line. Firing a boiler without sufficient water will damage the boiler and is dangerous.
5. No, steam will diffuse by itself throughout the system.

CHAPTER 19

1. They provide excellent protection against dirt or water hammer in a system to protect delicate trap parts.
2. It allows an alarm light or horn to be wired to alert the boiler operator to the fact that the burner is off due to low water.
3. They allow the air to escape, which encourages steam to enter the radiator.
4. Thermostatic, float and thermostatic, inverted bucket, and thermodynamic types.
5. The steam and condensate must use the same area.

CHAPTER 20

1. The low-water cutoffs will shut off the flow of water should it get below a certain level.
2. A compression tank is necessary to accommodate the volume changes in the system as the water heats up and cools down.
3. No.
4. Air.
5. Outdoor reset.

CHAPTER 21

1. One; two.
2. Yes.
3. As short as possible.
4. Balancing valves.
5. Direct return.

CHAPTER 22

1. Cast iron units were slow to heat up and slow to cool down. One advantage to this was the heat was gradually supplied to the space so that it was not very noticeable. The disadvantage was they were slow to respond to load changes.

2. No.
3. No.
4. Fan coil units consist of several horizontal tubes that allow the water to circulate back and forth across an air stream. The horizontal tubes are connected together with thin fins, usually made of aluminum. The aluminum fins give the coil the necessary surface area to dissipate heat into the building air. A fan is mounted to blow air across the coil. The fan and coil arrangements can be designed to be totally enclosed in metal housing that contains ductwork to mix outdoor air and filter the air as well. They can also be designed for use in large shop areas, where they can throw air for greater distances.
5. Yes.

CHAPTER 23

1. 80 percent.
2. Condensation.
3. Computerized control boards.
4. Humidifier.
5. Psychrometric chart.

CHAPTER 24

1. A solar collector receives radiation energy from the sun to heat water in the collector.
2. Insulated storage tank.
3. Cooler.
4. 100 percent.
5. Leaks in air conditioning ductwork systems can lower the efficiency of the system.

CHAPTER 25

1. Cavitation is when a gas bubble that has formed in the liquid rapidly collapses and as a result generates heat and removes material from the interior components of a pump.
2. Temperature and pressure.
3. Absolute pressure is the summation of the gauge's pressure reading plus the atmospheric pressure at the point of measurement.
4. 1 foot per second.
5. It will increase.

CHAPTER 26

1. No.
2. 8 to 10.
3. Centrifugal.
4. Pressure.
5. 25 feet.

CHAPTER 27

1. Pressure rise.
2. No.
3. No.
4. Rpm.
5. Quadruple.

CHAPTER 28

1. Contour line.
2. Center line.
3. An isometric sketch can show three sides of an object in one drawing.
4. A break line can be used whenever the current scale does not allow the drawing of a very long part and when there are no changes within the middle section, for example, a sewer pipe with no intersecting branches.
5. So he would know if there could be potential problems in venting.

CHAPTER 29

1. No.
2. No.
3. Different line types.
4. Yes.
5. Job title and name.

CHAPTER 30

1.

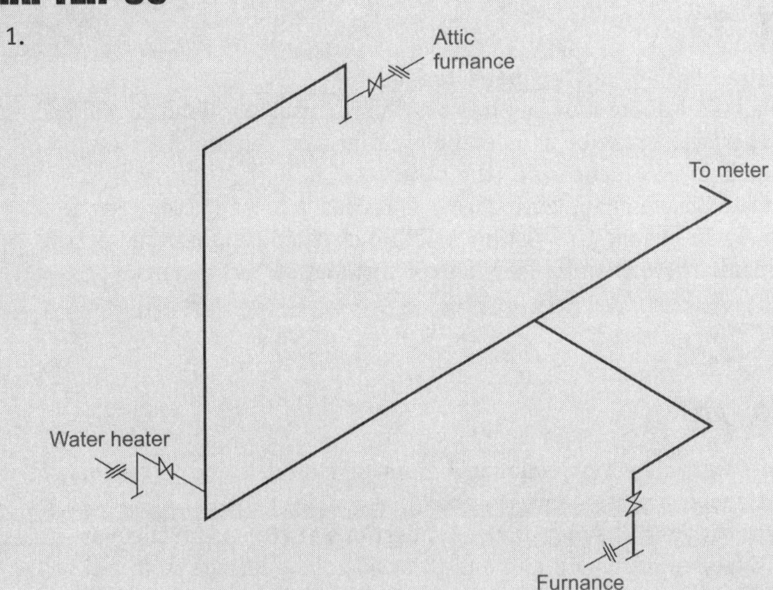

2.

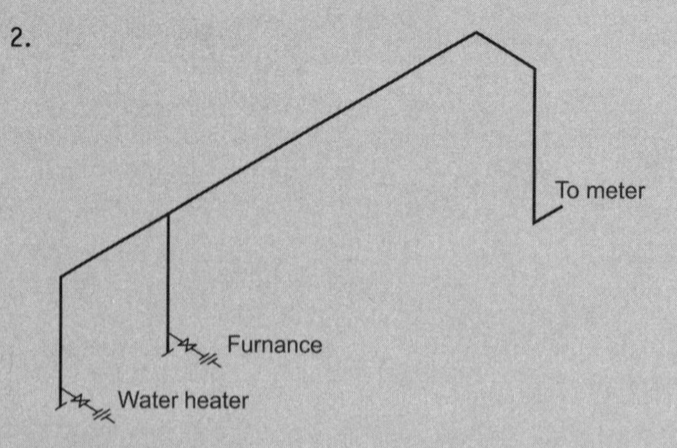

CHAPTER 31

No review questions for this chapter.

CHAPTER 32

1. Appeal is a process in which the installer requests a governmental authority to consider a practical, acceptable solution rather than mandating conformance to the code.
2. Clothes washer, water heater, dishwasher, and water softener.
3. Registered apprentice, journeyman plumber, master plumber, and plumbing contractor.
4. General regulations.
5. Shall.

CHAPTER 33

1. American Society for Testing Materials.
2. It is a brief clarifier provided to narrow down what the standard will address.
3. Costly mistakes can be avoided by ordering and installing the correct products in accordance with the codes.
4. Standards may change with each publication and may accept newer products or change the testing procedures of products covered by the standard. These changes may benefit installation technicians.
5. ASSE.

CHAPTER 34

1. Interceptors collect objectionable drainage materials and separators collect and transfer objectionable material to a containment vessel.
2. The objectives of having proper connections are to ensure adequate health and safety concerns are met and to avoid costly damage from leaks.
3. New cleaning equipment has provided greater drain cleaning abilities.
4. Connections between similar materials and dissimilar materials.

CHAPTER 35

1. *Accessible* describes a facility as being usable by persons having physical disabilities.

2. Bidet, lavatories, urinals, or water closets shall have at least 21 inches of clearance in front of the fixture lip.
3. A maximum water closet discharge of 1.6 gallons per flush is required.
4. Tub and shower valves shall be protected by an ASSE 1016 device and adjusted to a maximum temperature of 120°F.
5. A lavatory shall be protected by an ASSE 1070 device and adjusted to a maximum temperature of 110°F.

CHAPTER 36

1. When hangers fail in the drainage system, pockets could form in the line, hindering drainage and eventually resulting in obstructions.
2. Waste discharge flows provide lateral forces, which can be significant and damage a system.
3. The receptor discharge line shall not be less than the aggregate cross-sectional area of the indirect waste discharge pipes.
4. An air gap and an air break.
5. It is called a trapeze hanger.

CHAPTER 37

1. Hot water is defined as being equal to or greater than 110°F.
2. 80 psi under static conditions (no flow). Higher pressures are allowed at a sillcock.
3. No, they may provide a cross connection and are prohibited for underground installations by code.
4. The potable temperature shall be limited to 140°F.
5. Yes, the outlets shall be marked with signs identifying the water as dangerous and unsafe.

CHAPTER 38

1. A common vent serves two fixtures, which may be connected at the same level with a double sanitary tee or at two different levels with sanitary tees stacked.
2. Pipe slope creates force within the line to wash waste out of the drainage system. The desirable force is a velocity of 2 feet per second.
3. The increase in slope allows a drainage line to conduct a greater amount of DFU, translated to more fixtures.
4. This is the method used to avoid vent closure from frost building up in the vent terminal.
5. It retards the proper treatment of sanitary waste.

CHAPTER 39

1. All areas of healthcare facilities are of concern for life safety.
2. $1\frac{1}{2}$ inch pipe size.
3. No.
4. Medical gas piping systems.
5. To ensure backup.

CHAPTER 40

1. It is a method of testing the soil conditions to determine its ability to leach or absorb water away from the distribution fields.
2. Individual sewage disposal systems are required in the absence of a public sewer system.
3. Never, it is provided and conducted by the installer.
4. This enables the drain field to be closer to the terrain surface. Drainage water not only is absorbed by the soil for dispersion into the ground, but some water will be dispersed by evaporation.
5. They would not be gas tight during testing and the smoke would escape through the trap.

CHAPTER 41

1. Pitless adapters allow a connection from the submersible pump's drop pipe to the line supplying the water pressure storage tank.
2. Shallow-well jet pumps, deep-well jet pumps, and submersible pumps.
3. Bored, drilled, or driven, dependent upon the soil conditions.

CHAPTER 42

1. The unit is constructed to Department of Housing and Urban Development standards.
2. The individual dwelling unit may not have laundry facilities. The park's sanitary facilities are mindful of residents' laundry needs, similar to apartments. The one laundry for twenty units is the same for parks and apartments.
3. No, zoning may vary in some locations to allow them to be set on private property. The codes discussed in this chapter would then not be applicable.
4. No, newer HUD standards have units designed with independently trapped and vented sanitary systems.
5. This consideration dictates the underground installations of waste and water in the park based on where most units have their piping termination and water supply.

Index

Note: Italic page numbers indicate material in tables or figures.